PROCESS AND FORM IN GEOMORPHOLOGY

With the possible exception of molecular biology, no subject has been so transformed during the past forty years as that of the study of the landforms of the Earth. Central to that endeavour throughout that revolution has been Richard J. Chorley, whose achievement this volume honours. This book is not only about geomorphology: it is about how scientific change is effected and about how crucial people at critical times change the way we see the world.

The book is in two parts. The first presents state-of-the-art reports on fluvial, tectonic and climatic geomorphology by leading experts in their fields, all Chorley's colleagues and students. The second brings revisionary views to many aspects of the history of the discipline, on which Chorley is the world authority. Together they present not only new views on the landforms of the world but also incisive insights into how we have made sense of the environment around us from the early eighteenth century to the present day.

No student seeking to understand the complexity of landform development can afford to neglect these incisive reviews at the cutting edge of modern research; likewise, no historian of science can ignore this picture of revolution in action as paradigm change is acted out.

D.R. Stoddart is Professor of Geography, University of California at Berkeley.

PROCESS AND FORM IN GEOMORPHOLOGY

Edited by D.R. Stoddart

Routledge
Taylor & Francis Group

LONDON AND NEW YORK

First published 1997
by Routledge
2 Park Square, Milton Park, Abingdon, Oxfordshire OX14 4RN

Simultaneously published in the USA and Canada
by Routledge
711 Third Avenue, New York, NY 10017

First issued in paperback 2016

Routledge is an imprint of the Taylor and Francis Group, an informa business

Collection © 1997 D.R. Stoddart
Individual chapters © 1997 respective contributors

Typeset in Garamond by Solidus (Bristol) Limited

British Library Cataloguing in Publication Data
A catalogue record for this book is available from the British Library

Library of Congress Cataloguing in Publication Data
Process and form in geomorphology/edited by D.R. Stoddart.
p.   cm.
Includes bibliographical references and index.
1. Geomorphology.   2. Stoddart, D. R. (David Ross)
GB401.5.P743   1996
551.4′1–dc20   95-52607

ISBN 13: 978-1-138-98383-0 (pbk)
ISBN 13: 978-0-415-10527-9 (hbk)

CONTENTS

List of figures   vii
List of tables   xi
List of contributors   xiii
Preface   xv

Introduction

1 RICHARD J. CHORLEY: A REFORMER WITH A CAUSE   3
Robert P. Beckinsale

Part I On landforms

2 DRAINAGE DENSITY: PROBLEMS OF PREDICTION AND APPLICATION   15
Stanley A. Schumm

3 THE UNDERFIT MEANDER PROBLEM – LOOSE ENDS   46
George H. Dury

4 THE TROUBLE WITH VALLEYS   60
Barbara A. Kennedy

5 SUBSURFACE FLOW AND SUBSURFACE EROSION: FURTHER EVIDENCE ON FORMS AND CONTROLS   74
J.A.A. Jones

6 TECTONICS IN GEOMORPHOLOGICAL MODELS   121
Michael J. Kirkby

7 PROCESS AND FORM IN THE EROSION OF GLACIATED MOUNTAINS   145
Ian S. Evans

CONTENTS

8 LAND-USE CHANGES AND TROPICAL STREAM   175
   HYDROLOGY: SOME OBSERVATIONS FROM THE
   UPPER MAHAWELI BASIN OF SRI LANKA
   C.M. Madduma Bandara

9 PALAEOCLIMATOLOGY, CLIMATE SYSTEM   187
   PROCESSES AND THE GEOMORPHIC RECORD
   Roger G. Barry

10 ON THE LANDFORM HISTORY OF CHORLEY'S WEST   215
   SOMERSET
   Peter Haggett

Part II On theory and history

11 JAMES KEILL (1708) AND THE MORPHOMETRY OF   243
   THE MICROCOSM: GEOMETRIC PROGRESSION LAWS
   IN ARTERIAL TREES
   Michael J. Woldenberg

12 THEORY, MEASUREMENT AND TESTING IN 'REAL'   265
   GEOMORPHOLOGY AND PHYSICAL GEOGRAPHY
   *Keith Richards, Susan Brooks, Nicholas Clifford, Tim Harris and
   Stuart Lane*

13 OPEN SYSTEMS – CLOSED SYSTEMS: A PORTUGUESE   293
   VIGNETTE
   Robert J. Bennett

14 CHANCE AND NECESSITY IN GEOMORPHOLOGY   312
   Alan Werritty

15 A PLURALIST, PROBLEM-FOCUSED   328
   GEOMORPHOLOGY
   Olav Slaymaker

16 CARL SAUER: GEOMORPHOLOGIST   340
   David R. Stoddart

Epilogue

17 RICHARD J. CHORLEY AND MODERN   383
   GEOMORPHOLOGY
   David R. Stoddart

   Publications of Richard J. Chorley   400
   Index   406

FIGURES

Frontispiece: Richard J. Chorley

2.1	Relation between drainage density and mean annual precipitation (Gregory)	17
2.2	Location of study basins in California and Texas	19
2.3	Comparison of high and low drainage-density basins, Texas	24
2.4	Relation between drainage density and mean annual precipitation (Sortman)	24
2.5	Relation between drainage area and mean annual precipitation	25
2.6	Comparison of high and low drainage-density basins, California	26
2.7	Relation between active-channel drainage density and mean annual precipitation	27
2.8	Relation between valley drainage density and mean annual precipitation	28
2.9	Location of coal mines and Eccker study area	30
2.10	Relation between drainage density and mean annual precipitation (Eccker)	32
2.11	Drainage density as a function of relief ratio in the vicinity of the Dave Johnston Mine	35
2.12	Drainage density as a function of relief ratio in the vicinity of the Jim Bridger Mine	36
2.13	Drainage density as a function of relief ratio at McKinley Mine	37
2.14	Drainage patterns resulting from O'Brien's experiments	38
2.15	Sediment yield from drainage networks during Runs 1 through 4	40
3.1	Models for the interrelationship of incision rate and the passage of time	50
3.2	Exponential increase in incision rate through time on the Shoalhaven River, New South Wales	51

3.3	Constant net rate of incision through time on the Warwickshire Avon	53
3.4	Models for the interrelationship of increase in incision and increase in sinuosity	54
4.1	Maps to accompany Article VI on 'Géographie' in Buffon's 1749 *Histoire naturelle*	65
4.2	The principal trends of global relief, Noguès 1870	66
4.3	Two of Lyell's illustrations of valley formation	68
5.1	Runoff coefficients for hillslope processes	77
5.2	The trend in pipeflow runoff coefficients	78
5.3	Peak lag times and runoff rates for hillslope processes	79
5.4	Trends in peak lag times	80
5.5	Peak lag times for pipeflow and riparian seepage, Maesnant	81
5.6	Trends in peak runoff rates	82
5.7	The efficiency of drainage collection and yield for subsurface processes, Maesnant	85
5.8	Components of the Maesnant pipeflow model	86
5.9	The phreatic surface around a perennial pipe at baseflow	87
5.10	Partially collapsed ephemeral piping creating rills	92
5.11	The 'pipeflow streamhead', Maesnant	93
5.12	Partially collapsed perennial piping tributary to the Burbage Brook, Derbyshire	94
5.13	pH and aluminium concentrations, Maesnant	98
5.14	The response of pipeflow water quality to storm events, Maesnant	99
5.15	Electrical conductivity of topsoil extracts, Maesnant	100
5.16	The distribution of major plant associations around piping, lower Maesnant	102
5.17	The distribution of soil groups around piping, lower Maesnant	103
5.18	The frequency of piping in quadrat samples of major plant associations, Maesnant	104
5.19	Quadrat ordination showing distribution of piped quadrats in relation to major vegetation associations	105
5.20	Peat thickness around perennial piping, lower Maesnant	106
5.21	Piping in relation to the a/s index	108
5.22	Mean stormflow discharge and sediment yield compared with a/s index for Maesnant pipes	110
5.23	Distribution of piping in Britain in relation to soils	111
5.24	Catchment orientation and piping frequency in Britain	112
5.25	Distribution of piping in major soil groups in Britain	113
5.26	The topographic and climatic distribution of piping in Britain	114
6.1	Slope profiles generated for the process rates and types shown in Table 6.1	124

FIGURES

6.2	Final equilibrium relative denudation rates	130
6.3	Profiles in equilibrium with differing rates of constant downcutting	131
6.4	Schematic relationship between long-term rate of relative denudation and tectonic uplift rate	136
6.5	Equilibrium landforms associated with domino block extensional faulting	137
6.6	Example equilibrium profile obtained for a closed continental basin	138
6.7	Example simulation of closed continental basin evolution in the presence of extensional domino block faulting	139
6.8	Stream/slope profiles for erosion of a 1100 m plateau	140
6.9	Stream/slope profiles for same process rates and notation as in Figure 6.5	142
7.1	Calibration of cirques and troughs in the British Columbia Coast Mountains	158
7.2	A model of the contrast between classical armchair cirques and high-alpine 'cirques en van'	163
8.1	The upper Mahaweli basin	176
8.2	Annual rainfall trend at Nuwara Eliya	179
8.3	Changing runoff–rainfall ratios of the upper Mahaweli	180
8.4	Seasonal trends in streamflow at Peradeniya	181
9.1	Sub-gridscale processes	193
9.2	Zonally averaged surface temperatures for the Cretaceous	197
9.3	Annual mean precipitation for simulations with the GFDL global climate model	199
9.4	(a) Climate simulation for northern mid-latitudes at 18,000 BP and comparison with geological and palynological data; (b) estimated changes in summer temperature and in annual precipitation for 18,000 BP	202
9.5	Lake level status between 5°S and 45°N	204
10.1	West Somerset study area	217
10.2	Distribution of rocks of Devonian age in West Somerset	219
10.3	Cross-section of the Quantock Hills	222
10.4	Tectonic activity with downfaulted basins in the Minehead area, West Somerset	224
10.5	A.E. Frey's views of an alternative mechanism for forming the erosional 'flats' on the southern flanks of Exmoor	230
10.6	Diversion of the course of the Holford-Hodder stream, West Somerset	233
12.1	Correlograms of time series of downstream velocity components in a section of the River Sence, Leicestershire	272
12.2	The up-glacier contributing area on the hydraulic potential surface of the Haut Glacier d'Arolla, Valais, Switzerland	275

12.3   Three-dimensional computer maps of the Haut Glacier   276
        d'Arolla
12.4   A typology of river meander migration   282
12.5   Bed topography, bed material sizes, and bed flow directions at 284
        bankfull stage in an over-widened bend, Waireka Stream,
        Canterbury Plains, New Zealand
12.6   Depth-averaged flow vectors in a complex braided reach of   287
        the meltstream of the Haut Glacier d'Arolla
13.1   Sample catchment of Ribeira de Almádena and location in   294
        southern Portugal
13.2   Land-use change in the Ribeira de Almádena catchment,   304
        1951–1990
15.1   Classification of modes of enquiry   330
15.2   Geomorphology in the framework of the natural sciences   332
16.1   Kesseli's interpretation of five of Penck's slope development   350
        models
16.2   Sauer's only graphical essay in the Penckian mode   359
17.1   Professor Wooldridge in the field   385
17.2   Professor Steers as a gentleman fieldworker   387
17.3   Richard Chorley in 1962   392
17.4   Richard Chorley constructing one of his many works in the   395
        early 1960s, as viewed by a local wag

TABLES

2.1	Drainage basin data, Texas	21
2.2	Drainage basin data, California	22
2.3	Drainage basin data, Colorado	31
2.4	Experimental data	40
3.1	Terrace top heights and ages for the Warwickshire Avon near Evesham	52
5.1	Typical responses for hillslope drainage processes	76
5.2	Equations relating runoff parameters to catchment area	76
5.3	Comparison of sources of storm runoff in Maesnant Experimental Basin	83
5.4	Width–area relationships in soil pipes and open channels	90
5.5	Mean water quality parameters for rainfall, streamflow and pipeflow sites on Maesnant	97
6.1	File of slope process rates used for simulations shown in Figure 6.1	127
7.1	Mountain glacier erosion rates from sediment load of proglacial streams	147
7.2	Rock avalanches from cirque headwalls	168
8.1	Hydrological data for stream gauging stations in the upper Mahaweli basin	177
8.2	Land-use change in the Mahaweli basin above Peradeniya	181
8.3	Mean annual discharges of the Mahaweli Ganga prior to the development of large reservoir systems	182
9.1	GCMs commonly used in climatic studies	194
10.1	Classification and geomorphological impact of the Devonian rocks of West Somerset	220
10.2	Classification of the Permian and Triassic rocks of West Somerset	226
10.3	Peneplane remnants in West Somerset	228
10.4	Probable timing of Quaternary features in the West Somerset area	231
11.1	Values for R_A and R_L from Keill and Young	249

TABLES

12.1   A summary of the basic tenets of realism   267
12.2   Properties of downstream and vertical velocity series obtained   273
         in the flume and in the River Sence in Leicestershire
13.1   Association of surfaces and recent geological events in   296
         southern Portugal
13.2   Quaternary chronology for southern Portugal   297
13.3   Comparison of channel geometry above and below the   299
         limestone junction for four tributaries in the Ribeira de
         Almádena catchment
13.4   Land-use changes in the Ribeira de Almádena sample   305
         catchment, 1951–1990
13.5   Infiltration rate averaged over sample sites in the Ribeira de   305
         Almádena catchment
13.6   Flood discharge and channel characteristics in the Ribeira de   307
         Almádena catchment

CONTRIBUTORS

Dr Roger G. Barry Co-operative Institute for Research in Environmental Science (CIRES), University of Colorado, USA

Dr Robert P. Beckinsale Oxfordshire, United Kingdom

Professor Robert J. Bennett Department of Geography, London School of Economics and Political Science, United Kingdom

Dr Susan Brooks Department of Geography, University of Bristol, United Kingdom

Dr Nicholas Clifford Department of Geography, University of Hull, United Kingdom

Professor George H. Dury Suffolk, United Kingdom

Dr Ian S. Evans Department of Geography, University of Durham, United Kingdom

Professor Peter Haggett Department of Geography, University of Bristol, United Kingdom

Dr Tim Harris Department of Geography, University of Cambridge, United Kingdom

Dr J.A.A. Jones Institute of Earth Studies, University College of Wales, United Kingdom

Dr Barbara A. Kennedy St Hugh's College, Oxford, United Kingdom

Professor Michael J. Kirkby School of Geography, University of Leeds, United Kingdom

Dr Stuart Lane Department of Geography, University of Cambridge, United Kingdom

Dr C.M. Madduma Bandara University of Peradeniya, Sri Lanka

Dr Keith Richards Department of Geography, University of Cambridge, United Kingdom

Dr Stanley A. Schumm Department of Earth Resources, Colorado State University, USA

Professor Olav Slaymaker University of British Columbia, Vancouver, Canada

Professor David R. Stoddart Department of Geography, University of California at Berkeley, USA

Professor Alan Werritty Department of Geography, University of Dundee, United Kingdom

Dr Michael J. Woldenberg Department of Geography, University at Buffalo, State University of New York, USA

PREFACE

This book, written by his closest colleagues and his students, honours one of the most distinguished British geographers of this century, Richard J. Chorley, student of Robert P. Beckinsale at Oxford and Arthur Strahler at Columbia, University Demonstrator at Cambridge University 1958–1962, University Lecturer 1962–1970, Reader in Geography 1970–1974, Professor of Geography for twenty years from 1974, and now Emeritus Professor. Cambridge recognised his distinction with the degree of Sc.D. in the year he was appointed to his *ad hominem* chair. He served with distinction and indeed devotion as Head of the Department there for several years, and as Fellow of Sidney Sussex College.

Not all of his friends are represented in this collection: some had other duties or had moved on to other things, but all share the regard they have for him. The essays are in two groups, following his interests: on landforms, and on theory and history.

INTRODUCTION

1

RICHARD J. CHORLEY
A reformer with a cause
Robert P. Beckinsale

OXFORD

Richard John Chorley was born at 00.50 on 4 September 1927 and is always pleased to note that, discounting 'summer time', his birth date is 3.9.27. Three has always been his lucky number and it is understood that Plot 33 has been reserved for him at St Giles Cemetery, Cambridge. His birthplace was Minehead, a small town in Somerset only 20 miles west of the village of Pawlett, the home of Peter Haggett with whom later he was so happily linked. His father's name was Joseph and his mother's Mary, and he was their only child.

In 1937 he joined the local County (later Grammer (*sic*), as he persistently spelled it) School, where geography was to become his favourite, although not necessarily most successful, subject. His most influential teacher, Miss Lake, who was subsequently an Inspector of Schools, wrote on his reports: 'Has ability, but is not working consistently hard enough.... Both spelling and writing need improving. He can do well but his work is erratic.' One of his least successful subjects was French. As a 'credit' standard was then required for admission to university, he sat the School Certificate Examination in that subject on three occasions while at school, achieving the successive results of fail, pass, fail. As a result of this his knowledge of French grammar is very good even after almost half a century.

Between his volunteering for service in the army and his enlistment Germany surrendered, leading to his subsequent scepticism of simplistic correlations. There were to be two linked academic achievements during his almost three years of military service – the achievement of a credit in French and admission to Exeter College, Oxford. Without any further formal study of the language, Chorley presented himself again for examination and received the required credit standard. He has subsequently maintained that this success, the one of which he is most proud, was due to a combination of the language to which he was exposed in the army and to his release from formal education. However, this is not the whole story because, as has constantly happened in Chorley's career, luck played a strong part. He wrote the French examination in Manchester equipped only with an early ballpoint pen which, true to its trade name (Rollball), fell apart during the examination. Luck enabled Chorley to retrieve the minute ball from the dusty floor – not the first time he has fallen to his knees in the course of his scholarly career. The success in French, combined with the exemption from Latin allowed to ex-servicemen, made him eligible to sit the entrance examination for Exeter College, the *alma mater* of Charles Lyell (1797–1875), who became Britain's greatest geologist, internationally acclaimed as the master uniformitarian to whom, in the words of Charles Darwin, 'the science of geology is enormously indebted ... more so, as I believe, than to any other man who ever lived' (Chorley, Dunn and Beckinsale 1964: 190).

Chorley chose to read for an honours degree in geography, a subject which at Oxford had only recently blossomed into a separate entity. True, there had been temporary lecturers there on geographical topics during and since Tudor times but prolonged cementation into a unified educational subject did not materialise until after Halford J. Mackinder (1861–1947) was appointed Reader in Geography in 1887. However, it was not until twelve years later that Oxford University decided to establish a School of Geography which might grant a diploma to its successful students. Then, in the spring of 1899, Mackinder persuaded Andrew John Herbertson (1865–1913), who was lecturing at Herriot-Watt College, Edinburgh, to join him as his assistant at Oxford and the first examination there for a Diploma in Geography was held in 1901. Four years later, when Mackinder became Director of the London School of Economics, Herbertson succeeded him as Reader in Geography at Oxford and directed geographical affairs there until he died in 1915 at the early age of 50.

Thereafter, the ideas of Mackinder and Herbertson dominated thinking at the Oxford School of Geography and at most of the academies further afield to which its alumni radiated. The geographical approach was considered to be new because it had passed out of the encyclopaedic, data-listing bias common in early Victorian times to an approach which was more regional, more orientated towards topographical maps and strongly influenced by W.M. Davis's explanatory-descriptive method and cyclic concepts. H.O. Beckit, who succeeded Herbertson as Director of the Oxford School of Geography, actually joined Davis's European landform pilgrimage of 1911 and was a confirmed Davisian. Under his successor, Professor Kenneth Mason of Hertford College, a full honours course in geography was introduced at Oxford in 1930. However, the first half of the twentieth century was war-ridden and geographical progress here proved more material and organisational than intellectual, being dominated largely by pre-existing themes.

By chance, Chorley arrived at Oxford just at the wrong – or should we say the right? – moment for the anti-Davisian reformer. At first he was to find little co-operation in his recalcitrancy within British universities. In London, S.W. Wooldridge, a devout Davisian, ruled the roost; at Oxford landforms or geomorphology was considered a small part of a much wider geography, which was essentially a humane subject. In 1945 when the present writer chose to lecture on Rivers and Lakes the ruling powers altered the title to Rivers, Lakes and Man. In Chorley's undergraduate time the full honours three-year course in geography consisted mainly of a study of, and separate papers in, a preliminary examination and a final examination of a separate paper on: general physical geography (a mixture mainly of landforms, climate, hydrology and botany); general human geography; map work; British Isles; France; and each of three other regions of which Chorley selected Central and Southern Europe, India, and the USA. In addition, there was a thesis on a small area and a special subject, which was examined by two separate papers and was selected from a few specific topics each enlightened by a recommended bibliography. A few years before Chorley's arrival, enquiries on the selection of the Special Subjects had been circulated among the staff of the School of Geography and his future tutor, with an eye on the approaching International Hydrological Decade, suggested a study of European rivers. It was rejected and inexorably the Special Subject related to physical landscapes appeared as 'The Cycle in the Study of Landscapes' and was bolstered by a bibliography imbued with classic Davisiana. This topic was selected by both Chorley and his contemporary Yi-Fu Tuan, who went on to write a percipient paper on Penckian slopes (Tuan 1958) and later, under John E. Kesseli at the University of California at Berkeley, produced a fine doctoral thesis on 'Pediments in southeastern Arizona' (Tuan 1959).

By the age of 24 Chorley was beginning to feel his oats academically and was little inclined to conform to what he, perhaps misguidedly, considered to be the antiquated views of lecturers or examiners. Having realised the weaknesses of the uncompromisingly historical approach to geomorphology presented by the cyclic ideas of W.M. Davis and inflamed by its bombastic presentation by Professor Wooldridge during a guest lecture, Chorley asked a number of injudicious questions. These caused the then Professor at Oxford, famed more for the accuracy of his estimate of the height of Mount Everest than for his understanding of the depth of banality of much British geography of the period, to instruct him to refrain from the future questioning of distinguished visitors. The present writer was much more sympathetic and bears the responsibility of having been Chorley's sole tutor during his whole three years at Oxford.

Whereas most students would be expected to achieve high marks in their special subject, he obtained two of his lowest marks in his landform papers, in spite of an oral or viva lasting over fifty minutes. In the vernacular, he was batting on a sticky wicket. Ironically, had Chorley performed better, British geomorphology might well have fared worse, in that he was debarred from pursuing more conventional postgraduate studies in this country and was fortunate enough to gain a Fulbright Scholarship to study at Columbia University in New York City.

AMERICA

During the 1950s and 1960s many British geographers, human and physical alike, found a warm welcome in the USA, which, compared with war-depressed Europe, was a land flowing with milk and honey, abounding for earth scientists in real opportunities and progressive ideas. It was a further stroke of luck for Chorley that, having gone to Columbia to study under Armin K. Lobeck, he found that geomorphology there was the province of the youthful Arthur Strahler. At their first meeting Chorley was given an offprint of Strahler's recently published *American Journal of Science* paper in which ideas of quantitative dynamic geomorphology and systems analysis were applied to the classic problem of slope development (Strahler 1950). This publication had an effect on Chorley similar to that of Chapman's Homer on the poet Keats, and clarified for him his previous unique instinctive antipathy to the exclusively historical approach to geomorphology. These were vintage years for geomorphology in the Department of Geology at Columbia which have been described by Strahler (1992) in some detail, and it was a rare privilege and daunting challenge for the qualitatively/ historically trained Chorley to be exposed to the cutting edge of the new quantitative/dynamic thinking exemplified by such scholars as Strahler, Schumm, Melton, Broscoe and Morisawa. Schumm in particular became a lifelong friend and, along with Haggett, Stoddart, Barry, the present writer and others, combined such friendship with scholarly collaboration. After three years at Columbia, Chorley was appointed as Instructor in Geology at Brown University in Providence, Rhode Island, and spent a further very happy three years in that relatively small Ivy League university. In 1957 family matters prompted him to return to England, his father having died prematurely while Chorley was in the United States. Joseph's own father had been buried with two other paupers in an unmarked grave at the age of only 35, leaving a widow and seven children. Chorley's interest in genealogy was marked recently when, almost 100 years after his paternal grandfather's death, he inscribed a memorial to him in Locksbrook Cemetery, Bath.

Unemployed, Chorley returned to Oxford and re-established a contact with his former tutor which he never relinquished. During this rather difficult year he was fortunate to be befriended by George Dury, then one of the most radical and forward-looking of British geomorphologists. Under the mistaken apprehension that attendance at the annual meeting of the Institute of British Geographers would be both uplifting and career-enhancing, Chorley presented himself at Nottingham in January 1958. In a corridor Dury introduced him to Professor Wooldridge as someone who had been working with Strahler, which prompted the rejoinder: 'Next time you see Strahler tell him I spit at him!' This response was the product of what were becoming heady, emotional times for geomorphology and of course it was necessary that there should be some considerable figure of established authority, not quite so quick on the draw as of yore, towards whom young bucks took the long walk down Main Street. If Wooldridge had not existed it would have been necessary to invent him. Chorley, of course, recognised this and bore Wooldridge no lasting ill-will, writing more than thirty years later:

> Between the late 1920s and the late 1950s Wooldridge occupied a position of increasing pre-eminence in British geomorphology. By those who acknowledged this and followed his lead, he was rightly regarded as a brilliant researcher, an unequalled field teacher and a jovial colleague, ever ready to appear in amateur productions of Gilbert and Sullivan. To those who followed different precepts, especially as ill health set in during his later years, Wooldridge presented a rather different image.
>
> (Beckinsale and Chorley 1991: 285)

In a more proximate and less charitable frame of mind Chorley delivered at the meeting of the Institute of British Geographers held in Cambridge in 1959 (where he had in the interim been appointed to a University Demonstratorship), under the title 'The new geomorphology', a very explicit, ill-mannered but personally satisfying attack on the monolith of traditional British historical geomorphology – a house of cards which was to collapse with amazing rapidity. With this experience in mind, Chorley is perhaps the only British geographer not to be surprised by the rapidity of the collapse of the Soviet dictatorship, which bore such a similarity to the organisation of British geomorphology during the quarter of a century prior to 1959.

Chorley's employment at Cambridge was another occasion in which pure luck was to come to his assistance. In 1958 Professor Linton left Sheffield University and Chorley sent a curriculum vitae there enquiring whether there was a vacancy for a geomorphologist. Having a spare stamp, he also decided to send the one carbon copy of the c.v. to the Department of Geography at Cambridge in the faint hope that there might be the prospect of an opening there. The following day he was telephoned by Professor Alfred Steers and called for interview. Steers was looking primarily for a climatologist to join his strong team of geomorphologists, which included Bruce W. Sparks, A.T. (Dick) Grove and W. Vaughan Lewis. Steers, a wise and percipient director, was easily persuaded, if indeed he needed any persuasion, to appoint Chorley as a Demonstrator, in spite of the latter's obvious preference for landforms and deficiency as a climatologist. So the Oxford scholar embarked on weather and repaid this climatological debt when he and R.G. Barry collaborated in *Atmosphere, Weather and Climate* (Barry and Chorley 1968), which in expanded editions still flourishes (6th edition, 1992).

ROBERT P. BECKINSALE

CAMBRIDGE

The department which Chorley joined in October 1958 was remarkable for the composition of its youthful members – Peter Haggett had returned to a Cambridge Demonstratorship the previous year and the research student body included David Stoddart, David Harvey, Claudio Vita-Finzi, David Grigg, and others who were to change the course of world geography. From the outset the unison of the two geographers from West Somerset was fruitful, as was their friendship with David Stoddart. Together they used their influence in practical classes to substitute quantitative methods and locational analysis for the pre-existing arty cartography and laborious construction of map projections. In 1963 they began, in conjunction with the University Extra-Mural Board, the important Madingley Hall conferences for geography teachers which did so much to inject change into school geography and from which emerged *Frontiers in Geographical Teaching* (Chorley and Haggett 1965) and, especially, *Models in Geography* (Chorley and Haggett 1967). This innovation was described by the two in their preface to *Remodelling Geography* (Haggett and Chorley 1989), edited by Bill Macmillan.

In 1961 chance again took a hand in Chorley's career, this time in a macabre manner. Vaughan Lewis, who with Gus Caesar had been particularly welcoming to him, was killed in a motor smash in the United States. Prior to this Chorley had been warned that his job would terminate after five years, and it will be recalled that W.M. Davis had received a similar warning at Harvard seventy-nine years previously. The effect of the sudden death of this endearing and stimulating scholar was that Chorley was promoted to his vacant lectureship in 1963. Chorley wrote an affectionate reminiscence of Lewis in *The William Vaughan Lewis Seminars: No. 1. Tectonics and Geomorphology* (Chorley 1990).

The social changes of the 1960s coincided with a decade of intense professional activity for Chorley. The list of his research students who joined him between 1960 and 1970 reads like a roll call of innovation – M.J. Kirkby, H.O. Slaymaker, M.A. Carson, I.S. Evans, R.D. Hey, G.P. Chapman, J.A.A. Jones, B.A. Kennedy, R.I. Ferguson, C.M. Madduma Bandara, A. Werritty, M.G. Anderson, R.J. Bennett, R.S. Jarvis and K.S. Richards. As all these were highly motivated, first-rate scholars who worked on their own chosen topics, Chorley has never claimed a great deal of credit for their successes, but has never made any secret of his pride in being associated with them and of his gratitude for what they taught him. Two of Chorley's most important publications of about this time had to do with the application of the systems approach to geomorphology ('Geomorphology and General Systems Theory': Chorley 1962) and physical geography (*Physical Geography: A Systems Approach*: Chorley and Kennedy 1971).

In 1965 Chorley had the great good fortune to marry Rosemary More who, having graduated in geography at Cambridge before he arrived, pursued graduate work in irrigation at Berkeley and received a Ph.D. from Liverpool University, and became a Lecturer in Civil Engineering at Imperial College. She was to contribute important chapters both to *Models in Geography* (More 1967) and to Chorley's edited volume *Water, Earth and Man* (More 1969a, 1969b), as well as a book in her own right on *Water and Man* (More 1972). Her first encounter with her future husband was at the 1962 IBG dinner in Liverpool, when she enquired of her neighbour the name of the disruptive man who was throwing bread rolls! What modern social geographers disparagingly term the 'Kellogg's family' was soon completed by the advent of Richard (1966) and Eleanor (1968), the former now a research chemist and the latter a materials scientist.

THE HISTORICAL ASPECT

Chorley's researches into geomorphology naturally led him to consider the history of the study of landforms. Bolstered by the assistance of Antony J. Dunn and his former tutor, and by the information derived from direct contacts in the USA and elsewhere, a mountain of material was recast and reduced drastically to a sturdy, well-illustrated volume, *Geomorphology before Davis* (Chorley, Dunn and Beckinsale 1964). Scholars welcomed it and it has long been out of print.

But Chorley's pet theme, the inadequacy of the Davisian cyclic model, was now attacked full tilt, and in lectures at Madingley Hall he helped to shatter the Davisian dream. Published as part of *Frontiers in Geographical Teaching* (Chorley and Haggett 1965), his chapters on 'A re-evaluation of the geomorphic system of W.M. Davis' (Chorley 1965a) and 'The application of quantitative methods to geomorphology' (Chorley 1965b) had a strong influence on teaching in British educational establishments. Geomorphologists will especially note the suggestion that 'it is around the essentially non-cyclic model of Grove Karl Gilbert that much modern thought is centring' (Chorley 1965a: 24).

About this time Chorley, in collaboration with R.P. Beckinsale, embarked on a 'History of geomorphology' for the *Encyclopaedia of Geomorphology* edited by R.W. Fairbridge (Beckinsale and Chorley 1968). This general survey ends with the statement that since 1945 geomorphology has on the whole been progressing through its growing concern with measurement, the statistical analysis of data, and closer association with kindred sciences and practical problems.

The attraction of the historical angle was sustained when in July 1968 Charles C. Gillispie of Princeton University invited Chorley and Beckinsale to participate in a proposed *Dictionary of Scientific Biography* by advising on the choice of, and article-length appropriate to, earth scientists and by writing the relevant account of those who had not yet already been allocated to specialised authors. Among the persons the two dealt with were British geologists such as William Hopkins and Andrew Ramsay, German geographers and geomorphologists such as F. Ratzel, F. von Richthofen, and Albrecht and Walther Penck, French savants such as G.A. Daubrée and E. de Martonne, Swiss scholars such as J.A. de Luc, and the Americans D.W. Johnson and W.D. Johnson; Chorley himself was responsible for Daubrée, A. Heim, the two Johnsons and the younger Penck. This fine *Dictionary* was still in the course of publication in 1973 when the second volume of *The History of the Study of Landforms* appeared.

Entitled *The Life and Work of William Morris Davis* (Chorley, Beckinsale and Dunn 1973) and consisting of nearly 900 pages, this massive tome leaned heavily upon material collected by Chorley in the United States where he had met, among numerous other people who had been acquainted with Davis, the geomorphologist's eldest son, who supplied a cache of Davisiana. This detailed volume, which revealed an investigator who was saddled with an oversimplified, adolescent theme which had so engrossed him that his true contributions to geomorphology had been almost smothered, was acclaimed as a *tour de force*, the sort of biography that all would aspire to have.

During the late 1970s an opportunity arose to contribute to a professional salute to the work of G.K. Gilbert, who was perhaps always one of the geomorphologists Chorley admired most. In his clear assessment in *The Scientific Ideas of G.K. Gilbert* (Yochelson 1980) he concluded: 'Gilbert's legacy to contemporary dynamic geomorphology resides in his anticipation of the systems approach to the discipline' (Chorley and Beckinsale 1980: 140).

Chorley would regard his efforts to temper the monolithic, historically based, geological geomorphology of his youth with quantitative, dynamic, systems-based studies much more suited to a broad scientific base, the requirements of physical geography and to applied work of contemporary relevance as a particularly satisfying achievement. He has particular affection for two of his works involving collaboration with former research students – *Physical Geography: A Systems Approach* (1971) with Dr B.A. Kennedy and *Environmental Systems* (1978) with Professor R.J. Bennett.

RECENT YEARS

The decade since 1980 has been a very busy time for Chorley in both a literary and administrative sense, if indeed the two can be distinguished. By nature he is elephantine in memory, encyclopaedic in comprehension, Bactrian in stamina, and infectiously inventive and pleasant. In 1984 he collaborated with Stanley A. Schumm and David E. Sugden in the issue of a massive *Geomorphology* (Chorley, Schumm and Sugden 1984) which is richly illustrated and nicely up-to-date. Also he worked steadily, *inter alia*, on the history of the study of landforms, on which the available information accumulated at such an enormous rate that it soon outran the possibility of only one further single volume. So the decision was made to collect the more general aspects into a third volume and to elaborate on physical processes and most Quaternary landforms in a subsequent volume or volumes.

Volume 3, *Historical and Regional Geomorphology, 1890–1950* duly appeared early in 1991 (Beckinsale and Chorley 1991); it is well illustrated, carefully documented, very thorough, and in a scholarly and by no means uninteresting way points to the road ahead. One reviewer has warmly praised the authors' – it is collaboration – stamina but in any event the news of Chorley's threatened retirement has increased the possibility of the eventual completion of the whole vast history. In the meanwhile we close, in conjunction with all our readers, in wishing him and Rosemary a long and happy retirement blessed with the kind wishes of numerous friends and many decades of good fortune.

REFERENCES

Barry, R.G. and Chorley, R.J. (1968) *Atmosphere, Weather and Climate*, London: Methuen (6th edition, 1992, London: Routledge).

Beckinsale, R.P. and Chorley, R.J. (1968) 'History of geomorphology', in R.W. Fairbridge (ed.) *Encyclopaedia of Geomorphology*, New York: Reinhold.

——(1991) *The History of the Study of Landforms or the Development of Geomorphology*, vol. 3: *Historical and Regional Geomorphology 1890–1950*, London: Routledge.

Bennett, R.J. and Chorley, R.J. (1978) *Environmental Systems: Philosophy, Analysis and Control*, London: Methuen.

Chorley, R.J. (1962) 'Geomorphology and general systems theory', *U.S. Geological Survey Professional Paper*, 500–B: 1–10.

——(1965a) 'A re-evaluation of the geomorphic system of W.M. Davis,' in R.J. Chorley and P. Haggett (eds) *Frontiers in Geographical Teaching*, London: Methuen.

——(1965b) 'The application of quantitative methods to geomorphology', in R.J. Chorley and P. Haggett (eds) *Frontiers in Geographical Teaching*, London: Methuen.

——(1969) *Water, Earth and Man: A Synthesis of Hydrology, Geomorphology and Socio-economic Geography*, London: Methuen.

——(1990) 'William Vaughan Lewis', *The William Vaughan Lewis Seminars* (Cambridge: Department of Geography) 1: i–ii.

Chorley, R.J. and Beckinsale, R.P. (1980) 'G.K. Gilbert's geomorphology', *U.S. Geological Survey Special Paper* 183: 129–142.

Chorley, R.J. and Haggett, P. (eds) (1965) *Frontiers in Geographical Teaching*, London: Methuen.

——(eds) (1967) *Models in Geography*, London: Methuen.

Chorley, R.J. and Kennedy, B.A. (1971) *Physical Geography: A Systems Approach*, London: Prentice-Hall.

Chorley, R.J., Beckinsale, R.P. and Dunn, A.J. (1973) *The History of the Study of Landforms or the Development of Geomorphology*, vol. 2: *The Life and Work of William Morris Davis*, London: Methuen.

Chorley, R.J., Dunn, A.J. and Beckinsale, R.P. (1964) *The History of the Study of*

Landforms or the Development of Geomorphology, vol. 1: *Geomorphology before Davis*, London: Methuen.

Chorley, R.J., Schumm, S.A. and Sugden, D.E. (1984) *Geomorphology*, London: Methuen.

Haggett, P. and Chorley, R.J. (1989) 'From Madingley to Oxford: a foreword to *Remodelling Geography*', in B. Macmillan (ed.) *Remodelling Geography*, Oxford: Blackwell.

More, R.J.M. (1967) 'Hydrological models in geography', in R.J. Chorley and P. Haggett (eds) *Models in Geography*, London: Methuen.

——(1969a) 'The interaction of precipitation and man', in R.J. Chorley (ed.) *Water, Earth and Man*, London: Methuen.

——(1969b) 'Water and crops', in R.J. Chorley (ed.) *Water, Earth and Man*, London: Methuen.

——(1972) *Water and Man*, London: Collins.

Strahler, A.N. (1950) 'Equilibrium theory of erosional slopes approached by frequency distribution analysis', *American Journal of Science* 248: 673–696, 800–814.

——(1992) 'Quantitative/dynamic geomorphology at Columbia 1945–60: a retrospective', *Progress in Physical Geography* 16: 65–84.

Tuan, Yi-Fu (1958) 'The misleading antithesis of Penckian and Davisian concepts of slope retreat in waning development', *Transactions of the Indiana Academy of Sciences* 67: 212–214.

——(1959) 'Pediments in southeastern Arizona', *University of California Publications in Geography* 13: 1–140.

Yochelson, E.L. (ed.) (1980) 'The scientific ideas of G.K. Gilbert: an assessment on the occasion of the centennial of the United States Geological Survey (1879–1979)', *U.S. Geological Survey Special Paper* 183: i–viii, 1–148.

Part I
ON LANDFORMS

2

DRAINAGE DENSITY
Problems of prediction and application
Stanley A. Schumm

INTRODUCTION

From the ground one sees relief features, such as hillslopes, valley sides and scarps. From the air one sees rivers and drainage networks. In many areas the drainage networks are the dominant landscape feature. They are the arteries that move water and sediment through the fluvial system. With the advent of aerial photography and quantitative procedures for documenting landform characteristics (Horton 1945), it is no wonder that the drainage network and especially drainage density became a topic of great interest to geomorphologists and hydrologists. As a principal component of the landscape that could be investigated using maps and aerial photographs, drainage density (the ratio of total channel length to drainage area) was the subject of numerous studies. It was believed that if the independent variables that control drainage density could be related quantitatively to it, the results would be of great academic interest and of practical value. For example, it could lead to a quantitative climatic geomorphology.

Drainage density is clearly an important geomorphic variable (Gardiner and Gregory 1982) that has been related to climatic conditions (Gregory 1976; Chorley 1957; Chorley and Morgan 1962; Gardiner 1983), flood peaks (Carlston 1963; Gardiner and Gregory 1982), mean annual discharge (Hadley and Schumm 1961), and sediment yields (Schumm 1969). It has long been used to detect variations of rock type and structure by photogeologists (Chorley *et al.* 1984: 318) and to document the stage of erosional evolution of a drainage system (Ruhe 1950; Schumm *et al.* 1987).

Drainage density is also a significant factor for the reclamation of mined land in the USA (Toy and Hadley 1987). There is therefore a great need to improve the understanding of drainage density and to predict how altered conditions will affect it.

A federal law requires the restoration of a stable post-mining topography that approximates the original pre-mining contour. However, if a great thickness of coal is mined, which is the usual case, it may not be possible to design post-mining topography that will meet this requirement. In addition, if the pre-mining topography is very steep, the mining company may have difficulty conforming to the law. If the steep topography is re-established, it may not be considered to be a stable landform. Therefore, changes in drainage-basin topography, such as basin relief, slope, shape and drainage density can result from mining and reclamation design. However, according to the law, channel length and frequency, basin area and drainage density may be varied in order to provide a stable landform. Therefore, if the effects of changes in topography and vegetation on drainage density can be anticipated for a given lithology and within a given climate, the reclamation design can be improved.

The infiltration rate and permeability of reclaimed spoils and topsoil are often not the same as those of pre-mining soils and substrate. Mixing during the mining and reclamation processes causes the disruption of soil structure. Consequently, reclaimed spoils usually have an infiltration rate that is lower than the pre-mining rate until vegetation becomes re-established. However, when soils of low infiltration capacity are mixed with stony or sandy soils, the reclaimed spoils will have a higher infiltration rate. Vegetative type and cover also may change as the result of mining and reclamation. The resulting vegetative cover is typically less than that which existed prior to mining. All of these changes will influence drainage density. Another major concern is increased erosion and increased sediment yields from the reclaimed area. Mining companies are required to reclaim drainage basins and to re-establish stream networks so that they are stable over the long term in order to reduce erosion.

As one considers the literature, it is apparent that serious problems exist with regard to developing and using the empirical relations among drainage density and the numerous variables that affect it. The many controls that affect this important geomorphic variable are obviously a reason for controversy and the inability to predict with confidence. These controls include at least the following: mean annual precipitation, precipitation intensity, lithology, soil characteristics, relief, vegetation, human activity and stage of drainage-network development.

Climatic geomorphology has had a distinct European orientation as a result of the efforts of Tricart and Cailleux (1972), Büdel (1982) and Stoddart (1969). Their studies were primarily descriptive, in contrast to the quantitative studies of Peltier (1962, 1975), Melton (1957, 1958), Chorley and Morgan (1962), Carlston (1963), Williams and Fowler (1969), Abrahams (1972, 1984), Abrahams and Ponczynski (1985), Daniel (1981) and, of course, Gregory and Gardiner (1975), all of whom attempted to relate drainage density to climatic variables. In Peltier's (1962) early paper he showed that drainage density was a maximum in semi-arid regions, which also was the conclusion of the Gregory and Gardiner (1975) study (Figure 2.1). These studies demonstrated an increase of drainage density from humid to semi-arid, as vegetation cover decreases. Based upon the relation between sediment yield and precipitation in the USA, Schumm (1965) suggested that drainage density mirrors the

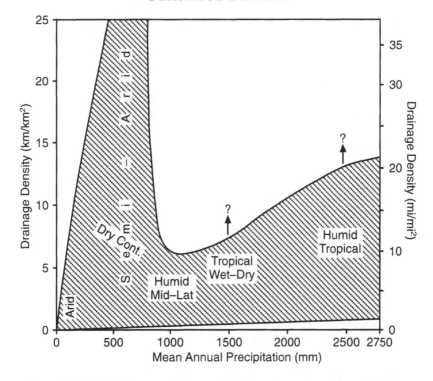

Figure 2.1 Relation between drainage density and mean annual precipitation
Source: After Gregory 1976

sediment–yield relation, reaching a maximum in semi-arid regions.

In semi-arid regions even microclimatic variations can affect drainage density. For example, south-facing hillslopes have a drier environment and less vegetation than north-facing slopes in the western USA. Therefore, if a drainage basin is small, its aspect, the direction in which it faces, can significantly affect drainage density and greatly increase the scatter of the data within an otherwise homogeneous geologic, climatic and geomorphic area. In addition, infrequent large floods may also produce a drainage density that is greater than that expected in a given region (Patton 1988).

In addition to climate and other natural controls of drainage density, there are two important aspects of human influences on drainage density. The first involves the modification of a drainage basin or drainage network. For example, timbering, overgrazing, urbanization or any activity which significantly decreases the vegetative cover or soil infiltration rates should cause an expansion of the drainage network and an increase of drainage density. Hence, human activity can increase drainage density and cause it to be out of balance with existing natural conditions.

The second human influence is not physical, but it involves knowing where to stop when measuring channel length. The measurement of drainage density is frequently subjective rather than objective, and there are different types of drainage density that can be recognised in one drainage basin (Gregory and Gardiner 1975). For example, there is a basic network, which is composed of perennial streams and which expands and contracts with runoff, or there is the active-channel network, which is composed of ephemeral, intermittent and perennial streams. In addition, the use of contour crenulations, as evidence of the presence of channels, will result in the inclusion of portions of valleys that do not contain active channels. My preference is to measure recognisable or active channels. Obviously, one must consider the objectives of the study when determining what should be measured, and therefore, what is measured by one investigator often cannot be compared with that of another. In any study of drainage density, areas that are significantly impacted by human activities must be avoided, and the investigator must describe how drainage density was measured.

A final problem relates to the scale of the maps or photographs that are used. Many first-order channels will not be visible on coarse-scale maps or photographs if the topographic texture is fine. This problem has been discussed by Chorley and Dale (1972) and Morisawa (1957).

It is necessary to consider drainage density in the development of estimates of sediment yield and runoff and in the reclamation of mined land, and it is important to develop a better understanding of drainage density and its controls. Five studies are therefore summarised in this chapter in order to test some of the assumptions about drainage density and to suggest procedures for designing stable drainage networks for the reclamation of disturbed lands in semi-arid western USA. Three of the studies were commenced in order to determine how mean annual precipitation and lithology affect drainage density (Eccker 1984; Sortman 1988; Levish 1992). One study was concerned with the effect of relief and slope on drainage density (Water Engineering & Technology 1985), and an experimental study was designed to determine how a drainage network responds to an artificial reduction of drainage density and how sediment yields are affected by such changes (O'Brien 1984).

EFFECT OF CLIMATE, LITHOLOGY AND RELIEF ON DRAINAGE DENSITY

In spite of the multiple variables acting to affect drainage density and other landform characteristics the author was convinced that a careful selection of field areas and a determined attempt to eliminate the effects of scale, relief and stage would yield a meaningful relation between drainage density and climatic variables. Therefore Vincent Sortman (1988) and Daniel Levish (1992) were set the task of finding relatively homogeneous rock types that occur through a wide climatic range. An example of this approach was

18

provided by Toy (1977), who measured hillslope characteristics developed on marine shales from Kentucky to Nevada along the 37th parallel. He showed that hillslopes become steeper and shorter as climate becomes drier. The first step was to locate Toy's field sites in order to obtain drainage-basin data at each. Unfortunately, although the sites could be used to study hillslope characteristics, many of the outcrops were too small for a drainage density study. Therefore, other sites were needed, and Sortman located a series of drainage basins developed on shale along an east–west line through Texas (Figure 2.2) while Levish located a series of drainage basins developed on granite along a north–south line in California (Figure 2.2). A third study by Eccker (1984) in a more limited area provided information on the effect of both climate and lithology on drainage density in northwestern Colorado.

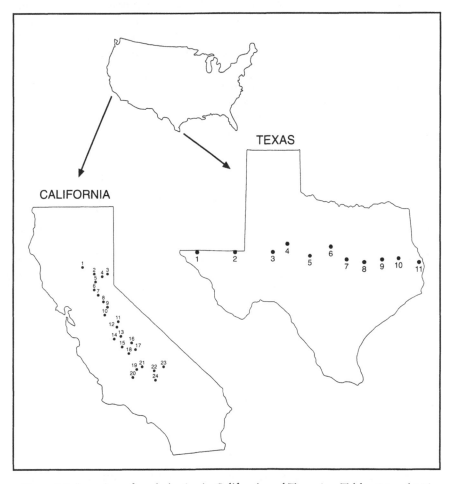

Figure 2.2 Location of study basins in California and Texas (see Tables 2.1 and 2.2)

Drainage density on shale in Texas

Sortman (1988) recognised that there is a great range of average annual precipitation across central Texas, from 1422 mm in the east to 178 mm in the west. This appeared to be an ideal situation to determine how climate affects drainage density. In order to evaluate the effects of climate the geologic variables of lithology and structure must be relatively constant. Therefore, the same rock type must underlie all of the selected basins. The rock should be easily eroded so that it is in equilibrium with the prevailing climate, and the soils should have a low infiltration rate in order to generate maximum runoff and to produce a well-developed drainage network. For these reasons it was decided that drainage basins developed on shale would be selected. Average permeability for the soils of these basins is 4 mm/hr (Table 2.1), therefore infiltration rates are very low. It was not possible to find a single shale formation that crossed Texas, but Sortman found drainage basins on shale throughout the climatic range (Figure 2.2), and he selected eleven for study. The effects of structure were minimal, as the dip of the beds was near zero.

It was not possible to select drainage basins with the same relief, but the stage of development of each network was comparable. This was determined by ensuring that each basin have a fully developed or stable drainage network (Figure 2.3). It was also decided that the basins should be small in order to facilitate measurement and to lessen precipitation and temperature variability. Finally, only fifth-order drainage basins were selected so that they would be morphologically comparable. Each basin was located near a weather station so that the climatic data were representative of the basin.

Sortman located outcrops of flat-lying shale using 1/250,000 geologic maps, and 1:24,000 scale topographic maps were used to locate specific drainage basins. Aerial photographs were then used to identify the actual drainage channels. Field visits permitted verification of the headward extent of active channels either by the existence of a headcut or the disappearance of the channel into a grassed hollow. Sortman measured only active channel lengths (Table 2.1).

In spite of the effort to eliminate confounding variables, problems remained. Although each basin was located on shale, there were differences among the lithologic units. For example, some units had a higher sand content or contained interbeds of limestone. In some cases a caprock maintained the main basin divide, but this appeared to have no significant effect on drainage density.

Results

Sortman's relation between drainage density and mean annual precipitation (Figure 2.4) supports the Gregory-Gardiner (1975) conclusion that drainage

Table 2.1 Drainage basin data, Texas

Location		Mean annual precipitation (mm)	Drainage density (km/km²)	Stream frequency (n/km²)	First-order stream frequency (n/km²)	Basin area (km²)	Total channel length (km)	Relief (m)	Average permeability (mm/hr)
1 Salt Flat	31° 38' N 105° 05' W	250	10.5	181	73	2.08	21.8	61	–
2 Paduca Breaks	32° 07' N 103° 37' W	355	9.28	130	74	2.93	27.2	43	–
3 Vealmoor	32° 36' N 101° 34' W	472	14.47	317	87	1.41	20.4	37	<1.5
4 Maverick Creek	32° 59' N 100° 48' W	596	10.89	285	106	1.92	20.9	61	<1.5
5 Bald Knob	32° 03' N 99° 52' W	677	10.0	194	66	1.75	17.5	55	3.3
6 Lone Camp	32° 45' N 98° 17' W	781	6.33	70	80	5.89	37.3	92	8.4
7 Aquilla	31° 48' N 97° 10' W	879	4.60	58	55	4.87	22.4	46	10.4
8 Bald Prairie	31° 15' N 96° 25' W	981	4.15	52	68	6.68	27.7	43	8.4
9 Grapeland	31° 25' N 95° 26' W	1073	3.87	38	40	5.42	21.0	46	3.3
10 Redland	31° 30' N 94° 39' W	1201	5.40	76	54	3.65	19.7	61	1.5
11 Fairmount	31° 11' N 93° 41' W	1351	4.57	51	46	4.60	21.0	76	1.5

Source: Sortman 1988

Table 2.2 Drainage basin data, California

Weather station name	Location	Elevation (m)	Mean annual precipitation (mm)	Basin area (km²)	Valley drainage density (km/km²)	Active-channel drainage density (km/km²)	Average permeability (mm/hr)
1 De Sabla	39° 52' N 121° 37' W	827	1660	4.5	5.84	2.18	79
2 Strawberry Valley	39° 34' N 121° 06' W	1160	2058	4.3	3.77	2.18	79
3 Sierraville	39° 35' N 120° 22' W	1516	675	4.0	3.15	1.80	330
4 Bowman Dam	39° 27' N 120° 39' W	1629	1699	4.7	4.75	2.34	102
5 Nevada City	39° 16' N 121° 01' W	768	1387	4.6	4.79	2.78	67
6 Auburn	38° 54' N 121° 04' W	393	875	2.8	5.40	2.23	102
7 Placerville	38° 44' N 120° 48' W	576	940	3.8	5.78	2.94	105
8 Tiger Creek	38° 27' N 120° 29' W	717	1146	5.4	3.19	2.23	42
9 Calaveras	38° 17' N 120° 19' W	1431	1340	3.9	4.09	2.08	33
10 Sonora RS	37° 59' N 120° 23' W	533	788	5.6	4.38	1.92	102
11 Yosemite Park HQ	37° 45' N 119° 35' W	1210	916	5.8	4.57	2.49	102

Station	Coordinates						
12 So. Entrance Yosemite	37° 30' N 119° 38' W	1560	1121	3.7	5.79	2.09	102
13 Auberry	37° 05' N 119° 30' W	652	606	5.4	5.97	2.08	95
14 Friant Gov. Camp	36° 59' N 119° 43' W	125	349	4.2	7.63	1.64	102
15 Orange Grove	36° 37' N 119° 18' W	131	326	5.1	7.69	2.56	95
16 Grant Grove	36° 44' N 118° 58' W	2011	1070	4.2	8.74	2.56	102
17 Ash Mountain	36° 29' N 118° 50' W	521	645	3.2	8.86	3.70	81
18 Lemon Cove	36° 23' N 119° 02' W	156	342	4.3	7.64	2.70	102
19 Glenville	35° 43' N 118° 42' W	957	460	3.1	7.80	1.52	67
20 Kern River	35° 28' N 118° 47' W	296	266	3.5	7.07	2.73	102
21 Kern River	35° 28' N 118° 47' W	824	323	5.5	11.78	3.92	102
22 Inyokern	35° 39' N 117° 49' W	744	106	5.0	9.01	4.48	77
23 Tronabla	35° 47' N 117° 23' W	517	100	5.7	10.78	10.78	77
24 Cantil	35° 18' N 117° 58' W	610	79	3.5	14.92	14.92	77

Source: Levish, 1992

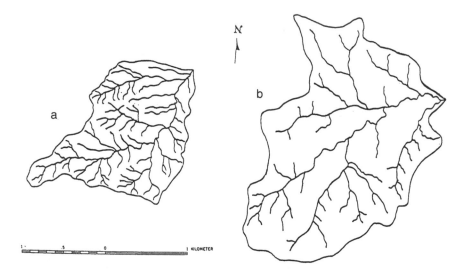

Figure 2.3 Comparison of high and low drainage-density basins: (a) Salt Flat basin,
Texas; (b) Fairmount basin, Texas
Source: Sortman 1988

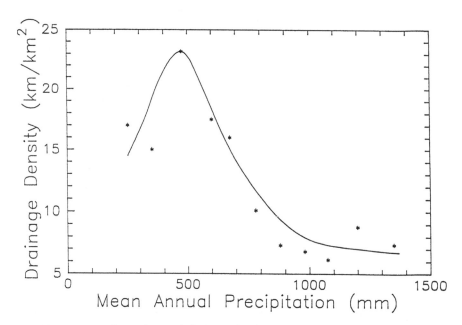

Figure 2.4 Relation between drainage density and mean annual precipitation
Source: Sortman 1988

density is at a maximum in a semi-arid climate. This was only because one drainage basin (Vealmoor, Table 2.1) plots high. If this is considered to be an anomaly one could interpret the scatter of points as indicating a relatively constant value of drainage density when annual precipitation is less than 700 mm. In fact, the fifth-order drainage basins abruptly increase in size above 700 mm of annual precipitation (Figure 2.5). This occurs at the change from brush and grass to open forest. Instead of a transition from small to large drainage basins (Figure 2.3) across the range of precipitation, the change of vegetative effectiveness is relatively abrupt. This may also explain the grouping of points on Figure 2.4, and the large difference between the values of drainage density and stream frequency between the Bald Knob and Lone Camp drainage basins (Table 2.1).

Sortman's study shows that active-channel drainage density in Texas is significantly affected by mean annual precipitation, but it is the effectiveness of vegetation in reducing runoff and its erosive potential that produces the shape of the drainage density curve (Figure 2.4).

Drainage density on granite in California

Levish (1992) selected 24 drainage basins for study, through an annual climatic range from 100 mm to 2100 mm, on granite in the Sierra Nevada and Mojave Desert of California (Table 2.2, Figures 2.2, 2.6). He identified

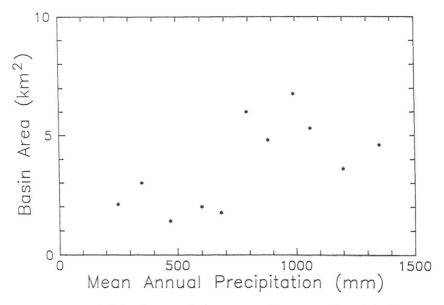

Figure 2.5 Relation between drainage area and mean annual precipitation
Source: Sortman 1988

25

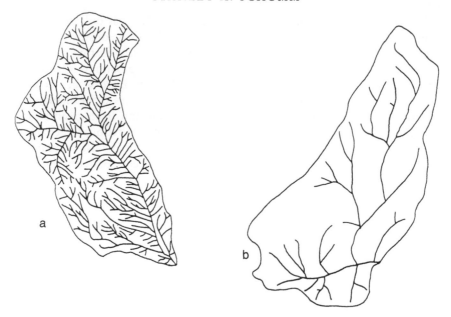

Figure 2.6 Comparison of high and low drainage-density basins: (a) Cantil, California; (b) Strawberry Valley, California
Source: Levish 1992

potential study areas on topographic maps, and then he determined the point at which first-order channels originate in the field. After the field visits, he identified the selected drainage basins on topographic maps and aerial photographs (1:63,000 and 1:20,000 scale). Levish measured the length of active channels with well-defined banks, and he also measured the length of unchannelled valley floors or hollows that formed the headward extent of some drainage networks. He could therefore calculate active-channel drainage density and valley drainage density.

Levish notes that 'it proved extremely difficult to locate basins of similar area and relief ... that were close to weather stations', which is a common complaint of quantitative climatic geomorphologists. By selecting granite drainage basins he assumed that he had eliminated lithologic and structural variations. However, Cooks (1983) has shown that the geotechnical properties of rocks, such as compressive and shear strength and elasticity also affect drainage density. Therefore the selection of a single rock type for study may not eliminate lithologic variability but it minimises it as much as possible. The average permeability of the weathered granite soils was 97 mm per hour, which is much higher than that for the shale soils of Texas.

Results

Unlike Sortman's results, Levish found that the maximum active-channel drainage density occurs under the driest climate conditions (Figure 2.7), and active-channel drainage density is relatively constant through a considerable range of precipitation. Only the two Mojave Desert basins plot well above the others (Table 2.2, Basins 23, 24). Valley drainage density shows a similar but more regular decrease of valley density with precipitation (Figure 2.8). In this figure the scatter of points is substantial, which suggests that in spite of Levish's best efforts, it was not always possible to select comparable basins and comparable climatic conditions. For example, Basin 3 plots very low and 16 high (Figure 2.8, Table 2.2). Upon checking Levish's data, it was determined that the permeability of the soil in Basin 3 was very high, and the intensity of precipitation was high in Basin 16, which was at a higher altitude

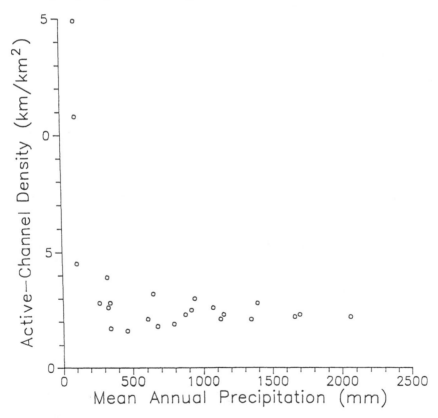

Figure 2.7 Relation between active-channel drainage density and mean annual precipitation
Source: Levish 1992

27

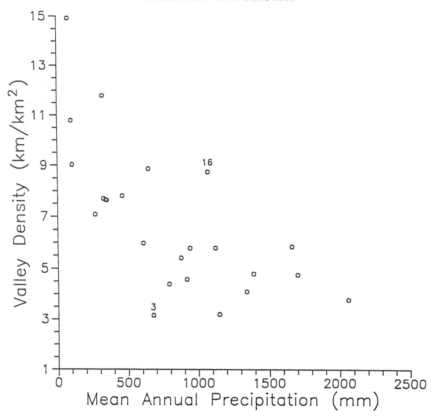

Figure 2.8 Relation between valley drainage density and mean annual precipitation
Source: Levish 1992

than many others. The position of these points on Figure 2.8 can be explained by extremes of soil permeability and precipitation intensity, and they could be eliminated from consideration. However, they provide valuable informa-tion about the development of the valley network. They demonstrate that, on the granites of the Sierra Nevada, the valley drainage network is responding to both climate and soil variables. In contrast to the valley network, active-channel drainage density remains relatively constant (Figure 2.7), except for the two most arid basins (Table 2.2, Basins 23 and 24), which suggests that it is relatively independent of climate above very low values of precipitation. In fact, these granitic basins, mantled by residual soil, may represent the permeable end-member of a continuum of basins in which the majority of runoff is transmitted by throughflow (Kirkby and Chorley 1967). Therefore, the active-channel density of these basins is related more to the character-istics of the weathered soil mantle than to precipitation. This is corroborated by the relatively low peaks and long duration of the hydrographs for similar granitic basins in the Sierra Nevada (Rantz 1970; Waananen 1973; Waananen and Crippen 1977).

Drainage density in north-western Colorado

In order to provide information on drainage density for mined-land reclamation in north-western Colorado, Eccker (1984) undertook to document drainage density changes with climate on two lithologic units. Ten fifth-order drainage basins were selected on the basis of accessibility, absence of structural control and absence of human influences. The field sites were located on two formations, Mesa Verde Group and Mancos Shale (Figure 2.9). These and similar rock types crop out over much of the western United States. Eccker therefore anticipated that her conclusions might have a general application for mined-land reclamation throughout the region.

In the field, channel longitudinal profiles and cross-sections were surveyed, channel morphology was described, channel and bank sediment was sampled, local vegetation was identified, and the percentage of ground cover was estimated. Morphometric measurements were made from aerial photographs (1:24,000) and topographic maps.

The Mancos Shale Formation is a distinctive lithologic unit that is composed predominantly of marine shale and sandy shale (Reeside 1924). The Mesa Verde Group is composed of marine, near-shore and terrestrial sediments. In north-western Colorado, it varies from interbedded fine-grained sandstone to shale with thin coal beds (Bass et al. 1955). The variability of each lithologic unit suggests that an attempt to develop relations between drainage density, climate and lithology may have little chance of success through a range of annual precipitation of only 6.5 inches. In order to resolve this problem, Eccker selected basins formed only on shale or on sandstone within each lithologic unit.

Unlike Sortman, Eccker measured total drainage density, which included active channels and grassed unchannelled valley floors. A first-order stream channel was thus defined as an unbranched linear depression with converging valley sides, which did not necessarily contain a distinct channel. This was done on the assumption that the grassed valley floor will eventually be incised, as commonly happens, and in order to be conservative regarding the length of channels required for the reclamation of mined land.

A problem faced by Eccker was the sparse nature of climatologic data. In order to determine mean annual precipitation for each of her drainage basins she had to extrapolate data from the distant weather stations. She did this by developing regression equations for the effect of elevation, exposure and distance from a north-western barrier and the Continental Divide.

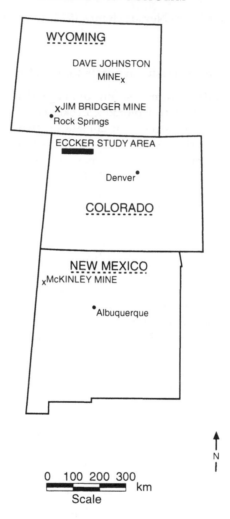

Figure 2.9 Location of coal mines and Eccker study area (black rectangle)

Results

Drainage basins that form on the Mesa Verde Group are large, circular and steep with a few widely spaced streams (Table 2.3), and on average they are larger than the basins formed on the Mancos Shale. The high infiltration capacity and permeability of the Mesa Verde Sandstone minimises runoff so that a drainage network of a given order generally occupies a larger watershed and drainage density is lower (Figure 2.10). The linear relation between drainage density (D) and mean annual precipitation (P) on this sandstone is as follows: $D = 91.4 - 0.14P$ ($r^2 = 0.79$).

Table 2.3 Drainage basin data, Colorado

Basin	Location	Elevation (m)	Mean annual precipitation (mm)	Drainage density (km/km²)	Relief (m)	Area (km²)	Total channel length (km)	First-order stream frequency (n)
Mesa Verde								
Upper Dunstan	T5N R89W	2411	483	21	241	0.62	13.0	137
Haunted	T6N R90W	2191	439	35	341	0.75	26.6	373
Coyote Eyes	T6N R85W	2156	483	26	210	0.72	18.5	247
Upper Winter Valley	T5N R98W	2036	345	38	82	0.26	9.9	170
Tumbling	T6N R94W	1941	351	48	73	0.28	13.5	219
Mancos								
Honey	T4N R87W	2224	442	34	91	0.21	7.1	139
Blister	T4N R89W	2202	467	41	235	0.57	23.5	307
Cross	T3N R92W	2127	437	36	82	0.54	19.5	262
Upper Boxelder	T5N R94W	1898	351	54	76	0.34	18.2	398
Forgotten	T4N R99W	1776	323	56	40	0.10	5.6	94

Source: Eccker 1984

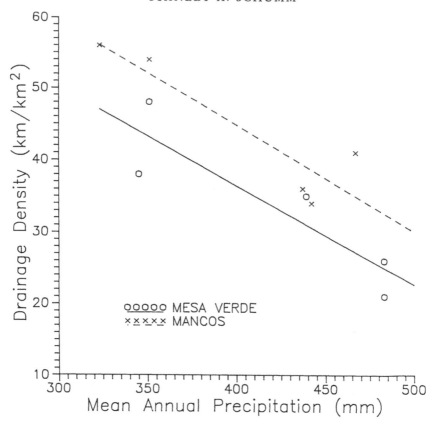

Figure 2.10 Relation between drainage density and mean annual precipitation from Mancos Shale and Mesa Verde Formation drainage basins (dashed line is best fit to Mancos Shale data, solid line for Mesa Verde Formation)

Source: Eccker 1984

Drainage basins on the relatively impermeable Mancos Shale tend to be smaller, with a gentle topography. The low infiltration capacity and permeability of the Mancos Shale and its soils promote more surface runoff than from the sandstone. Consequently, drainage density is higher (Figure 2.10). The relation between drainage density and mean annual precipitation on the shale is as follows: $D = 103.6 - 0.15P$ ($r^2 = 0.82$). The equations demonstrate the importance of both climate and geology on drainage-basin morphology in a limited area of western Colorado.

Summary

On shale, Sortman (1988) found a pattern of drainage density and climate (Figure 2.4) similar to Figure 2.1 and to the Langbein and Schumm (1958) sediment-yield curve, but Levish's results are very different (Figure 2.7). Shale represents the low-permeability high-runoff end-member of the Kirkby and Chorley (1967) continuum, whereas weathered granite represents their high-permeability low-runoff end-member. Soil permeability for Sortman's basins is in some cases two orders of magnitude less than that reported for the granitic basins. Surface runoff is responsible for the observed patterns of drainage density on shale but apparently it is less important on granite. The difference between the results of Sortman and those of Levish demonstrate that different lithologies can produce substantially different relationships between climate and drainage density, as a result of different geomorphic and hydrologic processes. The permeable weathered-granite soil mantle produces a low and relatively constant active-channel drainage density through a considerable range of annual precipitation. This is in contrast to the shale basins, where the infiltration rate is low and runoff high. This conclusion is supported by Day's (1980) observations of the extent and duration of flow in channels on both granite and sedimentary rocks in Australia.

Eccker's study of two sedimentary formations further emphasises the importance of rock type, infiltration capacity, and erodibility. Climate alone therefore cannot be used as an index of drainage density. The linear relations obtained by Eccker differ from the plots of Figures 2.4 and 2.7, but perhaps her relations are one segment of a curve similar to that of Figure 2.4. Because of cooler temperatures and better vegetative cover in Colorado the peak of the curve would occur at lower values of precipitation, as compared to the Texas curve of Figure 2.4.

SITE SPECIFIC STUDIES

A different approach to the drainage density problem was taken by the staff of Water Engineering & Technology (1985) in their study of three open-pit coal mines. Because each mine site has its own geomorphic, lithologic and climatic characteristics, extrapolation of data from other locations is not appropriate. Therefore, reclamation of a site can best be attempted if the local conditions are known in some detail. Three mines were selected in order to determine whether site-specific studies could provide the information on drainage density that was required for post-mining rehabilitation of the landscape.

The three mine sites selected for the study were: the Dave Johnston Mine, 16 km north of Glenrock, Wyoming; the Jim Bridger Mine, 40 km north-east of Rock Springs, Wyoming; and the McKinley Mine, 25 km north-west of Gallup, New Mexico. These sites were selected as being representative of different physiographic regions with various soil and climatic conditions (Figure 2.9).

Data on drainage density were obtained from USGS topographic maps,

33

high-altitude infra-red photography, and USGS orthophoto quadrangles. Elevations were determined from the topographic maps, whereas drainage area and channel-length measurements were taken from orthophoto quadrangles and high-altitude photography.

The Dave Johnston Mine (Figure 2.9) is located on the drainage divide between the North Platte, Powder and Cheyenne Rivers. The interbedded sandstones and shales of the Wasatch Formation dip gently to the east. Average annual precipitation is about 380 mm. Natural vegetation consists of grasses and sagebrush.

The Jim Bridger Mine is located on the east flank of the Rock Springs Uplift (Figure 2.9). The Wasatch Formation dips gently to the east, and beds of shale and sandstone control topography in a manner similar to the Dave Johnston Mine area. The topography is more variable, however, ranging from flat stream valleys to steep gullied badlands. Average annual precipitation is about 196 mm. Vegetation consists of sparse grasses.

At the McKinley Mine (Figure 2.9) relief is greater than at the other two sites. The outcrops of Mesa Verde Formation are composed of sandstone and shale, and the sandstone forms steep slopes in the headwaters of small basins. Vegetation includes grasses, shrubs and some juniper and piñon pines. Average annual precipitation is about 343 mm.

At each location, drainage basins were selected for measurement so that the effects of climate, lithology, and other morphologic variables, except basin slope, were minimised. For example, drainage basins were selected with a similar aspect. In the vicinity of a given mine climate can be assumed to be constant, and the geology appears generally similar within the area being studied. Drainage basins selected range from 0.13 to 3.9 square km.

Schumm (1956: 612) concluded that in areas of homogeneous lithology and similar stage of development, drainage density is a function of the relief ratio (ratio of basin relief to basin length). This measure of average basin slope provides an estimate of the erosive potential of a basin. Thus, the relationship between drainage density and relief ratio is positive. Experimental studies confirm these relations (Mosley 1974; Schumm et al. 1987).

Data collected from nineteen watersheds in the vicinity of the Dave Johnston Mine show a poor positive correlation between drainage density and relief ratio (Figure 2.11), probably because of lithologic variability. Hence, Figure 2.11 would not be a useful guide for reclamation. However, it may be that an upper-limit line could be useful in estimating the maximum drainage density that would develop on reclaimed areas for a given relief ratio (Figure 2.11). This, of course, is a conservative approach to mined-land reclamation, whereas the selection of average values from the regression line would be more cost-effective.

Data collected from thirty-four drainage basins in the vicinity of the Jim Bridger Mine shows a stronger correlation between drainage density and relief ratio (Figure 2.12). The scatter is large and again, for purposes of

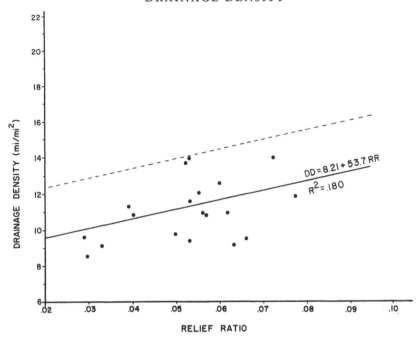

Figure 2.11 Drainage density as a function of relief ratio (ratio of drainage basin relief to length) in the vicinity of the Dave Johnston Mine (dashed line indicates maximum expected drainage density)

prediction, a worst-case maximum drainage density could be selected using the upper-limit line, or an average based upon the regression equations could be used.

Data from seventeen watersheds in the vicinity of the McKinley Mine show the best correlation between drainage density and relief ratio (Figure 2.13). The relief is greater in this area than at the other mine sites, and there is a larger range of relief ratio. In this case it would be appropriate to use the regression line to estimate an appropriate drainage density for the reclaimed landscape.

Summary

The data show a positive relation between drainage density and relief ratio. Additional study may explain the poor correlations, but it can be conjectured that local microclimatic and/or lithologic variability strongly influence the relations. In spite of the great care exercised in the selection of the drainage basins, the relations developed are disappointing. As stated previously, a key factor relating to the stability of a reclaimed area is whether an appropriate value of drainage density can be selected. If it is too low, channel extension

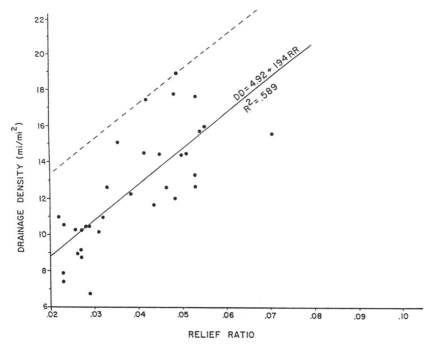

Figure 2.12 Drainage density as a function of relief ratio in the vicinity of the Jim Bridger Mine (dashed line indicates maximum expected drainage density)

and increased sediment yields will result. If it is too high, the channels may be stable, but too much money will be expended on the construction of the drainage network. Nevertheless, the relations of Figures 2.11, 2.12 and 2.13 provide a basis for estimating how drainage density will change at these mines if relief is altered during the reclamation process. For example, a 25 per cent decrease of relief ratio from .08 to .06 will cause a decrease of drainage density at the Dave Johnson Mine of about 14 per cent, a 20 per cent decrease at the Jim Bridger Mine, and a 12 per cent decrease at the McKinley Mine.

AN EXPERIMENTAL STUDY

In order to design a stable post-mining topography that blends into and complements the drainage pattern of the surrounding terrain, and that will minimise sediment yields, it should be designed using geomorphic principles (Stiller *et al.* 1980). Unfortunately, as demonstrated above, the variability of drainage density is great, which makes selection of an appropriate value difficult.

Experiments were performed to attempt to evaluate how changes of drainage density affect the stability of the drainage network and sediment production (O'Brien 1984). The experiments were performed in Rainfall

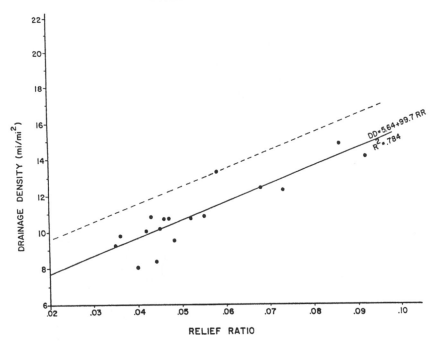

Figure 2.13 Drainage density as a function of relief ratio at McKinley Mine (dashed line indicates maximum drainage density)

Erosion Facility (REF) at Colorado State University, which is a 9.1 m by 15.2 m plywood box that was filled to an average depth of 1.5 m with sediment (median grain size = 0.2 mm). The actual watershed used to develop the drainage patterns for this study measured 7.1 m by 13.21 m (89.3 m² area). Simulated rainfall was applied to the surface from a set of seven sprinkler heads suspended from the ceiling. Precipitation was usually applied at the rate of 110 mm/hr.

After the initial emplacement and compaction of the sediment, the surface was graded to form two intersecting planes, which sloped towards the centreline at 0.7 per cent; the centreline sloped 3.5 per cent towards the outlet. After the surface was graded to the desired configuration, precipitation was applied, and a dendritic drainage pattern was permitted to develop (Figure 2.14a). Every two hours, the precipitation was stopped, the drainage network was mapped, and several cross-sections and the longitudinal profile of the main channel were measured. Application of precipitation was resumed, and the drainage network continued development. The run continued until the drainage network was visually judged to have ceased development (Figure 2.14a). At that time, a final set of data was collected, and the basin was prepared for the next run.

Two modifications were made to the original drainage pattern (Figure 2.14a) during the experiment. The first-order tributaries were obliterated by filling them with sediment (Run 2, Figure 2.14b), and all channels except the three main channels were obliterated (Run 3, Figure 2.14c). In addition, a third experiment was performed after the basin surface was graded to the original slope, and three artificial channels were cut into it (Run 4, Figure 2.14d). There was 2.5 cm of relief between the channels. In each case, precipitation was applied, and the experiment was continued until the drainage pattern was no longer adjusting.

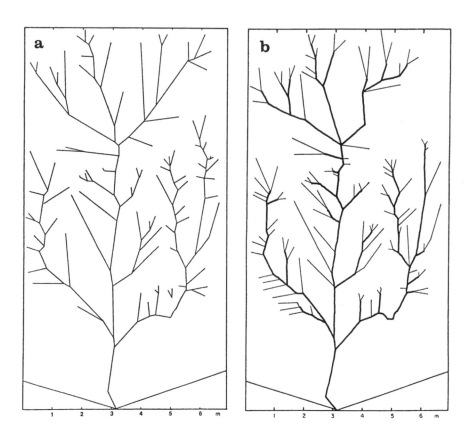

Figure 2.14 Drainage patterns resulting from O'Brien's (1984) experiments
(a) Final drainage pattern of unmodified network after Run 1
Source: O'Brien 1984

(b) Final drainage pattern of Run 2 after first order drainages were removed leaving only higher order channels (thick lines) of the Run 1 pattern
Source: O'Brien 1984

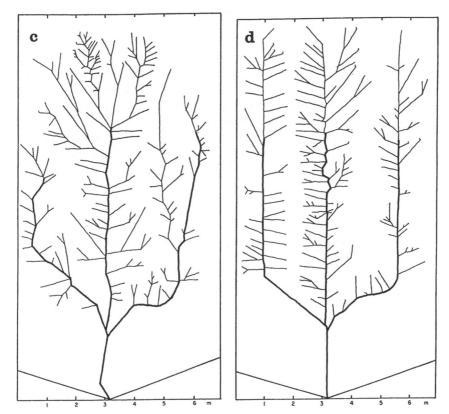

(c) Final drainage pattern of Run 3 after all channels except for the three main channels (thick lines) were removed from the Run 2 drainage pattern
Source: O'Brien 1984

(d) Final drainage pattern that developed from three artificial channels forming a 'pitchfork' pattern (thick lines) during Run 4
Source: O'Brien 1984

Results

A comparison of the unmodified drainage network (Figure 2.14a) with the drainage networks that developed, as a result of two modifications (Figures 2.14b, c), indicates that the modifications had a considerable effect upon drainage pattern development. When the precipitation was resumed, both patterns exhibited additional drainage network growth, which occurred primarily by the development of new tributaries. During Run 2 new first-order channels formed (Figure 2.14b). During Run 3 the network redevelopment was more elaborate, particularly in the upper basin (Figure 2.14c).

The drainage pattern that developed during Run 2 after obliteration of first-order tributaries had a slightly lower drainage density than the original pattern (Table 2.4). The drainage densities of the patterns that developed during Runs 3 and 4 were slightly greater than that of the unmodified pattern. For Run 4 this may be explained by the existence of the valley side slopes. The pattern of Figure 2.14d clearly shows this effect. Nevertheless, the results suggest that even when a drainage network reforms on identical materials, a plus or minus 14 per cent change of drainage density may occur.

Perhaps of more significance is the effect of the drainage pattern modification on sediment yield (Figure 2.15). The plot of sediment yield versus run time for Run 2, when the first-order tributaries were redeveloping, was equal to the sediment yield from the unmodified pattern (Run 1).

Table 2.4 Experimental data

Run number	Run time (hr)	Basin slope (%)	Drainage density (m/m²)	Mean sediment yield (g/s)
1	14	3.5	1.4	49
2	5	3.5	1.2	46
3	5.25	3.5	1.6	79
4	5.25	3.5	1.6	74

Source: O'Brien 1984

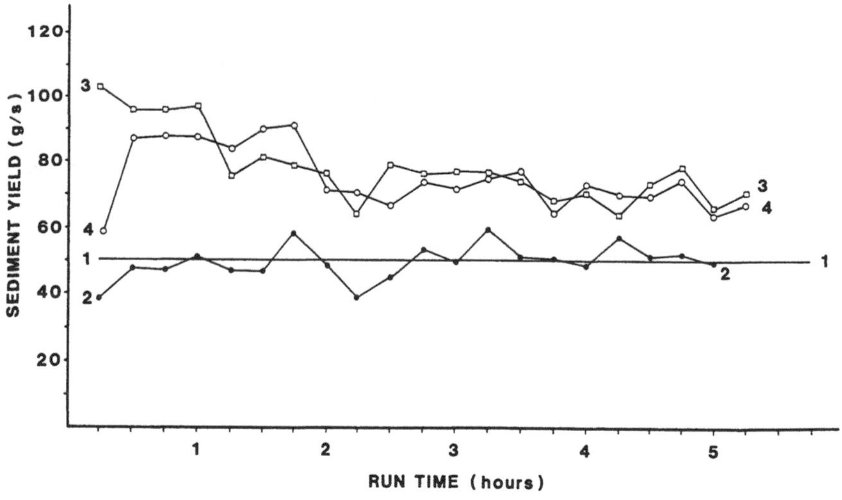

Figure 2.15 Sediment yield from drainage networks during Runs 1 through 4. Numbers indicate runs as listed on Table 2.4. The sediment yield for Run 1 is the average after full development of the drainage network
Source: O'Brien 1984

With this type of modification, most of the original drainage pattern was intact, and the regrowth of first-order channels produced little extra sediment that left the drainage basin. However, the adjustment to more significant modifications of the drainage pattern (removal of all channels except for the three main channels (Run 3) or replacement of the drainage pattern with three artificial channels (Run 4)), produced sediment yields that were higher than the sediment yield from the unmodified basin. This relationship was maintained even after tributary development ceased, but the sediment yields were declining slowly at the end of these runs (Figure 2.15).

Summary

The most important conclusion is that the most favourable drainage network modification was obliteration of first-order tributaries. The tributaries reformed, but sediment production did not significantly exceed that of the unmodified basin (Figure 2.15). This suggests that first-order channels can be eliminated from a reclaimed surface. If they form following reclamation their effects on sediment yields will be small. However, failure to replace the larger order channels will cause higher sediment yields as they develop. The greater the degree to which the drainage pattern is modified, or not restored, the greater will be the development of channels and sediment production. The results of the experiments suggest that selection of a maximum drainage density for reclamation (Figures 2.11, 2.12, 2.13) is not necessary because, even when the small-order tributaries form, the increase of sediment yields will be small (Figure 2.15).

CONCLUSIONS

The field and experimental studies reviewed here illustrate the problems associated with using climatic conditions, lithology or relief ratio to predict drainage density or vice versa. Nevertheless, Sortman's study (Figure 2.4) shows that, on readily eroded materials with low permeability, drainage density appears to reflect existing climate conditions in a way that supports the Gregory-Gardiner relation (Figure 2.1). On the other hand, rocks that weather to produce a soil with a high permeability produce drainage networks that are less closely linked to climate. Rather, limited surface runoff and the dominance of throughflow, which leads to groundwater sapping rather than surface erosion, permit little change of drainage density through a considerable range of climatic conditions (Figure 2.7). The valley network under these conditions nevertheless shows a progressive increase of drainage density as precipitation decreases (Figure 2.8). This relation resembles the right side of the Gregory-Gardiner curve (Figure 2.1).

In contrast to drainage basins formed on granite, the channels of the shale basins occupy most of the valley pattern so a comparison between shale active channels (Figure 2.4) and granite valley density (Figure 2.8) is reasonable. Obviously the shale basins will more readily reflect current climatic conditions than the granite basins because shale is much more responsive to change. Hence, it can be assumed that a large part of the valley pattern of the arid granite basins is a relic from a wetter climate. Therefore, the very different relations between drainage density and climate in the Texas and California studies suggest that perhaps past climatic conditions exert an important influence on lithologies that are resistant, because the drainage network cannot adjust readily to climate change.

What does drainage density actually represent on various lithologies? Many studies use contour-crenulation as a basis for obtaining drainage density, and it is assumed that this portrays the active channel drainage network. However, this assumption may not be true for granites and other rocks that produce soils of high permeability because the contour-crenulation drainage density may be the result of past climatic conditions.

With respect to reclaimed mines, O'Brien's (1984) study suggests that the stability of the imposed drainage pattern depends upon the degree to which the natural pattern is modified. The drainage density of a stable drainage pattern is dependent upon both the slope of the drainage basin (Figures 2.11, 2.12, 2.13), and the local side slopes immediately adjacent to the channel. Restoration of the original natural drainage pattern may therefore be inappropriate on surfaces that have undergone significant changes of relief ratio. Unless relief ratio has been significantly reduced, post-mining drainage networks, which have been restored with lower drainage densities than the natural pattern, will expand and develop additional drainage channels. If allowed to develop naturally, the exact position of these drainage channels will depend upon both the designed configuration of the surface and unplanned variations in microtopography. However, changes that affect only first-order drainages should have little effect on sediment yield, but more extensive changes of the drainage network may significantly increase sediment yields from the drainage basin (Figure 2.15).

This review has established what most of us intuitively know, that drainage density is not only a function of climate but also of lithology, relief, and perhaps past climatic conditions. Therefore, although a family of curves similar to Figures 2.1 and 2.4 may exist with erodible soils and rocks plotting high and resistant rocks plotting low, the form of the curves may also change (Figures 2.4, 2.7), as conditions change from the runoff-dominated to the throughflow-dominated end-members of the Kirkby and Chorley (1967) continuum.

ACKNOWLEDGEMENTS

I am very grateful for the careful reviews that were provided by Professor David Jorgensen and Professor W. Andrew Marcus. The research reported in this paper has been supported by several agencies as follows: Eccker by Office of Water Research and Technology (Project A-047-C020); Sortman and Levish by US Army Research Office (Project 24608-GS-VIR), and Water Engineering & Technology projects by the Office of Surface Mining.

REFERENCES

Abrahams, A.D. (1972) 'Drainage densities and sediment yields in eastern Australia', *Australian Geographical Studies* 10: 19–41.

——(1984) 'Channel networks: a geomorphological perspective', *Water Resources Research* 20: 161–188.

Abrahams, A.D. and Ponczynski, J.J. (1985) 'Drainage density in relation to precipitation intensity in the U.S.A.', *Journal of Hydrology* 75: 383–388.

Bass, N.W., Eby, J.B. and Campbell, M.R. (1955) 'Geology and mineral fuels of parts of Routt and Moffat Counties, Colorado', *U.S. Geological Survey Bulletin* 1027-D.

Büdel, J. (1982) *Climatic Geomorphology*, Princeton, N.J.: Princeton University Press.

Carlston, C.W. (1963) 'Drainage density and streamflow', *U.S. Geological Survey, Professional Paper* 422-C.

Chorley, R.J. (1957) 'Climate and morphometry', *Journal of Geology* 65: 628–638.

Chorley, R.J. and Dale, P.F. (1972) 'Cartographic problems in stream channel delineation', *Cartography* 7: 150–162.

Chorley, R.J. and Morgan, M.A. (1962) 'Comparison of morphometric features, Unaka Mountains, Tennessee and North Carolina and Dartmoor, England', *Geological Society of America Bulletin* 73: 17–34.

Chorley, R.J., Schumm, S.A. and Sugden, D.E. (1984) *Geomorphology*, London: Methuen.

Cooks, Johannes (1983) 'Geomorphic response to rock strength and elasticity', *Zeitschrift für Geomorphologie* N.F. 27: 483–493.

Daniel, J.R.K. (1981) 'Drainage density as an index of climatic geomorphology', *Journal of Hydrology* 50: 147–154.

Day, D.G. (1980) 'Lithologic controls of drainage density: a study of six small rural catchments in New England, N.S.W', *Catena* 9 v. 7: 339–351.

Eccker, S.L. (1984) 'The effects of lithology and climate on the morphology of drainage basins in northwestern Colorado', unpublished MS thesis, Colorado State University.

Gardiner, V. (1983) 'Drainage networks and palaeohydrology', in K.J. Gregory (ed.) *Background to Palaeohydrology*, New York: Wiley & Sons.

Gardiner, V. and Gregory, K.J. (1982) 'Drainage density in rainfall–runoff modeling', in V. P. Singh (ed.) *Rainfall–Runoff Relationships*, Littleton Colo.: Water Resources Publications.

Gregory, K.J. (1966) 'Dry valleys and the composition of the drainage net', *Journal of Hydrology* 4: 327–340.

——(1976) 'Drainage networks and climate', in E. Derbyshire (ed.) *Geomorphology and Climate*, New York: Wiley.

Gregory, K.J. and Gardiner, V. (1975) 'Drainage density and climate', *Zeitschrift für Geomorphologie* N.F. 19: 287–298.

Hadley, R.F. and Schumm, S.A. (1961) 'Sediment sources and drainage-basin characteristics in Upper Cheyenne River basin', *U.S. Geological Survey Water Supply Paper* 1531-B: 137–198.

43

Horton, R.E. (1945) 'Erosional development of streams and their drainage basins; hydrophysical approach to quantitative morphology', *Geological Society of America Bulletin* 56: 275–370.

Kirkby, M.J. and Chorley, R.J. (1967) 'Through flow, overland flow and erosion', *Bulletin of the International Association of Sci. Hydrol.* 12: 5–21.

Langbein, W.B. and Schumm, S.A. (1958) 'Yield of sediment in relation to mean annual precipitation', *Transactions of the American Geophysical Union* 39: 1076–1084.

Levish, D. (1992) 'The effect of climate on the morphology of granitic drainage basins, California', unpublished MS thesis, Colorado State University.

Melton, M.A. (1957) 'An analysis of the relationships among elements of climate, surface properties and geomorphology', *Office Naval Research Technical Report* 11.

——(1958) 'Geometric properties of mature drainage systems and their representation in E4 phase space', *Journal of Geology*, 66: 35–54.

Morisawa, M.E. (1957) 'Accuracy of determination of stream lengths from topographic maps', *Transactions of the American Geophysical Union* 38: 86–88.

Mosley, M.P. (1974) 'An experimental study of rill erosion', *American Society of Agricultural Engineers, Transactions* 68: 909–916.

O'Brien, M.J. (1984) 'An experimental study of drainage network modifications', unpublished MS thesis, Colorado State University.

Patton, P.C. (1988) 'Drainage basin morphometry and floods', in V.C. Baker, R.C. Kochel and P.C. Patton, *Flood Geomorphology*, New York: Wiley & Son.

Peltier, L.C. (1962) 'Area sampling for terrain analysis', *Prof. Geog.*, 14: 24–27.

——(1975) 'The concept of climatic geomorphology', in W. Melhorn and R. Flemal (eds) *Theories of Landform Development*, New York: Publications in Geomorphology, State University of New York, Binghamton.

Rantz, S.E. (1970) 'Runoff characteristics of California streams', *U.S. Geological Survey Water Supply Paper* 2009-A.

Reeside, J.B., Jr (1924) 'Upper Cretaceous and Tertiary formations of the western part of the San Juan Basin of Colorado and New Mexico', *U.S. Geological Survey Professional Paper* 134.

Ruhe, R.V. (1950) 'Graphic analysis of drift topography', *American Journal of Science* 248: 435–443.

Schumm, S.A. (1956) 'The evolution of drainage systems and slopes in badlands at Perth Amboy, New Jersey', *Geological Society of America Bulletin* 67: 597–646.

——(1965) 'Quaternary paleohydrology', in H.E. Wright and D.G. Frey (eds) *The Quaternary of the United States*, Princeton N.J.: Princeton University Press.

——(1969) 'A geomorphic approach to erosion control in semiarid regions', *American Society of Agricultural Engineers, Transactions* 12: 60–68.

Schumm, S.A., Mosley, M.P. and Weaver, W.E. (1987) *Experimental Fluvial Geomorphology*, New York: Wiley.

Sortman, V.L. (1988) 'The effect of climate on drainage network morphometry in central Texas', unpublished MS thesis, Colorado State University.

Stiller, D.M., Zimpfer, G.L. and Bishop, M. (1980) 'Application of geomorphic principles to surface mine reclamation in the semiarid west', *Journal of Soil and Water Conservation* 35: 274–277.

Stoddart, D.R. (1969) 'Climatic geomorphology: review and re-assessment', *Progress in Physical Geography* 1: 159–222.

Strahler, A.N. (1964) 'Quantitative geomorphology of drainage basins and channel networks', in V.T. Chow (ed.) *Handbook of Applied Hydrology*, San Francisco: McGraw-Hill.

Toy, T.J. (1977) 'Hillslope form and climate', *Geological Society of America Bulletin* 88: 16–22.

Toy, T.L. and Hadley, R.F. (1987) *Geomorphology and Reclamation of Disturbed Lands*, Orlando: Academic Press.

Tricart, J. and Cailleux, A. (1972) *Introduction to Climatic Geomorphology*, New York: St. Martin's Press.

Waananen, A.O. (1973) 'Floods from small drainage areas in California', *U.S. Geological Survey, Water Resources Division, Open File Report*.

Waananen, A.O. and Crippen, J.R. (1977) 'Magnitude and frequency of floods in California', *U.S. Geological Survey, Water Resources Investigation* 77-21.

Water Engineering & Technology (1985) 'Determination of drainage density for surface-mine reclamation in the western U.S.', unpublished report to Office of Surface Mining, Denver.

Williams, R.E. and Fowler, P.M. (1969) 'A preliminary report on an empirical analysis of drainage network adjustment of precipitation input', *Hydrology* 8: 227–238.

3

THE UNDERFIT MEANDER PROBLEM

Loose ends

George H. Dury

INTRODUCTION

My chapter title could equally well have been 'Unfinished business', except that this has become so over-used as to be hackneyed, and that anyway I propose to deal with questions that still remain open. When I served on Ph.D. committees at the University of Wisconsin-Madison, the final question to the aspirant was usually: If we pass you, what will you do next? Were the response 'More of the same', or words to that effect, the committee's collective heart would sink. To the credit of the aspirants, such a response was rarely forthcoming. It was usual for a successful dissertation to raise at least as many questions as it answered: that is to say, to point the way towards further fruitful enquiries. Loose ends were left dangling. I am encouraged by this circumstance to suppress my disappointment in what I have so far failed to do, and to suggest some directions in which future research might be directed.

My central research interest, during the last fifty years or so, has been underfit streams. We begin with manifest underfits – meandering streams contained within the far larger and usually ingrown meanders of winding valleys. In order to understand these, it seems essential to know the cause of meandering in the first place. But although a great deal has come to be known about the fluvial mechanics of meanders, and especially about controls on their essential shape, the fundamental cause of meandering still proves elusive (for multiple discussions see Elliot 1984). One hope for the future must surely be that that cause will be identified.

A parallel question to why so many rivers meander is the question why so many rivers, contained in meandering valleys, fail to meander. These are examples of the Osage-type underfit (Dury 1966), where pools and riffles are appropriately spaced at five to six bedwidths, but the channel does not swing from side to side. So far as I am aware, only one such stream has yet been analysed in detail – the Severn near Shrewsbury, where shear stress is

46

sufficient to deform the bed but not to deform the banks (Dury 1983). Many more tests are obviously required before this finding can be promoted to the status of a general hypothesis. On the other hand, Osage-type underfits do, by demonstration, deform their beds but not their banks.

The idea that, in the general case, manifest underfitness results from stream capture has long been laid to rest. The fact that we are dealing with climatically induced change in channel-forming discharge (although numbers of complicating factors, such as vegetation cover, can readily be envisaged), and that channel-forming discharge is $q_{1.58}$ on the annual series, namely, the most probable annual flood, now seem to be widely accepted (for possible complications, see Dury 1984). The early work on manifest underfits can, however, be seen as somewhat defective, concentrating as it did upon the disparity between the large meanders of ingrown valleys and the small meanders of existing streams. To some extent the defect here concerned relates back to the Davisian model of large meanders (supposed pre-capture) and small meanders (supposed post-capture). In another respect, however, the defect relates to the fact that the early work was performed on rivers of lowland England, where only the large former channels and the present sunken channels are readily identifiable. Among the earliest examples, only on the Cotswold Dorn is there a hint of a channel intermediate in size between the largest and smallest known (Dury 1958: Figure 6). For the alluvial fill in the valley of the Willow Brook in Northamptonshire, Sparks and Lambert (1961) identified disturbance, presumably erosional, in the bracket corresponding to the (mainland European) pollen zones VI and VII, namely Early Atlantic and Main Atlantic, that is, somewhere in the approximate range 7500–5500 BP. As will be seen, work on the European mainland has greatly elaborated the known sequence of episodes of high discharge subsequent to the last maximum.

In what now follows, all references to published works should be understood to imply *and references therein*.

THE LATEST DISCHARGE MAXIMUM AND ITS AFTERMATH

The dimensions of manifestly underfit streams that are the simplest to measure are the wavelengths of valley meanders and of stream meanders, averaged over the length of trains. It seemed an attractive proposition to appeal to the general equation $\ell \propto q^{0.55}$, where ℓ is wavelength and q is channel-forming discharge, and so obtain $Q/q = L/\ell^{1.82}$, where Q is former discharge, q is present discharge, and L, ℓ are respectively former and present meander wavelengths. Where, as is frequently the case, L/ℓ is about 10.0, this reckoning gives a value for former discharge about 50 to 60 times the present value. When, however, account is taken of confidence limits, it can be shown that, at the 95% confidence level, the retrodicted discharge lies somewhere between 15 and 655% of observed values (Dury 1985). Things are not much better with retrodiction from bedwidth (23 and 428%), which in any event can be difficult to determine for former channels. It seems likely that the dimensional equations retrodict discharges that are too great. Rotnicki (1991), in a careful study of the Prosna River in central Poland, for which the question of former roughness appears to have been resolved, concludes that the mean annual discharges in the Late Glacial were 3.6–6.5 times as great as they are today, and that they were greatest at the beginning of the Allerød climatic phase. For this phase, he finds that bankfull discharge was about five times mean annual discharge (his Table 20.5). In round values, then, former maximal bankfull discharge would appear to have been some 30 times present bankfull, comfortably within the 95% confidence limit of retrodiction from wavelength ratio, but still outside the 50% limit.

This information is needed here, because part of the future possible work to be suggested concerns the whole matter of the ingrowth of meandering valleys. There seems to be no evidence to suggest that the streams responsible for ingrowth were any larger, or any smaller, than the streams that flowed during the latest (i.e. Allerød) episode of maximal discharge. On the contrary; where former channel traces can be reconstructed, as for instance on the Warwickshire Avon and the middle Mosel, the dimensions of former (valley) meanders accord precisely with the dimensions of the ingrown bends.

The matter does not, however, appear to be as simple as a mere alternation between former maximal discharge, such as could have been sustained by mean annual precipitation twice that of today, and discharge at present levels. On the contrary, some of the work on the European mainland abundantly demonstrates repeated episodes of high former discharge, distinctly greater than that of today although less than that of the Allerød maximum. Thus, Rotnicki (1991) identifies for the middle Prosna River discharge peaks in the Younger Dryas phase (82% of the Allerød value: read off from his Figure 20.14, see also his Table 20.5), the pre-Boreal phase (62%), the Atlantic phase (53%), the sub-Boreal phase (31%) and the sub-Atlantic phase (21%). Bohncke and Vandenberghe (1991) also identify for the southern Netherlands five lesser episodes of high discharge subsequent to the latest maximum, but their proposed sequence does not appear to match well with that of Rotnicki, in terms of either radiocarbon dates, climatic phases, or spacing in time, except that both sequences involve a progressive decrease in discharge from one peak to the next. One can speculate here whether the quasi-cyclic representation of discharge variation in both papers might not be better shown as step-functional. The meticulous overview of Starkel (1991) suggests that the long-distance correlation may pose obdurate problems, notwithstanding that the Allerød climatic and hydrologic event in Europe may well have been matched in North America (see Broecker *et al.* 1960), due allowance being made for what is now known about the value of early radiocarbon dates. We may perhaps wonder whether episodes of maximal bankfull discharge, prior to the last, were not also followed by episodes of increased but still sub-maximal discharge. One thing does seem clear: that the schema of deep-sea stages is coarser than the post-Allerød record of fluvial activity.

MODELS AND INTERRELATIONSHIPS

The first full length discussion of underfit streams set up hypothetical models of underfitness (Dury 1964: Figure 4). Of the eight cases hypothesised, four have been identified in the field. The first and third are those of manifest underfits, with the large meanders ingrown, or with these contained in an open trough, as on the upper Evenlode (Dury 1958; other examples are cited in Dury 1964). The remaining four cases seem likely to be uncommon, if indeed they exist at all. As will be shown, it is possible to set up corresponding models for the development of ingrown meandering valleys, taking into account rates of incision and the relative rates of incision and the enlargement of bends.

The general problem here is that of dating. When work on underfits first began in England, pollen analysis failed to prove especially helpful, malaco- logical analysis was only slightly developed, and radiocarbon dating was effectively unavailable. Early work in the United States was also handicapped by a lack of dating techniques, particularly those concerning fossil pollen. It is now clear that radiocarbon dating is incapable of reaching back to the early episodes of meander ingrowth. Available techniques include uranium-series, thermoluminescence, and electron spin resonance dating, but these, as Bowen (1991) observes, have not yet been widely applied in Britain. The current main prospect of a general dating schema seems to reside in the oxygen isotope deep-sea stages identified by Shackleton and Opdyke (1973, 1976), namely, 22 stages of alternating high and low Northern Hemisphere ice volume during the last approximate 800,000 years. Numbers of workers on Pleistocene chronology, and in particular on alternating cut and fill within meandering valleys, hold that amino-acid geochronology can be used to correlate strandline deposits with the deep-sea stages, and fluvial deposits with those of the strandline, so that events in fluvial history can in effect be referred to appropriate deep-sea stages (Bowen et al. 1989; Bowen 1991). Some workers, however, remain sceptical, on the principle that the equatorial data used to define the deep-sea stages may not accurately reflect the fluvial events of middle latitudes (see for example West 1990, 1991). Field examples wherein fluvial events are correlated with deep-sea stages will be mentioned presently.

Simple models for the interrelationship of incision rate and the passage of time are presented in Figure 3.1. As shown, the rate of incision is conceived of as

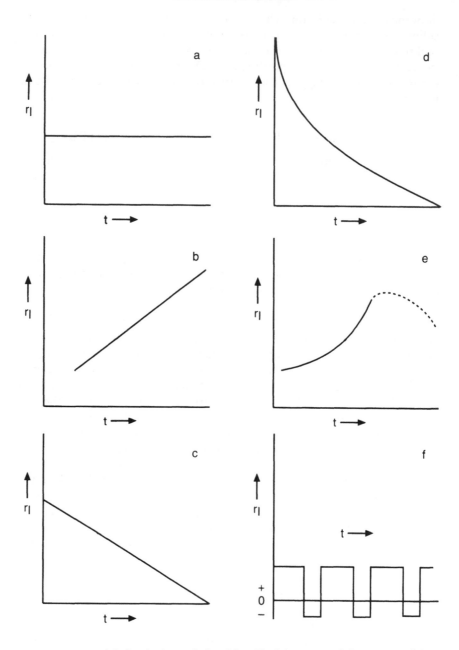

Figure 3.1 Models for the interrelationship of incision rate and the passage of time (for explanation, see text)

unchanging through time (Case *a*), linearly increasing (Case *b*), linearly decreasing towards an eventual zero value (Case *c*), exponentially decreasing, again becoming zero (Case *d*), and exponentially increasing (Case *e*). There must sooner or later be a reversal of this curve, as approach to base level will retard the rate of downcutting. In the meantime, the rising limb in Case *e* will be illustrated from the Shoalhaven River in New South Wales, Australia. It seems possible that a second instance is that of the well-known Goosenecks reach on the San Juan River, Utah. The final Case, *f*, is the only one where allowance is made for alternating cut and fill, although the other hypothetical cases could obviously be elaborated in similar fashion. The long-term effect for Case *f* is one of net incision, as will be illustrated for the Warwickshire Avon.

For the Shoalhaven River, J.F. Nott has obtained a series of thermolumi-nescence (TL) dates (kindly relayed by R.W. Young 1990) on terrace materials. They indicate an exponential increase in the incision rate, from 0.00054 mm/a at the beginning of the series, to 0.16 and 0.15 mm/a at the end. The two final values could possibly contain a hint that the rate of incision has ceased to increase. In Figure 3.2, rates of incision for five time brackets are

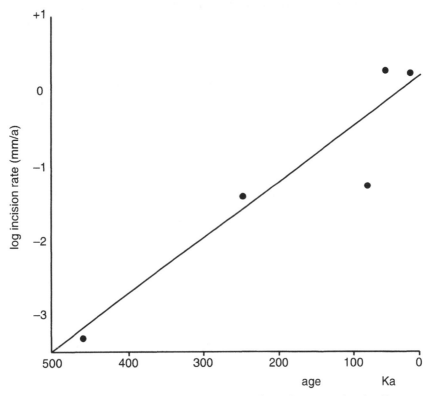

Figure 3.2 Exponential increase in incision rate through time on the Shoalhaven River, New South Wales, Australia

51

plotted against the terminal dates of each event. Regression analysis shows that $r = +0.935$, with $r^2 = 0.874$, and $0.02 > P$.

In a study of the Warwickshire Avon, Maddy et al. (1991) employ amino-acid analysis, among other means, to revise the terrace sequence into Terraces 5, 4, 3 and 2, in descending order of age and of height above the alluvium. For these four terraces, the history is one of increasing cold during aggradation. The sequence can be extended to Terrace 1, with the aid of the table presented by Barclay et al. (1992: Table 1), while a dummy terrace, 0, represented by the latest maximal channel, may be entered for the Allerød climatic phase. Using the height data for terrace tops near Evesham given by Tomlinson (1925), and the central values of possible age ranges kindly communicated by Maddy (1991), plus the data for Terraces 1 and 0, one can draw up a table (Table 3.1) of terrace top height against age (Tomlinson's non-metric height measurements have been retained for convenience).

The data indicate a linear rate of downcutting through time (Figure 3.3). The correlation between height and age is $+0.990$, with $r^2 = 0.980$, and $0.05 > P$. The net rate of downcutting is 0.162 m/ka, or 0.162 mm/a, similar to the two final rates determined for the Shoalhaven.

On the rough assumption that two-thirds of the lapsed time was accounted for by cutting, and one-third by filling, then the effective rate of cutting becomes about 0.243 m/ka = 0.243 mm/a.

Figure 3.4 presents hypothetical models of the relationship between increase in incision and increase in sinuosity. Case a is one end-member of the series, with no incision and no increase in sinuosity. It is readily illustrated from the Hungarian Plain (Lacsay 1977), where the effect of cutoff on river length is neatly balanced by the enlargement of existing loops and/or the development of new ones. Case b is the other end-member, with no change in sinuosity as incision progresses. If the initial value of sinuosity is 1.0, then the resulting feature will be a rectilinear slot. If the initial sinuosity value is, say, 1.57, then the result will be a train of vertically incised meanders; but despite the many references to entrenchment that extend back through the literature for well over half a century, this writer has yet to observe a truly entrenched meandering stream in the field. Case c, where sinuosity increases

Table 3.1 Terrace top heights and ages for the Warwickshire Avon near Evesham

Terrace	Top height (ft)	Age (ka)
5	150	295
4	90	191
3	50	130
2	40	65
1	10	25
0	0	12.5

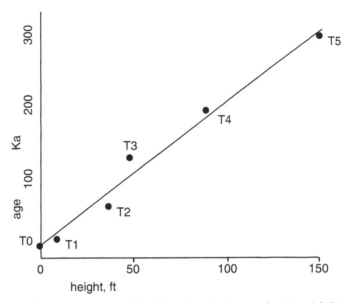

Figure 3.3 Constant net rate of incision through time on the Warwickshire Avon

in linear fashion as incision progresses, could be considered potentially widespread in nature, although sooner or later increasing sinuosity must lead to cutoff. The converse, Case *d*, where sinuosity progressively decreases (linearly or otherwise) may be considered highly improbable or even impossible. Case *e*, where the rate of increase in sinuosity falls off as incision progresses, is exemplified by the Goosenecks of the San Juan River, which, although they have been frequently described as entrenched, contain obvious slip-off slopes at the highest levels. Case *f*, with accelerating increase in sinuosity as incision progresses, appears likely, like Case *c*, to be widespread in nature, although once again a limit to the increase in sinuosity must be imposed by cutoff.

As yet, it has not been proved feasible to make dated reconstructions of past sinuosities to test the relative worth of Cases *c* and *e* in Figure 3.4. There does, however, appear to have been a widespread tendency for sinuosity to increase with increasing incision, such as is indeed implied in the term 'ingrown meanders'. The instance of the Warwickshire Avon is bedevilled by difficulties of Pleistocene chronology, and in particular by the correlation or even the existence of the Wolstonian glacial stage (see Ehlers and Gibbard 1991). Even when the Avon terraces are correlated with deep-sea stages, discrepancies occur between the ascriptions made by Maddy *et al.* (1991), and those made by Barclay *et al.* (1992). This writer's lack of success in tracking down comparable data for the middle Mosel, if such exist, is all the more regrettable, since the original work (Kremer 1954) lists seven terrace

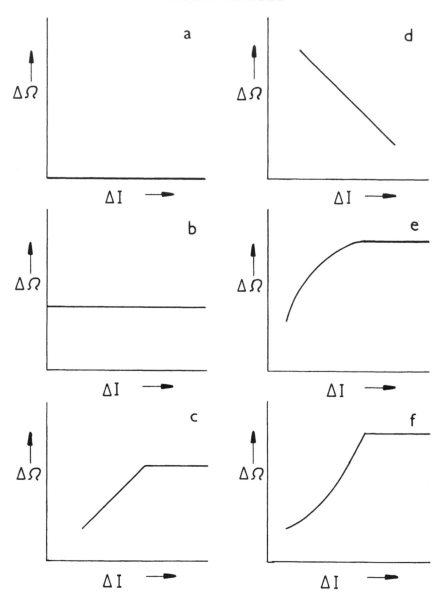

Figure 3.4 Models for the interrelationship of increase in incision and increase in sinuosity (for discussion, see text)

levels, or eight when allowance is made for the partial subdivision of the lowest terrace. If a dummy Terrace 0 is added, as for the Warwickshire Avon, and if the highest Mosel terrace is sunk at all below plateau top level, then the tally becomes 10, suggesting that the history of incision of the middle

Mosel may span a very large part of the last 800,000 years.

The increase in sinuosity with increasing incision can be conceptually reversed into decreasing sinuosity with decreasing incision, a fact that raises the whole question of the value of sinuosity when incision first began. The antique hypothesis that valley meanders are typically inherited from some unconformable cover, or alternatively from deep regolith on some planation surface, deserves to be abandoned, until such time as convincing instances can be adduced. The well-studied meandering gullies on spoil heaps in the old cement workings near Bishop's Itchington, Warwickshire, show conclusively that the gullies began as straight rills, and have developed their sinuosity during incision. Parallel cases are known from similar situations elsewhere. An admittedly crude analysis of ingrown valley meanders on natural streams, that uses relative depth of incision as a surrogate for dating, determining the sinuosity at 25, 50 and 100% of the depth of incision, suggests that, in all the small number of instances examined, sinuosity of the trains of ingrown meanders at plateau top level was 1.0: that is, that the former streams were initially straight (Dury 1969). (The title of this paper, 'Tidal stream action and valley meanders', appears in retrospect to be ill conceived: the paper is in actuality an attack on the peculiar notion that valley meanders were cut by tidal currents.)

SOME WIDER CONSIDERATIONS

If the findings of Maddy *et al.* (1991) for the Warwickshire Avon, namely, that infilling occurred during increasing cold, can be construed to imply that cutting occurred during increasing or increased warmth, and if, further, cutting is to be associated with increased discharge, as might be expected on hydrological grounds, then the analyses of Shackleton and Opdyke (1973, 1976) can be taken to imply eleven episodes of increasing/increased warmth, maximal stream discharge, and downcutting, during the last approximate 800,000 years, giving an average spacing between episodes of about 70,000 years. The average interval for the Avon Terraces is about 54.5 ka, while the actual intervals can be shown (by χ^2 test) to differ significantly from their common mean. That is to say, we are not dealing with a simple rhythm. No such rhythm would in fact be expected, if the deep-sea record reflects the effects of orbital perturbations. Furthermore, the Holocene record of stream discharge established in Europe suggests that something additional to those perturbations is in question.

The correlation of discharge fluctuations in middle latitudes with those in lower latitudes promises to be made in Australia, with its considerable range of climates and of latitude – an approximate 31° latitudinal range when Tasmania is included. There can be no doubt about the existence of underfit streams in humid Australian areas, or of the existence of former pluvial lakes in semi-arid to arid areas, any more than of the existence of formerly integrated drainage, now represented by discontinuous remnants, in the western half of the continent. It would be tempting to assume that the former large streams were reduced, and the former lakes dried out, at the same time. Evidence so far available, however, suggests that this may not have been so, and also that the record which is being established by TL and uranium-series dating is as yet not easy to match with the sequence of deep-sea stages.

The TL dates obtained by Nott for the Shoalhaven River indicate episodes of cutting that correspond in time to deep-sea stages 12 (possibly early in this stage), 7 (also possibly early), 5 (late), 3 (possibly mid-stage) and 2. Only three of the five episodes correspond to parts of odd-numbered stages. It could perhaps be that the Shoalhaven record is incomplete because of removal of some terrace material. Page *et al.* (1991), dealing with the TL dating of Quaternary sediments on the Riverine Plain of southeastern Australia, identify an episode of major fluvial activity from about 100 to 80 ka, which corresponds in time to (possibly part of) deep-sea stage 5, renewed fluvial activity in the north of the region from about 50 to 40 ka (corresponding to part of deep-sea stage 3), then a transition to mixed-load channels with discharges distinctly in excess of those of today, but still to be dated. Nanson *et al.* (1991) make comparative TL and uranium-series determinations on the deltaic sediments of the Gilbert River, Queensland, where an extensive sand body older than 85 ka and probably dating from about 120 ka represents a period of major fluvial activity not repeated subsequently. The sandy sediment corresponds in age to part of deep-sea stage 5. But the following episode of mud deposition, 110–50 ka, corresponds in age to the later part of deep-sea stage 5, all of stage 4, and the earlier part of stage 3. The next following deposit of muddy sand is dated at 50–40 ka, corresponding to part of deep-sea stage 3, while the final muddy deposit dated from 40 to 0 ka corresponds in age to the later part of deep-sea stage 3, plus all of stages 2 and 1. As so often happens, more information is clearly needed.

Again in the broad view, work in Australia may in time reveal how far underfit streams extend inside the tropic. They are already known to reach within the northern tropic. Brazil seems to offer a good prospect of discovering how nearly they approach the equator.

SITES POSSIBLY DESERVING EXPLORATION

This section is a kind of end note: it will name sites where subsurface exploration could possibly illuminate the chronology of underfit streams in the UK.

The inner valley of the Severn at Shrewsbury includes a cutoff valley bend (grid reference SJ 495150). The cross section presented by Morey and Pannett (1976) suggests the familiar combination of large former channel and much smaller later channel, although further drilling is desirable for the purpose of defining the precise form of the latter. The large channel is filled with silt, the small channel with silt, almost free of organic matter, at the base, and then with peat in which tree pollen, especially that of alder, is prominent. The silt in the small channel is taken to have accumulated in the cutoff after the ends had been dammed by levees. The existence of the smaller channel, which goes as deep as the large channel, suggests that cutoff occurred after the shrinkage to underfitness, that is, in post-Allerød and quite possibly post-Younger Dryas times.

In what now follows, sheet numbers and abbreviated (four-figure) grid references are cited for the Ordnance Survey 1:50000 Series. Some of the sites extend more widely than a 1 km^2 grid square, but the references should serve to locate them.

On the Evenlode upstream of the gorge (Sheet 163: SP 2719; Dury 1958), large former meanders that are not ingrown are cut into self-cemented limestone gravel that has been interpreted as periglacial sludge. As usual, the large former channel is present. The gravel is penetrable by screw auger. Downstream of the minor road bridge that is located roughly in the centre of the grid square cited, this gravel is in places underlain by peat, for which pollen analysis might prove profitable. The cutoff valley meander on the right bank of the Windrush at Asthall (Sheet 163: SO 2811) could be worth drilling out. The high-level cutoff valley meanders on the left side of the lower Wye at Redbrook and St Briavels (Sheet 162: SO 5094 and SO 5404) may not be particularly accessible, but could also be worth exploring. Finally, the Ancaster Gap in Lincoln Edge (Sheet 130: SK 9844) has been interpreted as a remnant of the former course of the proto-Trent. What, if any, fluvial deposits in the Gap have survived glaciation cannot of course be predicted, and exploratory work might be handicapped by the results of road and rail construction. Nevertheless, this site might perhaps prove significant indeed.

ACKNOWLEDGEMENTS

I am grateful to a large number of correspondents for their discussions of the ways in which fluvial morphology might move in the near future, and in particular, in the context of this paper, to D.Q. Bowen, D.J. Pannett, D. Maddy, R.G. West and R.W. Young for information and for the supply of reprints of pertinent papers.

REFERENCES

Barclay, W.J., Brandon, A., Ellison, R.A. and Moorlock, B.S.P. (1992) 'A Middle Pleistocene palaeovalley-fill west of the Malvern Hills', *Journal of the Geological Society of London* 149: 75–92.
Bohncke, S.J.P. and Vandenberghe, J. (1991) 'Palaeohydrological development in the southern Netherlands during the last 15,000 years', in L. Starkel, K.J. Gregory and J.B. Thornes (eds) *Temperate Palaeohydrology*, Chichester: John Wiley & Sons.

Bowen, D.Q. (1991) 'Time and space in the glacial sediment systems of the British Isles', in P. Ehlers, P.L. Gibbard and J. Rose (eds) *Glacial Deposits in Great Britain and Ireland*, Rotterdam: A.A. Balkema.

Bowen, D.Q., Hughes, S., Sykes, G.A. and Miller, G.H. (1989) 'Land–sea correlations in the Pleistocene based on isoleucine epimerization in non-marine molluscs', *Nature* 340: 49–51.

Broecker, W.S., Ewing, M. and Heezen, B.C. (1960) 'Evidence for an abrupt change of climate close to 11,000 years ago', *American Journal of Science* 258: 429–448.

Dury, G.H. (1958) 'Tests of a general theory of misfit streams', *Transactions of the Institute of British Geographers* 25: 105–118.

——(1964) 'Principles of underfit streams', *U.S. Geological Survey Professional Paper* 452-A.

——(1966) 'Incised valley meanders on the lower Colo River, New South Wales', *Australian Geographer* 10: 17–25.

——(1969) 'Tidal stream action and valley meanders', *Australian Geographical Studies* 7: 49–56.

——(1983) 'Osage-type underfitness on the River Severn near Shrewsbury, Shropshire, England', in K.J. Gregory (ed.) *Background to Palaeohydrology*, Chichester: John Wiley & Sons.

——(1984) 'Bankfull discharge through pool and riffle', in C.M. Elliott (ed.) *River Meandering*, New York: American Society of Civil Engineers.

——(1985) 'Attainable standards of accuracy in the retrodiction of palaeodischarge from channel dimensions', *Earth Surface Processes and Landforms* 10: 205–213.

Ehlers, J. and Gibbard, P.L. (1991) 'Anglian deposits in Britain and the adjoining offshore regions', in J. Ehlers, P.L. Gibbard and J. Rose (eds) *Glacial Deposits in Great Britain and Ireland*, Rotterdam; A.A. Balkema.

Elliott, C.M. (ed.) (1984) *River Meandering*, New York: American Society of Civil Engineers.

Kremer, E. (1954) 'Die Terrassenlandschaft der mittleren Mosel als Beitrag zur Quartär-geschichte', *Arbeit zur Rhein Landeskunde* 6: 1–100.

Lacsay, I.A. (1977) 'Channel changes of Hungarian rivers: the example of the Hernád River', in K.J. Gregory (ed.) *River Channel Changes*, Chichester: John Wiley & Sons.

Maddy, D. (1991) personal communication.

Maddy, D., Keen, D.H., Bridgland, D.R. and Green, C.P. (1991) 'A revised model for the Pleistocene development of the River Avon, Warwickshire', *Journal of the Geological Society of London* 148: 473–484.

Morey, C. and Pannett, D. (1976) 'The origin of the old river bed at Shrewsbury', *Shropshire Conservation Trust Bulletin* 35: 7–12.

Nanson, G.C., Price, D.M., Short, S.A., Young, R.W. and Jones, B.G. (1991) 'Comparative uranium-thorium and thermoluminescence dating of weathered Quaternary alluvium in the tropics of northern Australia', *Quaternary Research* 35: 347–366.

Page, K.J., Nanson, G.C. and Price, D.M. (1991) 'Thermoluminescence chronology of late Quaternary deposition on the Riverine Plain of south-eastern Australia', *Australian Geographer* 22: 14–23.

Rotnicki, K. (1991) 'Retrodiction of palaeodischarges of meandering and sinuous alluvial rivers and its palaeohydroclimatic implications', in L. Starkel, K.J. Gregory and J.B. Thornes (eds) *Temperate Palaeohydrology*, Chichester: John Wiley.

Shackleton, N.J. and Opdyke, N.D. (1973) 'Oxygen-isotope and palaeomagnetic stratigraphy of equatorial Pacific core V28-238', *Quaternary Research* 3: 30–55.

——(1976) 'Oxygen-isotope and palaeomagnetic stratigraphy of Pacific core

V28-239', *Geological Society of America Memoir* 145: 449–463.

Sparks, B.W. and Lambert, C.A. (1961) 'The postglacial deposits at Apethorpe, Northamptonshire', *Proceedings of the Malacological Society of London* 34: 302–315.

Starkel, L. (1991) 'Long-distance correlation of fluvial events in the temperate zone', in L. Starkel, K.J. Gregory and J.B. Thornes (eds) *Temperate Palaeohydrology*, Chichester: John Wiley.

Tomlinson, M.E. (1925) 'River terraces of the lower valley of the Warwickshire Avon', *Quarterly Journal of the Geological Society of London* 81: 137–169.

West, R.G. (1990) 'Global palaeoclimate of the Late Cenozoic', *Boreas* 19: 312.

——(1991) personal communication.

Young, R.W. (1990) personal communication.

4

THE TROUBLE WITH VALLEYS

Barbara A. Kennedy

Many of the valleys on the surface of the earth are formed by the mechanical agency of water; for, on examining them we find that their direction does not correspond to the internal structure of the stone; that the strata and beds on opposite sides of the valleys are similar, but intersected. All valleys, however, are not formed in the same manner; for many and very extensive valleys are formed by mountain groups disposed in a circular form ... others by the original inequalities of the crust of the earth; some by the unequal deposition of formations, and others by the widening of great rents.

It is also observed, that numerous rents and fissures, and the fall of great masses of mountains, take place during floods or wet seasons.

(Jameson 1808: 29)

INTRODUCTION

There can be little doubt that one of the major – and possibly most durable – of R.J. Chorley's contributions to geomorphology is to be found in the survey of the history of the discipline, undertaken in collaboration with R.P. Beckinsale (his Oxford tutor) and A.J. Dunn. Of the three volumes of *The History of the Study of Landforms* so far published (Chorley, Dunn and Beckinsale 1964; Chorley, Beckinsale and Dunn 1973; Beckinsale and Chorley 1991) it was the first which had the most dramatic impact, since it introduced a whole generation of physical geographers to the prehistory of the present discipline in a lively and extremely well-illustrated fashion. The succeeding thirty years has seen an enormous proliferation of other studies of the history of the earth sciences (most notably Davies 1969; Wilson 1972; Porter 1975; Greene 1982; Rudwick 1985; Tinkler 1985 and 1989; Gould 1987; Laudan 1987; and Huggett 1989) as well as several major studies of related aspects of the history of science (especially Mayr 1982 and Cohen 1985): in this light, volume 1 of *The History* is, perhaps, rather too Whiggish in its attempt to provide both a pedigree and a *raison d'être* for the process-dominated view of geomorphology which rose to prominence in the Anglo-

American tradition of the 1960s and 1970s (cf. Chorley 1962; Leopold, Wolman and Miller 1964; Schumm 1977; and see also Baker 1993). Nevertheless, Chorley *et al.*'s pioneering study remains an important landmark as well as a readable reminder that there is much to be gained by attempting to locate our present preoccupations against the backcloth of previous work.

In what follows, one recurrent theme in the study of landforms – the origin of valleys and their relationships to the fluvial network – will be examined in terms of both its historical development and our contemporary discussions. In particular, the deep-seated difficulties of interpretation which reflect problems in resolving views which derive from an emphasis upon the hydrophysical processes of runoff generation and stream network growth, with those studies whose focus is primarily upon the tectonic and structural bases of topography, will be examined in the context of the last 250 years.

THE TROUBLES WITH HUTTON AND PLAYFAIR

Whatever one's precise view about the role of James Hutton's 1785 (published 1788) *Theory of the Earth* as regards the initiation of the 'modern' view of geology (see Chorley *et al.* 1964; Davies 1969; Tinkler 1989; Gould 1987; Burchfield 1990), it *does* seem incontrovertible that John Playfair's *Illustrations of the Huttonian Theory of the Earth* (1802) can be considered as the starting-point for our present tradition of process geomorphology. No clearer statement concerning the dominance of water in earth sculpture can surely be found than the following:

> Here again water appears as the most active enemy of hard and solid bodies; and, in every state from transparent vapour to solid ice, from the smallest rill to the greatest river, it attacks whatever has emerged from above the level of the sea and labours incessantly to restore it to the deep. The parts loosened and disengaged by the chemical agents, are carried down by the rains, and, in their descent, rub and grind the superficies of other bodies.
>
> (Playfair 1802: 99, quoted in Chorley *et al.* 1964: 60)

From our present viewpoint, it seems so self-evident that this statement should be our starting-point that it would be easy to dismiss both the depth and the rationality of the obstacles which were raised to the acceptance of the view of the supremacy of presently observable exogenetic forces in earth sculpture. As Chorley *et al.* (1964) and Davies (1969) make plain, the objections were both reasonable and profound (and, of course, the reaction to Hutton's views covered a fair broader compass than merely the geomorphological). It is now generally conceded that there were two principal stumbling-blocks to the acceptance of Hutton and Playfair's interpretations of topography: first, the apparently infinite timescale which would be required for the work of rain and rivers to be effective (see Burchfield 1990, in particular); second, the almost overwhelming number of discrepancies which could be observed in northern and western Europe (as well as north-eastern North America) between the regularities postulated for fluvially eroded landscapes and the irregular legacy of – we now presume – repeated recent climatic changes involving not merely widespread glacial advances, but also wholesale oscillations of land and sea levels.

Whilst the intransigence of Lord Kelvin's insistence on the relative brevity (100 million years or even less) of the Earth's history prevented real progress on the estimation of denudation rates until after the discovery of radioactivity and the application of that discovery (by Rutherford in 1904) to provide a wholesale extension of the Earth's age (see Burchfield 1990), the role of glaciation in particular and climatic change in general was, in large measure, resolved by 1877 and the appearance of T.H. Huxley's *Physiography* (Chorley *et al.* 1964; Stoddart 1975). The story of the painstaking evaluation of the various roles of land ice, marine action, isostasy, eustasy and changing circumstances under which rain and rivers have operated which, in the end, were seen to vindicate the underlying truth of Hutton and Playfair's vision, is a fascinating one (see Chorley *et al.* 1964) and can be taken as a prime example of the truth of Kuhn's dictum that 'All historically-significant theories have agreed with the facts, but only more or less' (1962: 146).

To these well-studied stumbling-blocks to the acceptance of Hutton and Playfair's ideas has recently been added a third major strand: Gould (1987) has pointed to the problems which arose from the ostensibly ahistorical formulation of their views. The emphasis on the essentially timeless regularities of process seemed to provide no explanation whatsoever for what was, patently, historically unique topography. This reaction can be seen very clearly indeed in the attitude of the Wernerian, Robert Jameson, in his attack upon 'those *monstrosities* known under the name of *Theories of the Earth*'. Jameson continues:

> Almost all the compositions of this kind are idle speculations, contrived in the closet, and having no kind of resemblance to any thing in nature. Armed with all the *facts* and inferences contained in these visionary fabrics, what account would we be able to give of the mineralogy of a country, if required of us, or of the general relations of the great masses of which the globe is composed? Place one of these speculators in such a situation, and ... he cannot give a rational or satisfactory account of a single mountain.
>
> (Jameson 1808: 42)

I have argued elsewhere (Kennedy 1992) that both Gould and Jameson overlook the very important element of historical explanation contained in Hutton's ideas, partly because it was *implicit* and partly because it considered particular instances – such as the Jedburgh unconformity (Hutton 1795: 453–472; Chorley *et al.* 1964: 38) – as illustrative of *general* principles.

However, we should now be inclined to view such 'explanations' as Hutton's of the Jedburgh case as perfectly 'rational or satisfactory'; and, indeed, entirely comparable retrodiction of the historically specific has been regularly employed by geomorphologists, most notably by G.K. Gilbert in his famous study of the Henry Mountains (1877). Nevertheless, this apparent lack of interest on the part of both Hutton and Playfair in the detail of particular landscapes must be admitted as a further barrier to the acceptance of their views. Indeed, we can extend Gould's criticism to a fourth and closely connected field: the absence of links between Huttonian theories of earth sculpture and the profound regularities of the major structural lineaments of the globe. This, I consider, represents the fourth substantial stumbling-block to the ready acceptance of Hutton and Playfair's ideas, most especially in continental Europe during the Napoleonic era.

THE TROUBLE WITH MOUNTAINS

Although Playfair's *Illustrations* was not translated into French until 1815 (by Basset), it is clear that Hutton's *Theory* was quite widely known on the Continent well before that date. Cuvier, in his 1810 survey of the progress made in the natural sciences since the Revolution of 1789 makes very little direct reference to Hutton's empirical work (1810: 188) and then only with respect to his igneous theory for *vakes* (trap rock).

However, there seems little doubt that Cuvier has Hutton (amongst others) in mind when he launches a general attack upon the futility of large-scale theoretical schemes. (This distaste for what was seen as ill-founded speculation – rather than 'hard' fact-gathering – is interestingly echoed in Lyell's 1830 overview of the recent history of earth science given in volume 1 of his *Principles of Geology*). La Métherie, in his extensive survey of 'Systèmes sur la théorie de la Terre', given as the twelfth section of his geology course at the Collège de France (1816, vol. 3), not only includes Hutton alongside Descartes, Leibniz and Buffon, but also gives substantial attention to the topographic consequences of various viewpoints, including the Huttonian (to which we will return).

However, of all the contemporary responses to Hutton and Playfair, the most interesting is that of the Italian geologist Scipione Breislak (1750–1826: see Francani 1973). Although virtually unknown today, Breislak was an important figure during the Napoleonic period. He was, for example, one of the first to recognise the links between past and present volcanic phenomena and it was he who first drew attention to the evidence that the Temple of Serapis at Pozzuoli provided for *oscillations* of land and sea levels. Playfair picked up that observation (1802: 450) and, of course, Lyell then went on to employ the example as the frontispiece for his *Principles*, with an extensive discussion of its significance (1830: 449–459) in which the reference to Breislak's original observations are far from generous. What is of consequence here, however, is that Breislak would seem to be, potentially, one who should accept Huttonian views. Indeed, in his *Introduction à la géologie* (1812; translated from the Italian edition of 1811) his discussion of Hutton (1812: 112–118) is distinctly favourable in outline: he goes so far as to consider that the theory is 'lumineuse dans ses applications' (1812: 115) and yet he rejects it. Now the immediate reasons for the rejection given are various and concern discrepancies of observation, but the whole tenor of Breislak's work shows substantial and deep-seated unease (and, it has to be said, his own attempt to produce a coherent theory is far from successful, as the savage review in *Le Moniteur* (27 June 1812): quoted in Basset, 1815: 13–21, makes plain). After several re-readings, it seems to me that the underlying cause of Breislak's refusal to accept Hutton's views is their failure to account for the regularities of global topography. In this, Breislak is not alone, but Anglo-American discussions (to date) of the perceived difficulties with Hutton seem to have neglected this element in the situation although it appears to be quite as crucial as the other three general sources of difficulty.

The key section in which Breislak grapples with the joint problems of mountains *and* valleys comes in pages 292–329 of the 1812 French edition and the introductory sentence encapsulates his line of thought: 'Le troisième grand phénomène qui arriva lorsque la surface de la terre se consolida, fait la formation de quelques montagnes et vallées.' (In fact, as the succeeding list makes plain, he views the *majority* of the earth's mountain ranges as these primary features.) What Breislak and most eighteenth-century theoreticians – though not Hutton – were seeking was a device which would allow them to link, in one grand, Newtonian mechanism, the origins of strata *and* the principal lineaments of the Earth's surface. With the development of more-or-less modern atlases in the early eighteenth century, the geometric regularities of both the continental outlines and the location of mountain ranges – together with the observation that most active volcanoes occurred near coastlines – fired intense interest. An early example comes in Buffon's 1749 *Histoire naturelle* where Article VI on 'Géographie', is accompanied by two specially constructed maps – of Africa and Eurasia and of 'the New Continent' (see Figure 4.1) – whose grids are used to emphasise the regularities of the Earth's surface (which Buffon contrasts, p. 204, with the simple parallel latitudinal bands observed on Jupiter). The map projections emphasise not only the linked 'mirror-image' layout of the landmasses of the Old and New World (p. 209) but also the fact that their long axes correspond to areas of high relief (p. 210) which Buffon considers represent the oldest elements of the topography (the lower areas representing younger 'secondary' formations).

From this point onwards, the continental tradition devotes enormous and intense interest to both tracing-out the regularities of global structure and relating these to those forces which were deemed to have been responsible for fashioning the primordial surface. This tradition continues as late as 1852

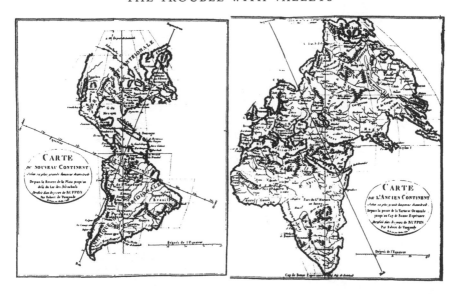

Figure 4.1 The two specially drawn and carefully oriented maps to accompany Article VI on 'Géographie' in Buffon's 1749 *Histoire naturelle*, pp. 209 and 210

with Elie de Beaumont (and see also Figure 4.2), but the major contributions were by Pallas (1778) on the formation of mountains, with especial reference to those of Russia; and Alexander von Humboldt's (1823) descriptions of the ranges of the Americas. However, the interest in – and excitement concerning – the regular form of mountain ranges was heightened by the widespread observation of the regular arrangement of strata on either side of the highest peaks: a crystalline centre, with inward-facing, successively softer and younger strata on each side. As Buffon (1749: 321–322) states: 'toutes les montagnes sont formées dans leur centres à peu près comme les ouvrages de fortification'.

One could multiply the examples of such discussions almost indefinitely from continental works of the period. It must be admitted, of course, that these observations did, indeed, point to some underlying principle at work; and the intense interest generated in the late eighteenth and early nineteenth centuries by the news of fresh measurements of mountain orientation and composition can fairly be compared, in my view, to that surrounding the successive discoveries of oceanic ridges and tectonic plate boundaries in the 1960s. *Surely* – Buffon, Pallas, La Métherie, Humboldt, Breislak and co. maintained – all these regularities *had* to mean something important? Surely the topographic trends *had* to have some crucial link to the most basic processes of the Earth?

Of course, where linear mountain ranges were, there, too, were the major valley systems.

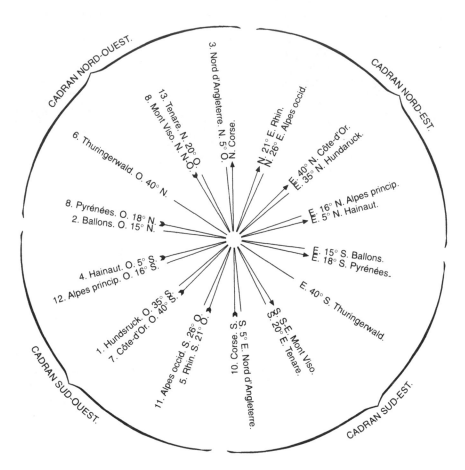

Figure 4.2 The principal trends of global relief as illustrated by Noguès 1870

Now there was, quite early on, a recognition that there was some kind of hierarchy in valleys (see Buffon, for example). There was also tremendous emphasis placed on Bourguet's early observation (1729: and see Ellenberger 1972) that interlocking spurs ('les angles saillans et rentrans') indicated that even large valleys had been *eroded* (since the same strata occurred on either side of the thalweg). This observation was, however, challenged by Saussure's later work in the Alps (1779–1796) and Breislak stresses the fact that Bourguet's apparently powerful generalisation was therefore highly suspect. Nevertheless, with one's eyes firmly fixed on large-scale topographic initiation, the question of valley formation *had* to be linked to that of mountains.

It is here, of course, that both Hutton and Playfair had nothing to say.

We can see the ramifications of this silence throughout the response to the

Huttonian vision. For those workers – notably McCullough and Scrope – whose focus is short term and/or relatively small scale – the absence of a grand topographical explanation in Hutton and Playfair's writings is almost irrelevant. Produce the high ground in any area as you will and *then* focus on the regularities created by fluvial dissection. (Interestingly, Hewitt (1989) draws attention to rather similar problems of 'grand designs' versus local topographic interpretations with respect to early work on the great Asian mountain massifs.) But given the significance of global structure in the eyes of many, and especially continental, authorities until well into the mid-nineteenth century, the *a-spatial* nature of Huttonian views was, to my mind, the fourth and in many ways final nail in their coffin.

How was this discrepancy overcome? Basically, by the switch from grand theory to more detailed observation and associated smaller-scale hypotheses and, ultimately, to the decoupling of studies of mountains and valleys which can be seen in Dana's *Manual of Geology* (1863) and which persists through to so recent a work as Summerfield (1991). We 'do' mountain massifs and grand drainage lines by the processes of plate tectonics; and then we 'do' the creation of local or, at best, regional channel networks by hydrophysical processes which derive firmly from the ideas of R.E. Horton (1945) which, in turn, were explicitly derived to provide a quantitative justification of Playfair's Laws. As so often in the history of science, a problem ostensibly disappears by redefining the focus of interest. But does it?

THE REAL TROUBLE WITH VALLEYS

If we look back to Jameson's 1808 outline of valley origin (given at the outset of this discussion) we see a clear recognition that both tectonic and fluvial processes can be involved in accounting for valley origin. Lyell – notoriously reluctant to allow a major role for rain and rivers in large-scale earth sculpture – goes further and implies (cf. 1830: 197 and Figure 4.3) that the two generally work together, with tectonism as the senior partner (see Kennedy 1992). And yet, this is still only part of the story: we have valleys created by mass movement (cf. Hack and Goodlett 1960), by collapse in karst or thermokarst regions, and so on and so forth. *Any* process which creates topographic irregularities will cause the subsequent concentration of any available surface moisture and, potentially, a stream-and-valley. Moreover, since valleys are exceptionally durable features – some networks in Australia are reckoned to be *c.* 500 million years old (Stewart *et al.* 1986) – we must face the fact that many networks will contain portions which owe both their ultimate origin and also their persistence to *different* processes. The problems recognised by Buffon, by Jameson and by Breislak have not gone away.

However, there are signs that some resolution of these difficulties may be at least within sight. If we accept that we must treat the topographic network

67

BARBARA A. KENNEDY

(a)

Figure 4.3 Two of Lyell's illustrations of valley formation: (a) a chasm caused by an earthquake in Calabria; (b) the Milledgeville gully, caused by anthropogenic accelerated erosion

of valleys as representing the potential network of functional channels (Kennedy 1978), then there are, at present, several proposals on the table which seek to link the 'fluvial' and 'other' to provide some *general* description of dissected topography. In order of radicalism in their departure from the Hutton–Playfair–Horton tradition these are: Dietrich's views about 'unchannelized basins' (cf. Dietrich *et al.* 1986; Montgomery and Dietrich 1992); Stark's (1991) invasion-percolation model which derives from a fractal view of landscape; and Shreve's (1966, 1967, 1975) probabilistic-topologic approach to drainage basin morphology.

Dietrich's ideas, most notably as expressed in the recent discussion based on Montgomery's 1991 doctoral thesis (Montgomery and Dietrich 1992), have the merit of trying to link the non-fluvial and fluvial portions of networks together in a more comprehensive fashion than Horton's original (1945) partial inclusion of 'mesh lengths' (see Kennedy 1978). However, my overwhelming reaction to this work is that, in large measure, it merely reinvents the wheel. The authors make twin claims: that 'landscape responses to changes in climate or land use depend on the corresponding changes in the thresholds of channelization'; and, that in consequence landscapes cannot be viewed as scale-independent (1992: 826). It seems first, that they simply restate the well-known relationships between channel density and the surface

68

(b)

force/resistance balance (see Melton 1958) and, second, that they ignore the very convincing case made by Church and Mark (1980) that drainage density relations are genuinely *iso*metric, not allometric. (However, later work (Dietrich *et al.* 1993) does go some way to combine mass movement and runoff as twin sources of channel development.)

Stark's work (1991), in contrast, is a more radical view, which attempts to link the observed fractal character of channel networks to physically plausible processes of network development. The underlying principle he develops is that of networks which 'branch and propagate at a rate that depends only on the local strength of the substrate. This model corresponds to the process of invasion percolation, with the added requirement of self-avoidance' (Stark 1991: 423). He goes on to show that – at least – strong regional tilt may be successfully incorporated in the model and that there are good reasons to hope that the entire 'dynamical link between surface flow and topography' (Stark 1991: 424) may be encompassable within the model. Whilst this approach is too new to accept wholeheartedly as the solution to the valley/stream network dilemma, Melton for one (personal communication) considers it to be extremely promising indeed.

In contrast, Shreve's probabilistic-topologic approach has been around for the best part of thirty years (Shreve 1966, 1967, 1975) and, although it has had some devoted adherents – most notably J.S. Smart – and although it has convincingly demonstrated both the generally low degree of influence of most structural controls and the ability to distinguish those cases where structure, rather than random processes, appears to determine stream network characteristics, it has been generally ignored as a solution to the link between structure and channel networks. In my view this rejection has been based in large measure on the gut feeling that, as runoff processes are demonstrably deterministic, so too must channel networks be. This view, to my mind, ignores the problem of multiple causation of valleys and, hence, stream nets. The feature of Shreve's approach which is its great strength – that it makes no assumptions about the *origins* of networks, merely allows one to describe and, hence, identify cases where some strong (i.e. non-random) control is operating – has indubitably been viewed as a substantial weakness. Just as Church and Mark's (1980) identification of isometric rather than allometric processes – again, an area where effective 'explanation' does not include 'deterministic' factors – has not been pursued, so Shreve's concepts have failed to find a key role in geomorphic thinking. This seems to be a question of temperamental bias on the part of the discipline: like Einstein, we are unhappy with any view which implies that 'the Almighty plays dice'.

However, whether Dietrich's, Stark's, Shreve's or some altogether new approach is ultimately accepted, it is encouraging that, at long last, we seem to be approaching a recognition that we must view topography in a *general* light which will allow both tectonics and fluvial processes (as well as all the other factors and processes, past and present, which contribute to the formation of dissected terrain and streams) to be viewed together, rather than separately.

CONCLUSION

The study of landforms has come a long way since Hutton and Playfair and the history of that study has also come a long way since Chorley, Dunn and Beckinsale wrote in 1964. What this discussion has tried to show is that a hitherto unacknowledged problem arising from the formulation of Hutton and Playfair's views – the divorce between the origin of large-scale topography and the generation of the regularities of fluvial networks – has remained problematic for the past two centuries. It is suggested, however, that there is now some growing recognition of the need to develop approaches which will allow us to view valleys and streams as a single, albeit multi-origin, set of phenomena. If we can once generate a convincing description for the mix of regular and irregular pattern which is the basic network of dissected terrain then, surely, we should be able to go on and generate further descriptions and, conceivably, predictions for the associated characteristics of both valleys and streams which empirical studies show us are intimately related to the form of networks (cf. Carter and Chorley 1961). If we can do that, then valleys will indeed cease to be troublesome.

ACKNOWLEDGEMENTS

I am indebted to the staff of the Rare Books rooms at the Universities of Cambridge and South Carolina, for their assistance; and to Martin Barfoot for his heroic efforts in the production of the illustrations.

REFERENCES

Baker, V.R. (1993) 'Extraterrestrial geomorphology: science and philosophy of Earthlike planetary landscapes', *Geomorphology* 7: 9–35.
Basset, C. (trans. and ed.) (1815) *Explication de Playfair sur la théorie de la Terre par Hutton . . .*, Paris: Bossange & Masson.
Beckinsale, R.P. and Chorley, R.J. (1991) *The History of the Study of Landforms*, vol. 3: *Historical and Regional Geomorphology, 1890–1950*, London: Routledge.
Bourguet, L. (1729) *Lettres philosophiques sur la formation des sels et des cristaux . . . Mémoire sur la théorie de la Terre*, Amsterdam: François l'Honoré.
Breislak, S. (1812) *Introduction à la géologie ou à l'histoire naturelle de la Terre*, trans. J.J.B. Bernard, Paris: J. Klostermann, Fils.
Buffon, J.M.L., Comte de (1749) *Histoire naturelle, générale et particulière . . .* vol. I, Paris: Imprimerie royale.
Burchfield, J.D. (1990) *Lord Kelvin and the Age of the Earth* (2nd edition), Chicago: University of Chicago Press.
Carter, C.S. and Chorley, R.J. (1961) 'Early slope development in an expanding stream system', *Geological Magazine* 98: 117–130.
Chorley, R.J. (1962) 'Geomorphology and General Systems Theory', *U.S. Geological Survey Professional Paper* 500-B: 1–10.
Chorley, R.J., Beckinsale, R.P. and Dunn, A.J. (1973) *The History of the Study of Landforms*, vol. 2: *The Life and Work of William Morris Davis*, London: Methuen.
Chorley, R.J., Dunn, A.J. and Beckinsale, R.P. (1964) *The History of the Study of Landforms*, vol. 1: *Geomorphology before Davis*, London: Methuen.
Church, M.A. and Mark, D.M. (1980) 'On size and scale in geomorphology', *Progress in Physical Geography* 4: 342–390.
Cohen, I.B. (1985) *Revolution in Science*, Cambridge, Mass.: Belknap Press.

Cuvier, G. (1810) *Rapport historique sur les progrès des sciences naturelles depuis 1789, et leur état actuel*, Paris: Imprimerie impériale.

Dana, J.D. (1863) *Manual of Geology*, Philadelphia: Theodore Bliss & Co.

Davies, G.L. [Herries] (1969), *The Earth in Decay: A History of British Geomorphology, 1578–1878*, London: Oldbourne, Macdonald Technical and Scientific.

Dietrich, W.E., Wilson, C.J., Montgomery, D.R. and McKean, J. (1993) 'Analysis of erosion thresholds, channel networks, and landscape morphology using a digital terrain model', *Journal of Geology* 101: 259–278.

Dietrich, W.E., Wilson, C.J. and Reneau, S.L. (1986) 'Hollow, colluvium, and landslides in soil-mantled landscapes', in A.D. Abrahams (ed.) *Hillslope Processes*, Boston: Allen & Unwin.

Elie de Beaumont, L. (1852) *Notice sur les systèmes des montagnes*, 3 vols, Paris: P. Bertrand.

Ellenberger, F. (1972) 'De Bourguet à Hutton: une source possible des thèmes huttoniens; originalité irréductible de leur mise en œuvre', *Compres rendus de l'Académie de Science de Paris* 275: 93–96.

Francani, V. (1973) 'Scipione Breislak', in C.C. Gillispie (ed.) *Dictionary of Scientific Biography*, II, New York: Charles Scribner's Sons.

Gilbert, G.K. (1877) *Report on the Geology of the Henry Mountains*, Washington: Government Printing Office.

Gould, S.J. (1987) *Time's Arrow, Time's Cycle: Myth and Metaphor in the Discovery of Geological Time*, Cambridge, Mass.: Harvard University Press.

Greene, M.T. (1982) *Geology in the Nineteenth Century: Changing Views of a Changing World*, Ithaca: Cornell University Press.

Hack, J.T. and Goodlett, J.C. (1960) 'Geomorphology and forest ecology of a mountain region in the central Appalachians', *U.S. Geological Survey Professional Paper* 347.

Hewitt, K. (1989) 'European science in high Asia: geomorphology in the Karakoram Himalaya to 1939', in K.J. Tinkler (ed.) *History of Geomorphology*, Boston: Unwin Hyman.

Horton, R.E. (1945) 'Erosional development of streams and their drainage basins; hydrophysical approach to quantitative morphology', *Bulletin of the Geological Society of America* 56: 275–370.

Huggett, R.J. (1989) *Cataclysms and Earth History: The Development of Diluvialism*, Oxford: Clarendon Press.

Humboldt, A. von (1823) *A Geognostical Essay on the Superposition of Rocks in Both Hemispheres* (trans. from the French), London: Longman, Hurst, Rees, Orme, Brown & Green.

Hutton, J. (1788) 'Theory of the Earth', *Transactions of the Royal Society of Edinburgh*, 1: 209–304.

——(1795) *Theory of the Earth, with Proofs and Illustrations*, vol. 1, London: Cadell, Junior & Davies.

Huxley, T.H. (1877) *Physiography: An Introduction to the Study of Nature*, London: Macmillan.

Jameson, R. (1808) *Elements of Geognosy*: vol. 3, part III of *System of Mineralogy*. Reprinted in facsimile, G.W. White (ed.) as *The Wernerian Theory of the Neptunian Origin of Rocks* (1976), New York: Hafner.

Kennedy, B.A. (1978) 'After Horton', *Earth Surface Processes* 3: 219–232.

——(1992) 'Hutton to Horton: sequence, progress and equilibrium in geomorphology', *Geomorphology* 5: 231–250.

Kuhn, T.S. (1962) *The Structure of Scientific Revolutions*, Chicago: University of Chicago Press.

La Métherie, J.-C. de (1816) *Leçons de géologie, données au Collège de France*, 3 vols, Paris: Courcier.

Laudan, R. (1987) *From Mineralogy to Geology: The Foundations of a Science, 1658–1838*, Chicago: University of Chicago Press.

Leopold, L.B., Wolman, M.G. and Miller, J.P. (1964) *Fluvial Processes in Geomorphology*, San Francisco: W.H. Freeman.

Lyell, C. (1830) *Principles of Geology*, I, London: John Murray.

Mayr, E. (1982) *The Growth of Biological Thought: Diversity, Evolution and Inheritance*, Cambridge, Mass.: Belknap Press.

Melton, M.A. (1958) 'Geometric properties of mature drainage systems and their representation in an E4 phase space', *Journal of Geology* 66: 25–54.

Montgomery, D.R. and Dietrich, W.E. (1992) 'Channel initiation and the problem of landscape scale', *Science* 255: 826–830.

Noguès, M.A.F. (1870) *Traité d'histoire naturelle ... 4ème année: géologie appliquée*, Paris: Victor Masson & Fils.

Pallas, P.S. (1778) 'Observations sur la formation des montagnes et les changements arrivé au Globe, particulièrement à l'égard de l'Empire de Russie', *Acta Academiae Scientiarum Imperialis Petropolitanae*, Part I 1777: 21–64.

Playfair, J. (1802) *Illustrations of the Huttonian Theory of the Earth*, London: Cadell & Davies. Reprinted in facsimile, G.W. White (ed.), 1964, New York: Dover Publications Inc.

Porter, R. (1975) *The Making of Geology: Earth Science in Britain 1660–1815*, Cambridge: Cambridge University Press.

Rudwick, M.J.S. (1985) *The Great Devonian Controversy*, Chicago: University of Chicago Press.

Saussure, H.-B. de (1779–1796) *Voyages dans les Alpes, précédés d'un essai sur l'histoire naturelle des environs de Genève*, Neuchâtel: L. Fauche-Borel.

Schumm, S.A. (1977) *The Fluvial System*, New York: John Wiley.

Shreve, R.L. (1966) 'Statistical law of stream numbers', *Journal of Geology* 74: 17–37.

——(1967) 'Infinite topologically random channel networks', *Journal of Geology* 75: 178–186.

——(1975) 'The probabilistic-topologic approach to drainage-basin geomorphology', *Geology* 3: 527–529.

Stark, C.P. (1991) 'An invasion percolation model of drainage network evolution', *Nature* 352: 423–425.

Stewart, A.J., Blake, D.H. and Ollier, C.D. (1986) 'Cambrian river terraces and ridgetops in central Australia: oldest persistent landforms?', *Science* 233: 758–761.

Stoddart, D.R. (1975) '"That Victorian Science": Huxley's *Physiography* and its impact on geography', *Transactions of the Institute of British Geographers*, 66: 17–40.

Summerfield, M.A. (1991) *Global Geomorphology*, Harlow: Longman Scientific and Technical.

Tinkler, K.J. (1985) *A Short History of Geomorphology*, London: Croom Helm.

——(ed.) (1989) *History of Geomorphology*, Boston: Unwin Hyman.

Wilson, L.G. (1972) *Charles Lyell. The Years to 1841: The Revolution in Geology*, New Haven: Yale University Press.

5

SUBSURFACE FLOW AND SUBSURFACE EROSION

Further evidence on forms and controls

J.A.A. Jones

Thirty years ago shallow subsurface water movement was considered of little importance by most geomorphologists and hydrologists. Even in 1968, Alfred Jahn failed to include subsurface water flow in his list of erosional processes to be monitored.

It is very different today. Numerous recent reviews indicate the quantum leap that research has made during the last two decades, to which the seminal paper of Kirkby and Chorley (1967) gave a timely fillip. The collection of papers in Higgins and Coates (1990) provides an excellent overview of the links between landforms and subsurface water, and many papers in Anderson and Burt (1990a) reveal the progress made from a hydrological viewpoint. Substantial progress has been made in understanding the chemical aspects of soil stability, from Aitchison et al. (1963) to the syntheses of Sherard and Decker (1977) and beyond, that were incorporated, for example, in Jones's (1981) review of piping and Gerits et al.'s (1990) review of surficial erosion. Marked progress has also been made in studying the role of subsurface flow in the development of surface channel networks, as reviewed by Dunne (1980, 1990), Jones (1987a), Baker et al. (1990) and Higgins (1990a) and modelled by Howard (1990), as well as its role in stimulating mass movement (Coates 1990).

Nevertheless, a number of important questions remain only partly answered, and this paper focuses on a few of these. Progress in studies of subsurface erosion ultimately depends upon establishing a firm understanding of subsurface flow processes, yet major gaps remain in our knowledge of their rates and distribution. How important are subsurface processes for the development of surface channel networks or valleys, and are the resultant landforms uniquely distinguishable? How widespread is subsurface erosion and what factors determine its distribution? And what implications does subsurface flow have for environmental response?

COMPARATIVE HYDROLOGY OF HILLSLOPE DRAINAGE PROCESSES

There is still considerably less quantitative information on subsurface flow than surface flow. Probably the greatest difficulty in comparing surface and subsurface responses lies in the lack of information on the area contributing drainage to a given subsurface point. There is also a dearth of catchment scale assessments of contributions to streamflow from subsurface sources. It is particularly regrettable that still only one catchment scale investigation has studied pipeflow in the context of streamflow and the other sources – diffuse throughflow, overland flow and groundwater movement – and provides sufficient data to allow quantitative assessments of both its hydrological contribution to the stream and the size of its own source areas. Wide variations in the mix and relative efficiency of processes are to be expected (cf. Whipkey and Kirkby 1978), and pipeflow may often be absent or unimportant. However, the evidence from Gardiner (1983) suggesting a minor contribution in deep blanket peat (Burt *et al.* 1990) should be treated with caution, since he failed to monitor the larger pipes in his catchment.

Published flow velocities collated by Jones (1987b) show pipeflow ranking first, averaging 300 mm s^{-1}, followed by overland flow at 50 mm s^{-1}, macropore flow (3 mm s^{-1}), and finally matrix throughflow at 0.65 mm s^{-1}. These rates suggest that overland flow would take less than an hour to reach the stream channel in the average British basin, whereas diffuse throughflow would take over 100 hours and therefore arrive too late to contribute to the stormflow hydrograph. However, field observations in Britain suggest that overland flow is more likely to be saturated than Hortonian, and therefore restricted spatially to areas of convergent flow and often partly fed by throughflow. Even where conventional infiltration measurements suggest that Hortonian flow will be common, in practice large macropores and pipes can cause subsurface flow to dominate (Jones 1990). Nevertheless, overland flow may still dominate in discharge terms through having a larger cross-sectional area of flow.

Table 5.1 compares three key parameters of discharge for subsurface flow and Hortonian overland flow. The throughflow and overland flow values are based on data collated by Dunne (1978). In order to facilitate comparison, all values have been interpolated for a range of catchment sizes, using regression equations (Table 5.2). The equation for throughflow peak lag time has been calculated as Dunne (1978), by omitting the evidence from Weyman's (1970) and one of Knapp's (1974) measurements. Their figures deviate considerably from the general trend and make the regression non-significant: are catchment sizes perhaps underestimated?

For pipeflow, the maximum dynamic contributing area (DCA) recorded in the Maesnant programme (Jones and Crane 1984; Jones 1987a) has been

Table 5.1 Typical responses for hillslope drainage processes

Basin area (km²)	Peak runoff rate (mm h⁻¹)			Peak lag times (h)			Runoff coefficient		
	HOF	TF	PF	HOF	TF	PF	HOF	TF	PF
10	24	(0.1)	(0.3)	0.7	(25.3)	(9.6)	0.18	(0.1)	(0.11)
1	36	0.15	(0.7)	0.4	14.3	(6.7)	0.21	(0.1)	(0.17)
0.1	54	0.2	1.6	0.3	8.0	4.6	0.25	0.1	0.25
0.01	81	0.3	3.6	0.2	4.6	3.2	0.29	0.1	0.38
0.001	122	0.4	8.4	0.1	2.6	2.2	0.35	0.1	0.56

Notes: HOF – Hortonian overland flow, TF – throughflow, PF – pipeflow
Brackets indicate extrapolation beyond the range of data
Overland flow and throughflow calculations based on data from Dunne (1978)

Table 5.2 Equations relating runoff parameters to catchment area

Process	Peak runoff rate	Peak lag time	Runoff coefficient
Hortonian overland flow	$Q_n = 35.88\,\mathrm{Area}^{-0.15}$	$L_n = 0.42A^{0.20}$	Not significant
Throughflow	Not significant	$L_n = 14.3A^{0.25}$	Not significant
Pipeflow	$Q_n = 0.6815A^{-0.36}$	$L_n = 6.65A^{0.16}$	$RO_{coeff} = 0.1683A^{-0.175}$

Note: All equations significant at 5%. Overland flow and throughflow equations based on data from Dunne (1978)

used as the best available estimate of subsurface catchment size. DCAs were calculated as

$$DCA = \frac{\text{total storm discharge}}{\text{total storm rainfall}}$$

for perennially flowing pipes and

$$DCA = \frac{\text{total storm discharge}}{\text{total storm rainfall before end of pipeflow}}$$

for ephemerally flowing pipes. Each regression is based on 15 different sized 'catchments', nesting pipe sub-catchments together in order to get the three largest catchments. Alternative ways of determining pipeflow catchment areas, such as (1) using water tracers, (2) surface topography or (3) using the average calculated DCAs, were rejected either because of impracticality (1), inadequacy (2) or in the case of (3) a circular argument that would guarantee a runoff coefficient of one. There is no corresponding circular argument in using maximum DCA to calculate peak runoff rates, because the latter are instantaneous measurements whereas all DCAs calculated by the above formulae are averages for the whole storm.

Runoff coefficients

Throughflow has very low coefficients that are insensitive to catchment size. In contrast, pipeflow coefficients are comparable to those for overland flow. The average runoff coefficient for a microcatchment drained by a perennial pipe at an average depth of 500 mm in the Maesnant basin was 0.35. Amongst the shallow ephemeral pipes, running at a depth of only 150 mm, the average was 0.43. The scatter in the coefficients for overland flow is so great that both pipeflow and throughflow map within its range (Figure 5.1). Pipeflow is the only process to show a clear trend with catchment size (Figure 5.2; Table 5.2).

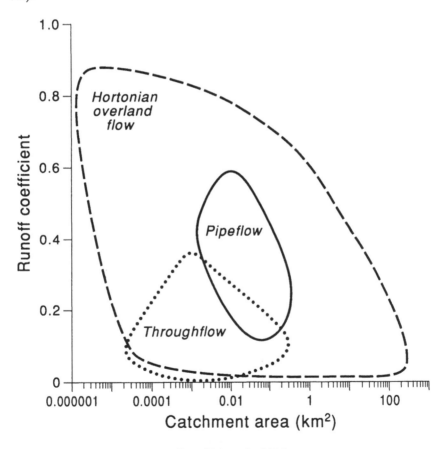

Figure 5.1 Runoff coefficients for hillslope processes
Source: Based on data from Dunne (1978) and the author

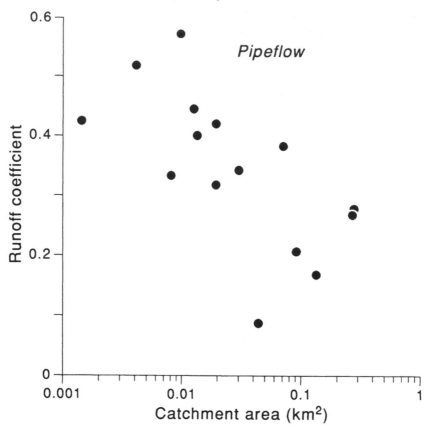

Figure 5.2 The trend in pipeflow runoff coefficients

Peak lag time

Lag times in Tables 5.1 and 5.2 are based on the time between a 'burst of intense rainfall' (Dunne 1978) and peak discharge. Table 5.1 suggests that lag times are 30–40% shorter for pipeflow than for normal throughflow. Figure 5.3 shows that pipeflow maps into the graph produced by Anderson and Burt (1990b) just below throughflow and some way above saturation overland flow, perhaps becoming closer to throughflow in extremely small catchments (Figure 5.4). However, they are still 15–20 times longer than Hortonian overland flow. Taking the runoff coefficients into consideration, this suggests that in multiple source stormflow pipeflow is more likely to peak after overland flow and to create a more gradual recession curve in the stormflow hydrograph. However, where overland flow is absent, pipeflow will tend to provide a runoff peak before throughflow, as noted in the East Twins

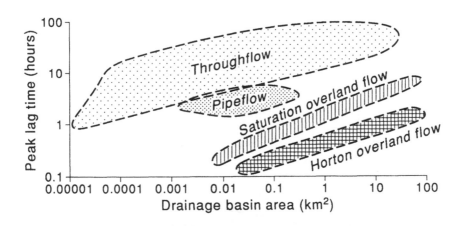

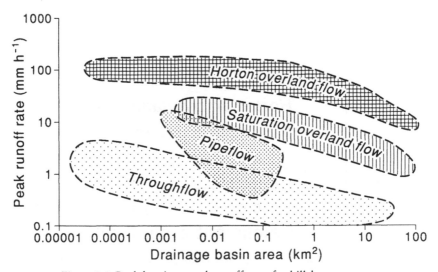

Figure 5.3 Peak lag times and runoff rates for hillslope processes
Source: Based on collations by Dunne (1978), Kirkby (1985), Anderson and Burt (1990b), Burt (1992) and the author

catchment near Bristol by Weyman (1970), Stagg (1974), Finlayson (1977) and Burt (1992).

On Maesnant, where the stream has an average peak lag time of 4.4 h, the mean peak lag time at the streamside outfalls of perennial pipes was 4.7 h. For ephemeral pipes feeding the stream directly it was 3.6 h, although Figure 5.5 shows that most ephemeral pipes have lag times nearer to 2 h. Piped riparian seepage zones average a 3.1 h lag. Since 70% of pipes feeding the stream (and 50% of perennial pipes) peak before the stream, and since pipeflow accounts

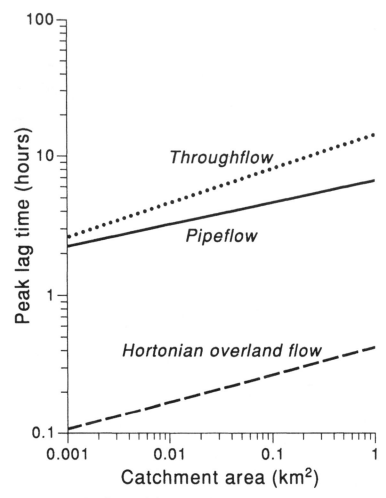

Figure 5.4 Trends in peak lag times. (Based on equations in Table 5.2)

for 49% of storm runoff in the stream, it is clear that the streamflow peak is substantially driven by pipeflow.

Peak runoff rates

Peak runoff rates for pipeflow once again lie between throughflow and overland flow (Table 5.1). In catchments of less than 0.01 km^2, pipeflow occupies approximately the mid-point in terms of proportionality, roughly 15–20 times greater than throughflow and 15–20 times less than Hortonian overland flow. Rates get closer to throughflow as catchment size increases (Figure 5.6). Mapping pipeflow into the graph produced by Anderson and

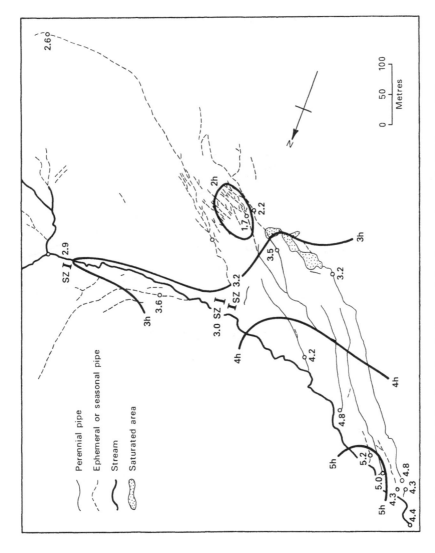

Perennial pipe

Ephemeral or seasonal pipe

Stream

Saturated area

Figure 5.5 Peak lag times for pipeflow and riparian seepage on the hillslopes of Maesnant

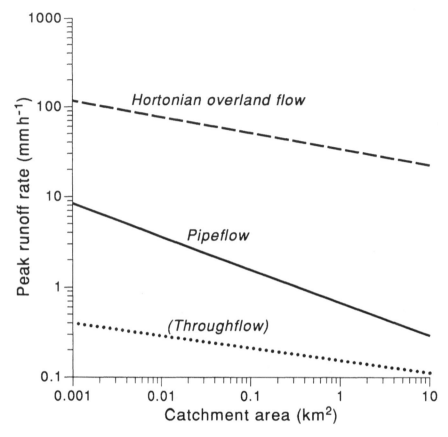

Figure 5.6 Trends in peak runoff rates
Source: Based on equations in Table 5.2

Burt (1990b) in Figure 5.3 shows that pipeflow rates can overlap with saturation overland flow in catchments of less than 0.01 km², and with throughflow in slightly larger catchments.

In terms of absolute values, however, the dominance of overland flow remains unquestionable wherever it is able to operate. Kirkby (1985) would be correct in inferring that this should mean that runoff models should be biased towards the accurate assessment of the occurrence of overland flow, were it not for three facts: (1) that Hortonian overland flow is so rarely observed in well-vegetated, humid landscapes (Chorley 1978), (2) that pipeflow bridges the gap between saturation overland flow and throughflow, and (3) that the latter three processes involve at least an element of contribution from subsurface flow. Even though pipeflow itself is a far from ubiquitous process, subsurface flow in general can contribute to stream runoff in such a variety of ways that it would be folly to ignore it. Not least of these is the concentration of subsurface flow in seepage lines and the generation of saturation overland flow in shallow swales or seepage lines by return flow.

Stormflow yields

Table 5.3 lists processes in the Maesnant basin in order of average yield per square metre of effective catchment area feeding each source, defined as the maximum recorded DCA for each process. Average storm rainfall throughout was approximately 30 mm.

This reveals the relative efficiency of shallow, ephemeral pipes as a means of draining a slope (column 3). Like the pipes described by Gilman and Newson (1980) in the nearby Nant Gerig basin, the ephemeral pipes on Maesnant are clearly developed from desiccation cracks in the shallow organic horizon of the aquod soils on the midslopes. The dense network of cracks allows rapid and almost total infiltration around these pipes. Even conventional measurements of infiltration capacity, avoiding obvious macropores, reveal infiltration capacities around these pipes in the range 1×10^{-4} to 1×10^{-3} mm s^{-1}, which coincides with mean rainfall intensities in 86% of the 180 storms monitored, whereas away from the pipes infiltration capacities are around 1×10^{-6} mm s^{-1} or lower.

In contrast, the deeper-seated perennial pipes yield approximately two-thirds as much storm runoff per square metre of catchment. One factor is the delay caused by percolation to three times the depth through an organic horizon that is less cracked. This will dampen stormflow response, which alone can result in delayed outflow not being recognised as stormflow, and cause greater losses to soil water storage. But possibly more importantly, the maximum DCAs for the perennial pipes tend to be much larger in relation to the average DCA than for ephemeral pipes, averaging 3.8 times as large

Table 5.3 Comparison of sources of storm runoff in Maesnant Experimental Basin

Process	Contribution in an average storm (m³)	Mean contributing area per pipe or zone (m²)	Average yield per m² of maximum source area (mm)
Ephemeral pipeflow direct to stream	100	1808	12.6
Perennial pipeflow	2148	11877	8.2
Piped seepage zones	242	4786	6.7
Diffuse riparian seepage zones	389	160[1]	3.7
Other sources	702	–	–

Note: [1] per metre of streambank

compared with only 2.3 times for the ephemerals. This means that there are much larger areas of the estimated micro-catchments that remain inactive or non-contributing during the average storm amongst the perennial pipes, and it follows from the fact that perennial pipe networks are more capable of major expansion and tapping large areas of the hillside during heavier storms.

The behaviour of the bankside seeps was estimated on the basis of three selected sections of bankside flush areas, identified by *Juncus* and *Sphagnum* vegetation and monitored by continuous logging equipment (Jones and Crane 1984). Trenches were dug down to the impermeable clay till substrate and lined with heavy waterproofing polythene sheet. Although this method also collects saturation overland flow, the surface collecting areas are small, *c.* 30 m².

The two lower bankside seeps (SZ in Figure 5.5) are partly fed by shallow ephemeral pipes. The average stormflow yield per square metre of source area given in Table 5.3 for these 'piped seepage zones' is about half that of ephemeral pipes alone and essentially reflects the mixture of deep through-flow and shallow pipeflow. The section of diffuse bankside seep at the upper site yielded only half as much stormflow per square metre of catchment as the perennial pipes. Average contributing areas for this, the most common type of riparian seep, were calculated to be 160 m² per metre length of streambank. A photogrammetric map of the basin, combined with a ground truth survey, suggests that there are 297 m of similar bankside seepage in the basin. Hence, the average storm yield from diffuse bankside seepage zones given in Table 5.3 has been calculated on the assumption that the upper site is representative of this bank area. Tentative calculations of the residual contribution from other, unmonitored sources suggest that these yield *c.* 1.2 m³ per metre of streambank, which would imply contributing areas of about 40 m² per metre of bank in the average storm. Most of this must be provided by deep throughflow and groundwater flow direct to the stream-bank, since some twenty strategically placed crest stage gauges for overland flow picked up very little.

Figure 5.7 expresses the relative effectiveness of each source of subsurface flow in terms of average size of contributing areas and the efficiency of yield per square metre. The more effective a process is as a source of stormflow in the stream, the nearer it is to the top right-hand end of the graph.

Sources of pipeflow – 'old' water or 'new'?

Two recent investigations into the sources of pipeflow introduce the possibility of conflicting evidence. Monitoring and modelling of pipeflow and phreatic surface movement around the main pipe network on Maesnant by Jones *et al.* (1991) and Connolly (1993) confirms the conclusion of Jones and Crane (1984) that the rising water table is primarily responsible for pipe

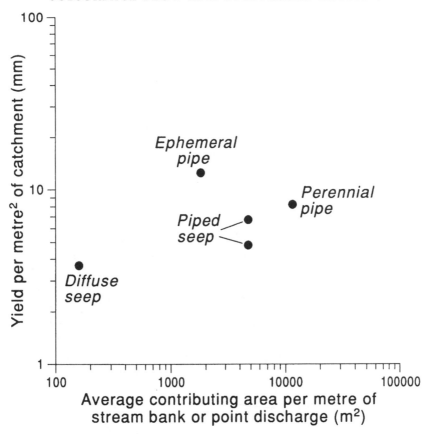

Figure 5.7 The efficiency of drainage collection and yield for subsurface processes on Maesnant

stormflow in the perennial pipes, but that the rapidity of the response is caused by bypass infiltration through macropores feeding the saturated zone, aided by direct inflow from overland flow via pipe blowholes.

Potentially conflicting evidence comes from hydrogen isotope data collected by Sklash *et al.* (in press a and b) from the nearby Nant Gerig, which suggest a preponderance of 'old' water even in ephemeral pipeflow.

The evidence of Jones and Connolly derives partly from the 'good fit' between pipeflow data and the predictions of a computer model based upon three sources of runoff (Figure 5.8). The sources are: (1) an upslope supply, which consists of (a) captured overland flow for the ephemeral pipes and (b) this plus a supply of return flow and 'channel precipitation' from the mid-slope bogs for the perennial pipes, and (2) groundwater supply via flow orthogonal to both types of pipe. The latter is limited to the width of the pipe swales, *c.* 5 m either side of perennial pipes or *c.* 1 m for ephemerals, as

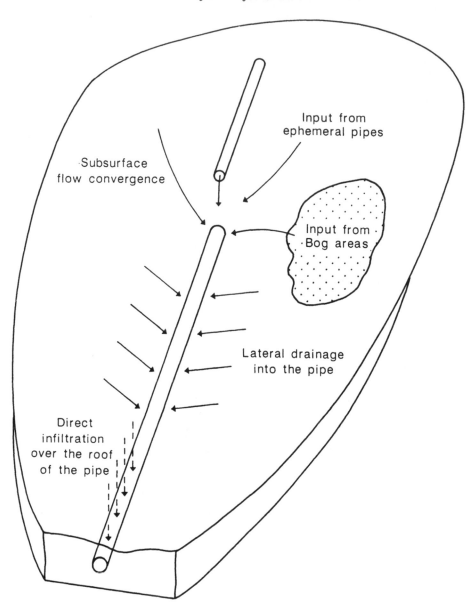

Figure 5.8 Components of the Maesnant pipeflow model
Source: After Connolly 1993

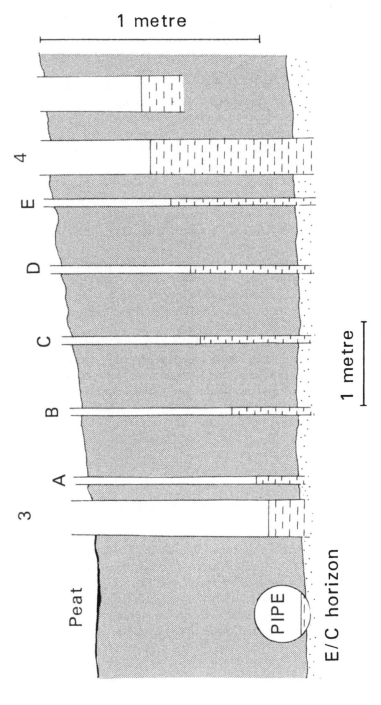

Figure 5.9 The phreatic surface around a perennial pipe at baseflow on Maesnant as shown by piezometers (narrow tubes) and dipwells

Source: Based on Connolly 1993

indicated by cross-profiles of the water table (Figure 5.9). The case for bypass flow is supported by measurements of infiltration rates and hydraulic conductivity in the ephemeral pipe zone which suggest that rainwater would take more than a day to percolate to the phreatic surface by diffuse flow (Jones 1990), whereas pipe responses throughout the network are far more rapid (Jones 1988).

The 'obvious' conclusion is that the pipes are draining new rainwater, albeit water that enters the pipes by way of the upper part of the saturated zone. This is supported by the calculation that the total volume of quickflow in the Maesnant averages 68% of storm rainfall, and by the high runoff coefficients of the pipes (see above). Mosley (1979) drew a similar inference regarding macropore flow in forest soils.

Similarly, a number of chemical analyses of pipeflow and streamflow have shown that pipeflow typically lies between peat matrix water and overland flow, and appears to be a mix of the two tending to become more dilute during stormflow (Cryer 1980; Roberts et al. 1984; Reynolds et al. 1986; Hyett 1990).

Sklash et al. (in press a and b) conclude that the ephemeral pipes on Nant Gerig are draining displaced groundwater, and that the rapid response indicates the presence of piston flow. In the same area, Neal and Rosier (1990) conclude from analyses of chlorine and oxygen isotopes that streamflow owes little to current rainfall.

Piston flow could well be part of the answer. Certainly, the author has observed displaced soil water with exceptionally high dissolved salts during the first ephemeral pipeflow after the summer dry period on Maesnant. Preliminary results from the Maesnant model, combining field measurements of hydraulic conductivity and soil profile observations with known pipe sizes, suggest that piston flow *could* account for a large part of the response. However, the detailed water quality analyses of Hyett (1990) suggest that the average response involves a substantial proportion of 'new' water (see below).

DISTINGUISHING SEEPAGE LANDFORMS

Interest in this long-standing problem has been revived by attempts to interpret the Martian landscape (Higgins 1982, 1984; Baker 1982, 1985; Howard 1986a) and by recent concern over the role of subsurface processes in riverbank erosion (Jones 1989; Hagerty 1991a, 1991b, 1992). The problem has two facets: (1) are the landforms created directly by subsurface erosion or exfiltration distinctive, and (2) are the indirect products caused by exposure of subsurface landforms distinguishable from the products of surface erosion?

Surface landforms created by subsurface processes

Most interest has been focused on the development of linear drainage features. However, Higgins (1990b) has widened the issue to consider broad-scale cliff recession.

Features generally regarded as evidence of subsurface origin mainly comprise: (1) 'sharp-edged' features, especially gullies with broad, flat bottoms, steep sides and uniform width; (2) gully networks with stubby tributaries or blind valleys meeting with large junction angles; (3) drainage networks that show strong structural control or may be markedly elongated and roughly equispaced; (4) partial collapse features such as bridges and short tunnels; (5) haystack hills. Features (1) and (2) seem to have been amply confirmed by Howard's (1986b) simulation model. Haystack hills are a common feature of badlands containing piping, although the relative importance of subsurface processes is still open to question. But perhaps the greatest doubt centres on the question of structural control raised, for example, by Dunne (1990). Howard (1986a), Laity and Malin (1985) and Campbell (1973) have linked uniform spacing with guidance by master joints in the Colorado Plateaux. Conversely, Howard's (1986b) model of seepage erosion produced equispaced channels due to competition for drainage water. Clearly, neither of these controls is unique to subsurface erosion, but evidence from soil pipe networks at least suggests that (1) structural controls may be more marked and (2) that competition may be less important than for surface networks. Jones (1981: 136ff.) noted that these networks tend to exhibit more 'primitive' features than stream channels, such as greater clustering of junctions, triple junctions, more anastomosing channels and the persistent influence of cracking. Jones (1978b) found greater clustering of junctions compared with stream networks in pipe networks from Arizona to Wales. N.O. Jones (1968) detected the effect of polygonal cracking in the largely collapsed pipe-gully networks of the San Pedro valley, Arizona. Pipe networks in Wales show the persistence of desiccation cracking in the pattern of ephemeral pipes (Gilman and Newson 1980; Jones 1982) and subsidence cracking in perennials (Jones 1978a). Similarly, evidence that many British pipe networks are transmitters of water from higher slopes rather than onslope collectors (Weyman 1974; Gilman and Newson 1980) would seem to suggest that local midslope competition may be less important. The even spacing of perennial pipes on Maesnant seems to be largely governed by discharge from the groundwater bogs which lie transversely across the midslope at their heads. However, analysis of trans- and cis-link frequencies in the collapsed Arizonan networks shows a more 'normal' pattern, suggesting a degree of competitive adjustment weeding out close same-side tributaries (Jones 1981: 147).

It seems unlikely that many of the traits of small-scale subsurface networks will persist long after exhumation and merging with larger surface networks. Gullies formed by the coalescence of a number of adjacent pipes, as described by Higgins (1990a) in the case of Kutz Canyon, New Mexico, are bound to lose most of the characteristics of individual pipes. Not only will surface processes tend to remove evidence of former provenance, but many gullies seem to have mixed origins from the outset (e.g. Blong *et al.* 1982). In addition, the exhumed features may not be so different in form by the time they are sufficiently developed to collapse.

There is evidence that the hydraulic geometries of pipes approach those of local surface channels as they mature (Jones 1981: 119ff., 1987b). In lieu of discharge measurements being available for comparison, Table 5.4 is based on calculated relationships between the maximum width, W, and capacity cross-sectional area, A. The table suggests that the more 'mature' perennial pipes discharging directly into Burbage Brook have geometries closer to normal open channels than do disconnected ephemeral pipes on the upper hillslopes of the Wye. Similarly, the higher exponent for partially collapsed pipes issuing into semi-arid gullies in Arizona is reflected in open channels in the semi-arid south-western USA. In each case, higher exponents indicate greater lateral corrasion in more mature channels.

Despite such difficulties in identifying subsurface origins, there is abundant evidence of the importance of subsurface erosion in the initiation of gullies, and latterly of rills also. Most of this evidence comes from the drylands, where Parker and Higgins (1990) state that piping is 'a major mode

Table 5.4 Width–area relationships in soil pipes and open channels

Downstream width–area relationship	Pipe form	Location	Source
$W \propto A^{0.33}$	Ephemeral (multiple)	Upper Wye, UK	Based on data from Morgan (1977)
$W \propto A^{0.61}$	Perennial (multiple)	Maesnant, UK	Author's unpublished data
$W \propto A^{0.43}$	Perennial (single)	Burbage Brook, UK	Jones (1981)
$W \propto A^{0.62}$	Perennial (single)	Burbage Brook, UK	Jones (1981)
$W \propto A^{0.76}$	Ephemeral (single)	San Pedro River, Arizona	Based on data from Jones (1968)

Comparative relationships for open channels (deduced from width–area–discharge formulae of sources)

$W \propto A^{0.48}$		Pennsylvania, USA	Wolman (1955)
$W \propto A^{0.56}$		Humid midwest, USA	Leopold and Maddock (1953)
$W \propto A^{0.63}$		Semi-arid SW USA	Leopold and Miller (1956)

of gully development', although it may occur so rapidly that traces of subsurface origin are not preserved. De Ploey (1974) regarded piping as a dominant process in Tunisia and responsible for the advance of frontal gullies. Dardis and Beckedahl (1988) state that it is a 'major cause' of gullying in African drylands. Other recent descriptions of gullying initiated by pipe erosion come from Rooyani (1985) in Lesotho, Jacobberger (1988) in Mali and Stocking (1981) in Zimbabwe. In a less arid environment, Higgins (1990a) concludes that 'many, if not most gullies' in an area near Davis, California, derive from piping, and in San Mateo County, California, Swanson (1983) explained gully extension in terms of sapping or piping. Numerous other descriptions from humid environments, e.g. Galarowski (1976) in Poland, are reported in Jones (1981). In addition, rilling can be initiated by diffuse sapping in rill walls, as found by Savat and De Ploey (1982), and by pipe collapse, as indicated by field observations in loessial soils near Louvain, Belgium (Figure 5.10).

A number of reports refer to interesting 'symbiotic' sequences whereby cycles of piping and gullying succeed each other within the same network, e.g. linking discontinuous gullies (Heede 1976; Higgins 1990a), and even within the same channels, causing cyclical lowering of the beds (Aghassy 1973; Hamilton 1970). Pipe-gully erosion typically proceeds by stepwise slumping of steep headwalls or the collapse of gullyhead tunnels. Not uncommonly in the British uplands slumping in headwater hollows takes the form of a series of broad 'stepped crescents', which indicate undermining by seepage or piping. The role of landslide mechanics in this process recalls the observations of seepage-induced channel networks by Dietrich et al. (1986) in California, and the concentration of landslides in piped hollows in Honshu (Tsukamoto et al. 1982).

Dunne (1990) quite rightly points out, however, that steep headwalls are not universally associated with subsurface erosion and that pointed head-forms can occur. Howard's (1990) computer simulation experiment seems to resolve the apparent conflict between the blunt-ended forms commonly created by sapping in non-cohesive beach material (Higgins 1984) and narrower pointed channels created in other landscapes in terms of the magnitude of critical discharge needed for erosion, which increases with cohesion. In cohesive materials only a small number of active seepage nodes have sufficient discharge to undergo erosion.

Piping can have a similar effect, creating spatially restricting erosion. Jones (1981: 82) observed miniature pointed channel networks developing on mud-flats in the Fal estuary, Cornwall, where the coarse beach material has been covered by a thin cohesive overlayer of 73% silt and 20% clay. The miniature channels emerge from open piping, which has formed at the interface between the cohesive overlayer and the underlayer with a 37% gravel content. Figure 5.11 illustrates the head region of the northern branch of Maesnant, which is plainly a streamhead that is derived from the collapse of

91

Figure 5.10 Partially collapsed ephemeral piping creating rills on De Ploey's
experimental plot at Huldenberg, Louvain, Belgium

piping. Perhaps this case offers some of the 'proof' that Dunne (1990)
believes is lacking for Jones's (1971) hypothesis that such tunnels develop
into stream channels. The evidence of McCaig (1983, 1984) that most erosion
in the Slithero Clough, Yorkshire, derives from a network of streamhead
pipes adds further support, as do Carling's (1986) and Newson's (1980)
reports of gullies actively extending along perennial pipes. Figure 5.12
illustrates another case from Burbage Brook, showing a tributary in the
making, where piping has eroded to water-table level and is approaching the
local baselevel determined by the stream channel. Even so, the vast majority

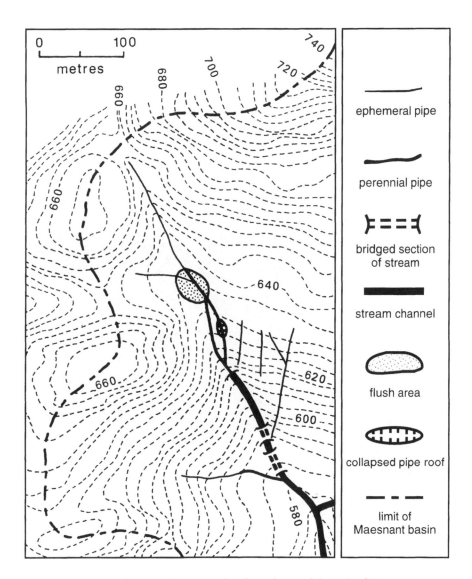

Figure 5.11 The 'pipeflow streamhead' on the north branch of Maesnant

of pipes in the British landscape are not undergoing active collapse, because the landsurface as a whole is not undergoing rapid erosion, and hence, as Dunne (1990) and Jones (1987b) suggest, they exist in quasi-equilibrium with the surface channels.

(a)　　　　　　　　　　(b)

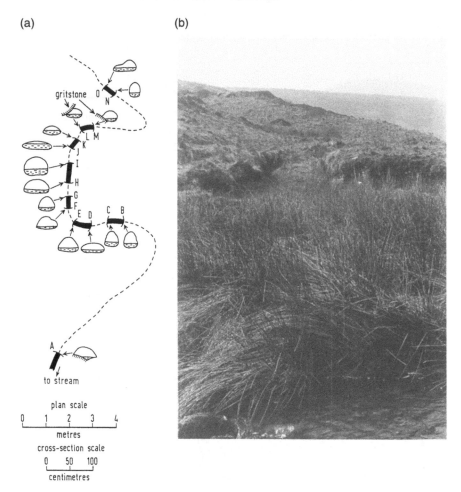

Figure 5.12 Partially collapsed perennial piping tributary to the Burbage Brook, Derbyshire: (a) plan and cross-sections, (b) view from stream

Longevity of subsurface landforms

It is commonly assumed that piping landforms are relatively short-lived; that either soil creep or blockage eventually causes infilling, or the roofs collapse. Dunne (1990) states that tunnels are ephemeral landforms, which are either self-destructive or destroyed by surface erosion, although his qualification 'on a geological time scale' removes much of the force from the observation – what landforms are not? But how long do they survive and is it possible that pipes might remain in equilibrium with slope retreat and not be destroyed? Many landscapes appear to be predominantly eroding subcutaneously rather than surficially, as evidenced by numerous New Zealand writers (e.g. Cumberland 1944; Gibbs 1945). Indeed, the lack of evidence of Hortonian overland flow in humid, temperate regions (Chorley 1978) means that landsurface degradation must be achieved by subsurface or mixed-source flow, a conclusion supported by McCaig's (1984) Gerlach trough experiment. For subsurface landforms to survive surface subsidence, the collapse must be gradual and widespread, creating swales or general slope retreat rather than 'sharp-edged' features, or else the landforms must undergo cyclical renewal.

Understandably, assessing the age of pipe networks is extremely difficult. Most observations of development have been limited to unstable and rapidly evolving systems. Dating pipes according to the age of the surrounding peat, as done by Rasmussen (1972) in the Faeroe Islands, begs a major question. On Maesnant, it seems evident that the perennial piping owes a lot to stream rejuvenation which cut a terrace into the solifluced till and created the necessary hydraulic gradient (Figure 5.21b). This could have occurred during the Hypsithermal when Wales may have experienced 16% more rainfall (Lamb 1977), but man-induced deforestation since the Mesolithic (Taylor 1980) and higher rainfall during the Early Mediaeval Optimum could also have increased stream power. Whichever was most important, these perennial pipes seem to be many centuries, if not millennia, old (cf. Jones 1987a). In contrast, the ephemeral pipes are prone to short-term change associated with desiccation cracking in years of extreme drought (Gilman and Newson 1980).

There is more conclusive evidence of very old pipes in America. But these are fossil pipes in clay horizons of buried soils developed during Pleistocene pluvials in New Mexico (Gile and Grossman 1968), or in Tertiary deposits in North Dakota which show an active period of piping during the Eocene and renewed activity in the Holocene associated with a semi-arid climate (Bell 1968). There are probably many other fossil pipes in the drylands that have simply been bypassed by the foci of erosion.

THE ENVIRONMENTAL IMPACT OF PIPING: A BRITISH PERSPECTIVE

In global terms, piping is commonly associated with accelerated soil erosion. It appears to be a normal feature of badlands (Bryan and Yair 1982; Harvey 1982; Gutierrez-Elorza and Rodriguez-Vidal 1984) and to be frequently associated with environmental degradation following deforestation or excessive agricultural practices (Jones 1981: 44ff.), such as over-irrigation (Garcia-Ruiz *et al.* 1986; Harris and Fletcher 1951), overgrazing (Parker 1963; Neboit 1971) and careless ploughing on steep slopes (Aghassy 1973). Masannat (1980) reports that almost 50% of agricultural land in the San Pedro valley near Benson, Arizona, has been destroyed by erosion initiated by piping. The experiments of the Soil Conservation Service of New South Wales offer the most extensive case histories of attempts to control such runaway development (Newman and Phillips 1957; Floyd 1974; Crouch et al. 1986; Boucher 1990).

Impacts are more subtle in Britain, where piping is probably a *victim* of the plough rather than a product. Recent work in the Maesnant basin has focused on its effects on acid runoff (Jones and Hyett 1987; Hyett 1990), and on solute movement, soil development and wetland ecology (Jones et al. 1991; Richardson 1992).

Acid rain

Jones and Hyett (1987) suggested that piping exacerbates problems of acid runoff in upland Britain by reducing buffering effects as a result of (1) more rapid transmission and reduced residence times, and (2) directing through-flow through upper organic horizons and reducing contact between storm-water and weathering mineral surfaces. Drainage and aeration of large sections of the hillside can also encourage the release of sulphates and organic acids from peaty soils, which increases the acidity of runoff. Consequently, the perennial pipes tend to issue more acidic water during storms and thereby contribute to 'acid flushes' in the stream, which ecologists believe may be the main source of acid rain impact on fish and microfauna (Gee and Stoner 1989; Edwards et al. 1990).

The most significant acidic contributions come from those perennial pipe networks with more flashy responses, which tend to be those with more extensive networks of tributary ephemeral pipes. However, in basins where significant contributions of runoff are made by ephemeral pipes issuing directly to the stream, the effects could be even more marked. Detailed monitoring by Hyett (1990) on Maesnant revealed average yields from ephemeral pipes that were more acidic than the rainfall, pH 3.8–4.2 against 4.8 for rainwater. Table 5.5 clearly demonstrates (1) the more acidic discharge of the ephemeral pipes, (2) the higher acidity of pipeflow than either rainfall or streamflow, (3) the greater acidity of stormflow in the large perennial pipe (number 4), and (4) an increase in acidity between the heads and the outfalls of the perennial pipes. The higher pH at the heads of the perennial pipes (Figure 5.13) is due to mid-slope bogs which are fed by deep groundwater that has had longer periods of residence in contact with weathering mineral surfaces.

Figure 5.14 illustrates another salient feature of Table 5.5, namely, the concentrations of dissolved aluminium, which closely follow the pattern of pH. Aluminium seems to be the main cause of fish-kills as it leads to asphyxiation. All pipe sources here are generally above minimum toxicity levels. Stream aluminium seems to be predominantly driven by pipeflow, with groundwater having near-zero concentrations. Again, the greatest risk

Table 5.5 Mean water quality parameters for rainfall, streamflow and pipeflow sites on Maesnant

	Rainfall	Stream	Perennial pipe outfalls			Heads of perennials		Ephemeral pipe outfalls		Pipe average (n = 63)	Storms only
			pipe 2	pipe 3	pipe 4	pipe 5	pipe 9	pipe 13	pipe 14		pipe 4
pH	4.84	5.16	4.90	4.48	4.52	5.20	5.50	4.26	4.10	4.58	4.27
Conductivity (μS cm^{-1} at 20°C)	39.5	34.1	35.8	41.9	40.0	39.9	33.4	57.4	47.0	41.5	47.0
Dissolved aluminium (mg l^{-1})	–	0.211	0.162	0.238	0.295	0.104	0.208	0.524	0.370	0.271	0.206
Dissolved organic carbon (mg l^{-1})	–	2.69	4.11	4.95	3.19	1.60	1.40	2.20	15.60	3.81	3.48

Source: Based on data presented by Jones and Hyett (1987). Pipe reference numbers as used in Jones (1987a)

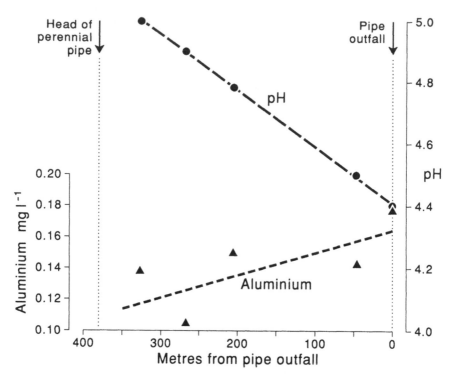

Figure 5.13 pH and aluminium concentrations down the largest perennial pipe on
Maesnant
Source: Based on data presented by Jones and Hyett 1987

to stream fauna is likely to occur in basins with high runoff contributions
from ephemeral pipes, in which both total aluminium and the more toxic
monomeric form are highest.

The role of pipeflow as a source of both discharge and aluminium seems
to have important implications for 'source area' protection as a method of
reducing the impact of acid rain. It suggests that efficient targeting of liming
should concentrate on the pipeflow contributing areas. This extends the
proposal that target source areas should be identified using the a/s index,
which was put forward after bankside forest clearance by the Forestry
Commission and treatment of a 10 m wide riparian swath failed to affect
stream pH in part of the Welsh Acid Waters experiments (Gee and Stoner
1989; Edwards *et al.* 1990). It also suggests that 'no-go areas' for conifer
planting in the uplands should be extended to cover pipeflow contributing
areas, in order to avoid feeding the system with waters of even higher acidity.

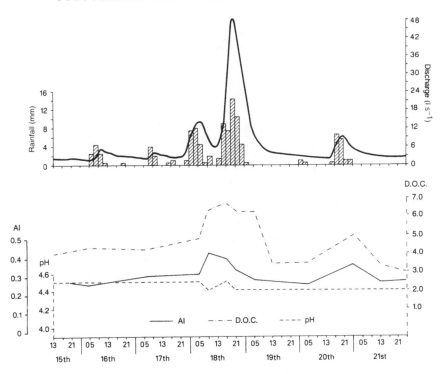

Figure 5.14 The response of pipeflow water quality to storm events on Maesnant in
November 1986. Note the increase in aluminium concentrations and the
perturbation in pH associated with a major storm
Source: Based on data presented by Jones and Hyett 1987

Plant ecology and soil development

The latest Maesnant surveys show that the pattern of swale microtopography
created by piping (Jones 1987a) is replicated in soil profile development and
vegetation associations.

Measurements of nutrient concentrations in the top 150 mm of soil in 234
quadrats in the $0.5 \, km^2$ basin by Richardson (1992) suggest that the
ephemeral pipes play an important role in the redistribution of nutrients.
Figure 5.15 shows two main areas of high topsoil solute levels: (1) the two
streamhead blanket peat areas, and (2) the outfall area for the main group of
ephemeral pipes at lower right. The latter seem to be depleting solutes
upslope and depositing them on the surface below their resurgences. The
pattern of pipeflow enrichment is also seen in maps of potassium, phosphor-
ous and magnesium. Earlier unpublished work by Richardson suggests that
the depressions around the lower sections of the perennial pipes have
similarly elevated nutrient levels at around pipe depth (0.5 m).

Drainage and aeration of the hillslopes by the pipes, together with material

99

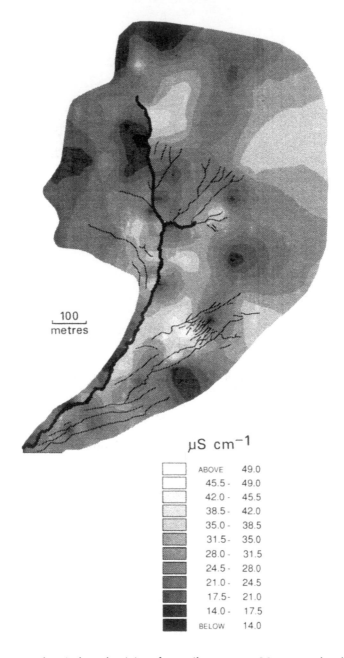

μS cm⁻¹

ABOVE	49.0
45.5 -	49.0
42.0 -	45.5
38.5 -	42.0
35.0 -	38.5
31.5 -	35.0
28.0 -	31.5
24.5 -	28.0
21.0 -	24.5
17.5 -	21.0
14.0 -	17.5
BELOW	14.0

Figure 5.15 Electrical conductivity of topsoil extracts on Maesnant related to the
pipe network
Source: Based on Richardson 1992

and solute transport, demonstrably affect the environment for plant growth and the differentiation of soil profiles (Jones *et al.* 1991), and this is borne out in Figures 5.16 and 5.17. The pipes run predominantly beneath shallow, grass-covered swales separated by interfluves covered by heath vegetation. Figure 5.18 shows the strong association with grassland: 92% of perennial pipes and 75% of ephemeral pipes occur under grassland; the higher concentration among perennial pipes may indicate their greater control on the micro-environment.

Quadrat ordination suggests that most pipes tend to have a drier than average environment, and to occur in the more mesotrophic areas. Figure 5.19 plots the piped quadrats in terms of the most significant axes of ordination. Axis I is clearly a wetness axis; axis II is less clear, but includes an important element of nutrient status. Pipes are noticeably absent from the drier region near the origin, despite 27 quadrats being measured in this range, and similarly absent from the wetter top right of the graph, despite 67 quadrats in this region. Their propensity for the 'middle ground' seems likely to result from a balance between opposing tendencies, namely, (1) the need for water flow to maintain the pipes and (2) their draining effect on the micro-environment.

Similarly, the marked banding in soil types (Figure 5.17) and its correspondence with the pipe network suggest that this must be the product rather than a cause of soil piping. Drainage of the upper slope (top right) by ephemeral pipes seems to reduce the accumulation of peat and to develop an eluvial, podzolic horizon and a shallow loamy ferric stagnopodzol profile. On the lower slope, aeration and erosion around the perennial pipes has caused thinning of the peat (Figure 5.20) and created oligo-amorphous humified peat soils.

Monitoring of groundwater levels around selected perennial and ephemeral pipes over a two-year period has shown that both categories of pipe drain phreatic water, and that stormflow is only initiated in the ephemeral pipes when the phreatic surface intersects the floor of the pipe (Connolly 1993). At the beginning of a storm, however, hydraulic gradients can become reversed immediately around the pipe due to influent seepage. Dipwells nearest to the pipes could overflow onto the surface before any reaction was observed in dipwells on either side. They also tended to mimic pipeflow stage levels very closely, whereas those in the heath vegetation on the micro-interfluves failed to exhibit the detailed fluctuations (Jones *et al.* 1991). Data logger records of the behaviour of the phreatic surface, taken at 10-minute intervals at 5 cross-profiles spaced down the swale surrounding the main pipe network, suggest that the microtopographic depressions channel waves of throughflow parallel to the pipes during storms. These waves are propagated through a zone of higher hydraulic conductivity immediately around the pipes (0.0030 mm s^{-1} against 0.0018 on the edge of the swale), which has probably been created by a combination of oxidation of the peat and erosion

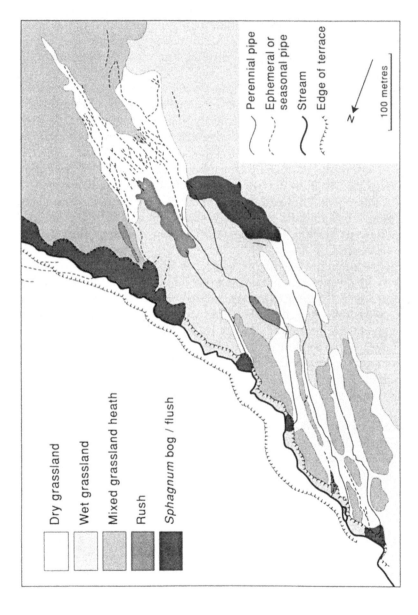

Figure 5.16 The distribution of major plant associations around piping on the lower Maesnant slope
Source: As presented by Jones et al. 1991

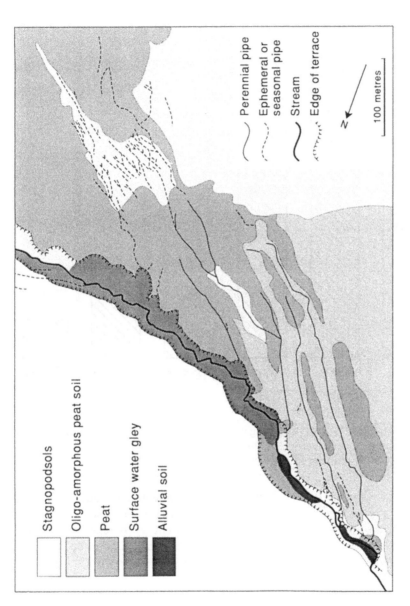

Stagnopodsols

Oligo-amorphous peat soil

Peat

Surface water gley

Alluvial soil

Perennial pipe

Ephemeral or seasonal pipe

Stream

Edge of terrace

100 metres

N

Figure 5.17 The distribution of soil groups around piping on the lower Maesnant slope
Source: As presented by Jones *et al.* 1991

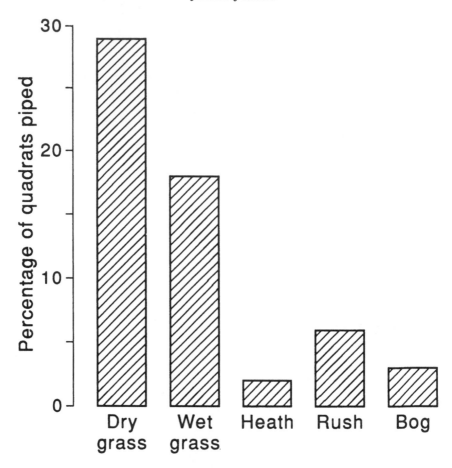

Figure 5.18 The frequency of piping in quadrat samples of major plant associations on Maesnant
Source: Based on the data of Richardson 1992

caused by rapid hydraulic drawdown. Consequently, although the actual walls of the pipes tend to be depleted in nutrients, because of prolonged leaching, the swales in general have their nutrient levels repeatedly replenished during storms. These nutrients may be channelled from up to 3.56 ha of hillside into the swale at the lower end of the largest network.

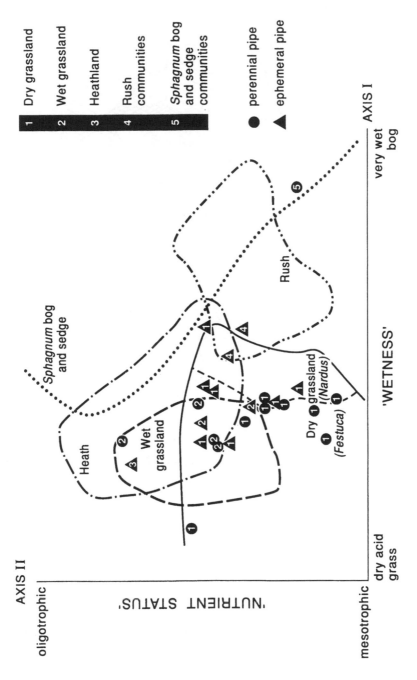

Figure 5.19 Quadrat ordination showing distribution of piped quadrats in relation to major vegetation associations

Source: Simplified from Richardson 1992

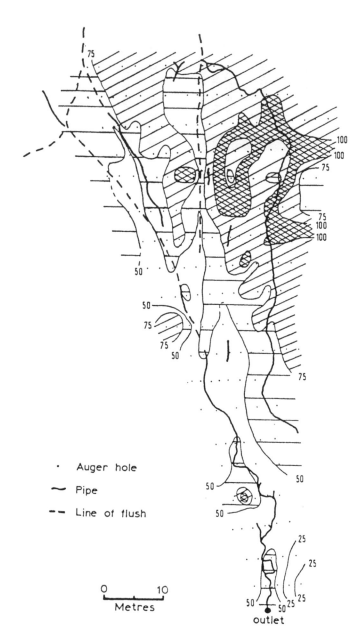

Figure 5.20 Peat thickness (cm) around perennial piping in the lower Maesnant

TOPOGRAPHIC AND CLIMATIC CONTROLS ON THE DISTRIBUTION OF PIPING IN BRITAIN

Intrabasin distribution

Whilst piping plainly influences pedogenesis on Maesnant, its distribution is also subject to a number of pre-existing topographic and pedologic controls. Piping is not encountered within eutro-amorphous peat soils, where conductivity is too low, or in the extremely shallow rankers on the higher slopes (udorthents and lithic histosols), which are too coarse, or the stony alluvial soils along the streambanks, which lack sufficient hydraulic gradient (Jones *et al.* 1991).

Figure 5.21 explores the relationship between the a/s index of water concentration and the occurrence of pipes. The area drained per unit contour length, a, should control the degree of concentration for sheetwash and diffuse throughflow, whereas lower slope, s, should increase saturation (Carson and Kirkby 1972). The maps indicate a broad correspondence at the catchment scale, but also areas of marked discord. Ephemeral pipes begin around a/s = 0.25–1.0 km on the southern slopes, but around a/s = 2.6 km in the eastern headwater area and at a/s = 0.1 km in the stream source area (Figure 5.21a). Perennial pipes begin around a/s = 2 km on the southern slopes but only 0.5 km in the headwaters. These figures can be compared with a critical a/s of 0.5 km for perennial overland flow calculated by Kirkby (1978) for similar high rainfall areas. Figure 5.21b shows the breakdown of the relationship in detail on the southern slopes, as discovered by Jones (1986). The storm discharge and sediment yield graphs in Figure 5.22 reflect the weak association, in which the only statistically significant relationship depends on the inclusion of the largest perennial pipe with the best developed surface micro-catchment.

The implication is that piping is partly controlled by factors that are antipathetic to a/s, e.g. (1) subsurface erosion is encouraged by high slopes, (2) desiccation cracking tends to occur on exposed, even convex, hillslopes, (3) pipes generate a subsurface collecting network that may owe more to in-profile variations than to surface topography, and (4) they can persist as allogenic channels through 'unfavourable' areas.

National distribution

The first attempt at a comprehensive survey of piping in Britain was undertaken by Richardson (1992), using published reports and a questionnaire survey of field scientists. Figure 5.23 shows the distribution of 74 piped catchments in relation to upland soils in the Soil Survey's Winter Rainfall Acceptance Potential (WRAP) groups 2 and 5 (Farquharson *et al.* 1978). Gilman and Newson (1980) deductively predicted that piping should

(a)

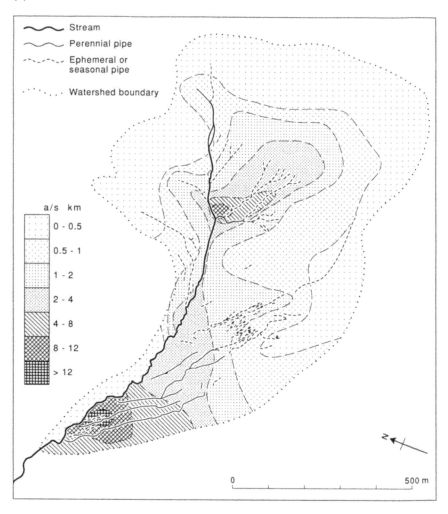

Figure 5.21 Piping in relation to the a/s index: (a) general map of Maesnant, based on 4 m contour resolution 1:7500 photogrammetric map; (b) detail of lower slopes, based on 1 m contour resolution 1:1000 ground survey plan

predominate in the uplands and in class 5 soils and diffuse throughflow in the more permeable class 2 soils. Figure 5.23 confirms that piping occurs primarily in the impermeable and poorly drained soils of WRAP class 5, and that it is relatively widespread in the uplands.

Abstraction of topographic, pedological and climatic data for these

(b)

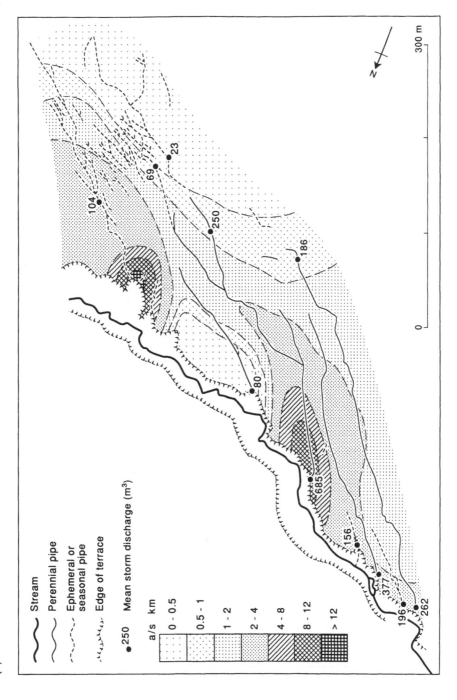

Stream

Perennial pipe

Ephemeral or
seasonal pipe

Edge of terrace

●250 Mean storm discharge (m³)

a/s km

	0 - 0.5
	0.5 - 1
	1 - 2
	2 - 4
	4 - 8
	8 - 12
	> 12

N

0   300 m

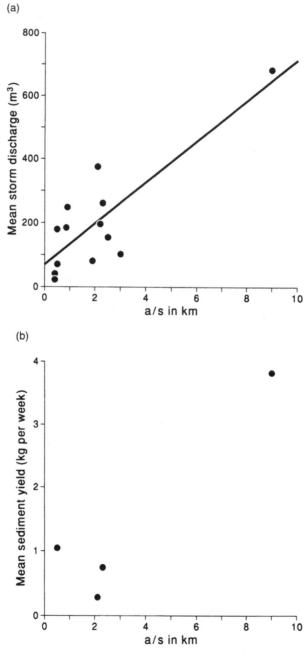

Figure 5.22 Mean stormflow discharge (a) and sediment yield (b) compared with the a/s index for Maesnant pipes. Only (a) shows a statistically significant relationship ($r = 0.84$, $\alpha = 1\%$), but this is due solely to the largest pipe

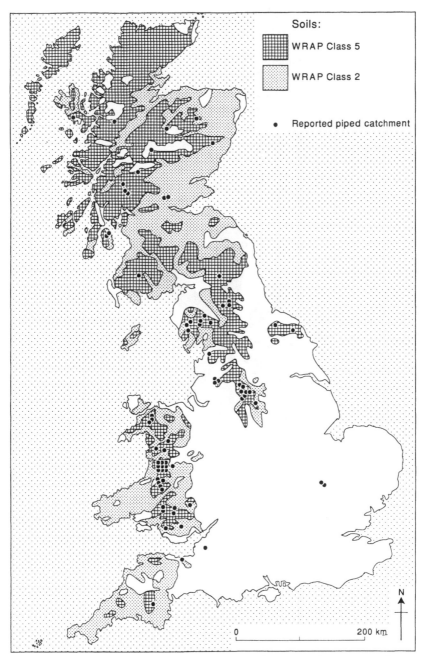

Figure 5.23 Distribution of piping in Britain in relation to soils in WRAP classes
5 and 2 in the uplands above 150 m OD

Sources: Compiled from Richardson 1992 and Gilman and Newson's 1980 generalisation and
extrapolation of soil classes from Farquharson *et al.* 1978

catchments revealed a number of consistent patterns. Figure 5.24 shows a predominance of south-facing catchments (significant at 5%), suggesting an important role for desiccation cracking in the initiation of pipes. Aspect has not been seriously considered as a factor in Britain, in contrast to observations in Victoria and New Zealand (Jones 1981: 53) and the quantitative support for the effect of desiccation from Hughes (1972).

Figure 5.25 shows a distinct concentration of piping in peats and podzols, i.e. histosols and aquods. These soils offer a number of favourable properties, such as high shrinkage coefficients, stable roofing, erodible subsoils and impeded drainage in mid-profile (Jones 1981). They also tend to be located

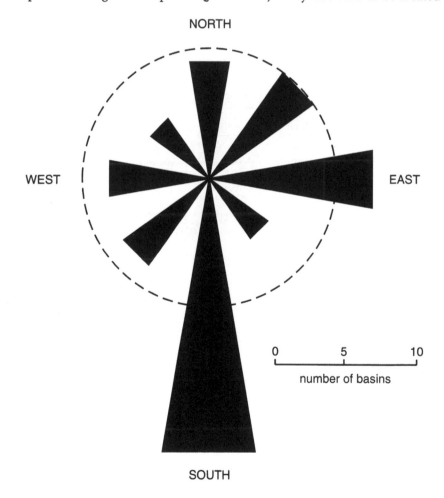

Figure 5.24 Catchment orientation and piping frequency in Britain. Unshaded bars indicate the null hypothesis of even distribution
Source: After Richardson 1992

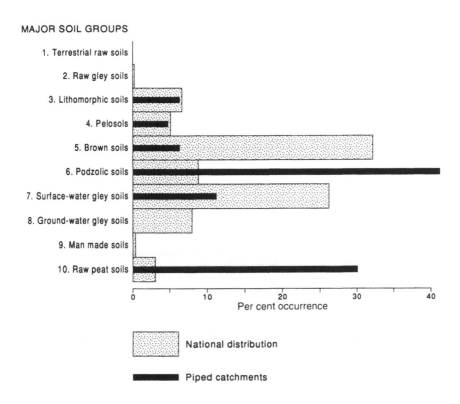

MAJOR SOIL GROUPS

Figure 5.25 Distribution of piping in major soil groups in Britain compared with the national distribution of soils according to Avery (1980)
Source: After Richardson 1992

in environments with high water balance surplus and steep slopes.

Figure 5.26 compares the results of Richardson's survey with an estimate of national frequency distributions based on data used in the UK Flood Studies Report (NERC 1975). Piping peaks somewhat above the national average in all respects – higher rainfall, altitude and mainstream slope. The distribution seems to indicate the importance of the factors that control 'subsurface stream power', i.e. the erosive potential. It is possible that the distribution has been partially curtailed in the lowlands by agricultural activity, but this seems unlikely to have been very marked. A survey in the New Forest, a lowland area protected from agriculture for almost a millennium, revealed no clear evidence of piping (Jones 1978a), and Conacher and Dalrymple (1977) report only short sections of small pipes in uncultivated soils on hills in Berkshire.

113

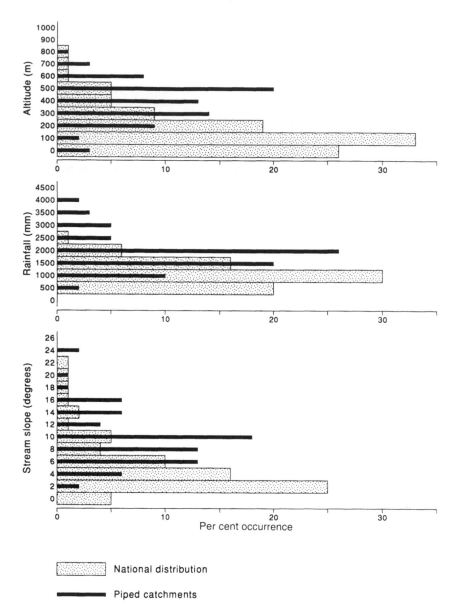

Figure 5.26 The topographic and climatic distribution of piping in Britain compared with national distributions of catchment parameters derived from the UK Flood Studies Report (NERC 1975) by Richardson (1992)

CONCLUSIONS

Research over the last quarter century has dramatically expanded our appreciation of the importance of subsurface processes in creating or influencing surface form. There are still many aspects that need further research, especially controls on the distribution of subsurface processes, and there will always remain problems of identifying the origins of many surface products, especially where they are multi-source landforms. Equally, it is important to recognise the potential effects of subsurface drainage processes on surface water quality and ecology.

ACKNOWLEDGEMENTS

Research at Maesnant has been conducted with the assistance of Natural Environment Research Council grants GR3/3683, GR3/6792 and GT4/86/AAPS/52, and support from the University of Wales. Francis Crane, Glyn Hyett, Mark Richardson, Liam Connolly, Lindsay Collin, Neil Chisholm and Peter Wathern all made valuable contributions. This is Publication Number 254 of the Institute of Earth Studies.

REFERENCES

Aghassy, J. (1973) 'Man-induced badlands topography', in D.R. Coates (ed.) *Environmental Geomorphology and Landscape Conservation*, vol. III: *Non-urban*, Stroudsberg: Dowden, Hutchinson & Ross.

Aitchison, G.D., Ingles, O.G. and Wood, C.C. (1963) 'Post-construction deflocculation as a contributory factor in the failure of earth dams', *Proceedings of 4th Australia–New Zealand Conference on Soil Mechanics and Foundation Engineering*, Adelaide: 275–279.

Anderson, M.G. and Burt, T.P. (eds) (1990a) *Process Studies in Hillslope Hydrology*, Chichester: John Wiley.

——(1990b) 'Subsurface runoff', in M.G. Anderson and T.P. Burt (eds) (1990a) *Process Studies in Hillslope Hydrology*, Chichester: John Wiley.

Avery, B.W. (1980) 'Soil classification for England and Wales', *Soil Survey of England and Wales Monograph* 14.

Baker, V.R. (1982) *The Channels of Mars*, Austin: University of Texas Press.

——(1985) 'Models of fluvial activity on Mars', in M.J. Woldenberg (ed.) *Models in Geomorphology*, London: Allen & Unwin.

Baker, V.R., Kochal, R.C., Laity, J.E. and Howard, A.D. (1990) 'Spring sapping and valley network development', *Geological Society of America Special Paper* 252: 235–265.

Bell, G.L. (1968) 'Piping in the badlands of North Dakota', *Proceedings of the 6th Annual Engineering Geology and Soils Engineering Symposium*, Boise, Idaho: 242–257.

Blong, R.J., Graham, O.P. and Veness, J.A. (1982) 'The role of sidewall processes in gully development: some N.S.W. examples', *Earth Surface Processes and Landforms* 7: 381–385.

Boucher, S.G. (1990) *Field Tunnel Erosion: Its Characteristics and Amelioration*, Clayton, Victoria: Monash University.

Bryan, R.B. and Yair, A. (eds) (1982) *Badland Geomorphology and Piping*, Norwich: GeoBooks.

Burt, T.P. (1992) 'The hydrology of headwater catchments', in P. Calow and G.E. Petts (eds) *The River Handbook*, vol. 1, Oxford: Blackwell.

Burt, T.P., Heathwaite, A.L. and Labadz, J.C. (1990) 'Runoff production in peat-covered catchments', in M.G. Anderson and T.P. Burt (eds) *Process Studies in Hillslope Hydrology*, Chichester: John Wiley.

Campbell, I.A. (1973) 'Controls of canyon and meander form by jointing', *Area* 5: 291–296.

Carling, P.A. (1986) 'Peat slides in Teesdale and Weardale, Northern Pennines, July, 1983: description and failure mechanism', *Earth Surface Processes and Landforms* 11: 193–206.

Carson, M.A. and Kirkby, M.J. (1972) *Hillslope Form and Process*, Cambridge: Cambridge University Press.

Chorley, R.J. (1978) 'The hillslope hydrological cycle', in M.J. Kirkby (ed.) *Hillslope Hydrology*, Chichester: Wiley.

Coates, D.R. (1990) 'The relation of subsurface water to downslope movement and failure', *Geological Society of America Special Paper* 252: 51–76.

Conacher, A. and Dalrymple, J.B. (1977) 'The nine unit landsurface model: an approach to pedogeomorphic research', *Geoderma* 18: 1–154.

Connolly, L.J. (1993) 'Modelling natural pipeflow contributions to the stormflow hydrograph', unpublished Ph.D. thesis, University of Wales, Aberystwyth.

Crouch, R.J., McGarity, J.W. and Storrier, R.R. (1986) 'Tunnel formation processes in the Riverina area of N.S.W., Australia', *Earth Surface Processes and Landforms* 11: 157–168.

Cryer, R. (1980) 'The chemical quality of some pipeflow waters in upland mid-Wales and its implications', *Cambria* 6: 1–19.

Cumberland, K.B. (1944) *Soil Erosion in New Zealand*, Wellington: Soil Conservation and River Control Council.

Dardis, G.F. and Beckedahl, H.R. (1988) 'Drainage evolution in an ephemeral soil pipe-gully system, Transkei, Southern Africa', in G.F. Dardis and B.P. Moon (eds) *Geomorphological Studies in Southern Africa*, Rotterdam: Balkema.

De Ploey, J. (1974) 'Mechanical properties of hillslopes and their relation to gullying in central semiarid Tunisia', *Zeitschrift für Geomorphologie*, N.F. 21: 177–190.

Dietrich, W.E., Wilson, C.J. and Reneau, S.L. (1986) 'Hollows, colluvium, and landslides in soil-mantled landscapes', in A.D. Abrahams (ed.) *Hillslope Processes*, Winchester, Mass.: Allen & Unwin.

Dunne, T. (1978) 'Field studies of hillslope processes', in M.J. Kirkby (ed.) *Hillslope Hydrology*, Chichester: John Wiley.

——(1980) 'Formation and controls of channel networks', *Progress in Physical Geography* 4: 211–239.

——(1990) 'Hydrology, mechanics, and geomorphic implications of erosion by subsurface flow', *Geological Society of America Special Paper* 252: 1–28.

Edwards, R., Gee, A. and Stoner, J. (1990) *Acid Waters in Wales*, Monographiae Biologicae 66, Dordrecht: Kluwer.

Farquharson, F.A.K., Mackney, D., Newson, M.D. and Thomasson, A.J. (1978) 'Estimation of runoff potential of river catchments from soil survey', *Soil Survey of England and Wales Special Survey* 11: 1–29.

Finlayson, B.L. (1977) 'Runoff contributing areas and erosion', *University of Oxford School of Geography Research Paper* 18.

Floyd, E.J. (1974) 'Tunnel erosion: a field study in the Riverina, New South Wales', *Journal of Soil Conservation Service, N.S.W.* 30: 145–156.

Galarowski, T. (1976) 'New observations of the present-day suffosion (piping) processes in the Bereznica catchment basin in the Bieszczady Mountains (the East Carpathians)', *Studia Geomorphologica Carpatho-Balcanica* 10: 115–122.

Garcia-Ruiz, J.M., Lasanta-Martinez, T., Ortigosa-Izquierdo, L. and Arnaez-Vadillo, J. (1986) 'Pipes in cultivated soils of La Rioja, Spain: origins and evolution', *Zeitschrift für Geomorphologie*, N.F., Supplement Band 58: 93–100.

Gardiner, A.T. (1983) 'Runoff and erosional processes in a peat-moorland catchment', unpublished M.Phil. thesis, Huddersfield Polytechnic.

Gee, A.S. and Stoner, J.H. (1989) 'A review of the causes and effects of acidification of surface waters in Wales and potential mitigating techniques', *Archives of Environmental Contamination and Toxicology* 18: 121–130.

Gerits, J.J.P., De Lima, J.L.M.P. and Van den Broek, T.M.W. (1990) 'Overland flow and erosion', in M.G. Anderson and T.P. Burt (eds) *Process Studies in Hillslope Hydrology*, Chichester: Wiley.

Gibbs, H.S. (1945) 'Tunnel-gully erosion in the Wither Hills, Marlborough, New Zealand', *New Zealand Journal of Science and Technology* 27, Section A(2): 135–146.

Gile, L.H. and Grossman, R.B. (1968) 'Morphology of the argillic horizon in desert soils of southern New Mexico', *Soil Science* 106: 6–15.

Gilman, K. and Newson, M.D. (1980) *Soil Pipes and Pipeflow – a Hydrological Study in Upland Wales*, Norwich: GeoBooks.

Gutierrez-Elorza, M. and Rodriguez-Vidal, J. (1984) 'Fenomenos de sufosion (piping) en la depresion media del Ebro', *Cuadernos de Investigacion Geografica (Logrono)* 10: 75–83.

Hagerty, D.J. (1991a) 'Piping/sapping erosion: I Basic considerations', *Journal of Hydraulic Engineering* 117: 991–1008.

——(1991b) 'Piping/sapping erosion: II Identification–diagnosis', *Journal of Hydraulic Engineering* 117: 1009–1025.

——(1992) *Identification of Piping and Sapping Erosion of Streambanks*, Final Report Contract HL-92, US Army Engineer Waterways Experiment Station, Vicksburg, Mississippi.

Hamilton, T.M. (1970) 'Channel-scarp formation in western North Dakota', *U.S. Geological Survey Professional Paper* 700-C: 229–232.

Harris, K. and Fletcher, J.E. (1951) 'Report of the so-called soil piping, Yolo Ranch, Yoma Country, Arizona', unpublished report, Soil Conservation Service, Arizona State Office, Phoenix, Arizona.

Harvey, A. (1982) 'The role of piping in the development of badlands and gully systems in south-east Spain', in R.B. Bryan and A. Yair (eds) *Badland Geomorphology and Piping*, Norwich: GeoBooks.

Heede, B.H. (1976) 'Gully development and control: the status of our knowledge', *US Department of Agriculture, Forest Service Research Paper* RM-68.

Higgins, C.G. (1982) 'Drainage systems developed by sapping on Earth and Mars', *Geology* 10: 147–152.

——(1984) 'Piping and sapping: development of landforms by groundwater outflow', in R.G. La Fleur (ed.) *Groundwater as a Geomorphic Agent*, Boston, Mass.: Allen & Unwin.

——(1990a) 'Gully development', *Geological Society of America Special Paper* 252: 139–156.

——(1990b) 'Seepage-induced cliff recession and regional denudation', *Geological Society of America Special Paper* 252: 291–318.

Higgins, C.G. and Coates, D.R. (eds) (1990) 'Groundwater geomorphology: the role of subsurface water in earth-surface processes and landforms', *Geological Society of America Special Paper* 252: 1–368.

Howard, A.D. (1986a) 'Groundwater sapping on Mars and Earth', in A.D. Howard,

R.C. Kochel and H.E. Holt (eds) *Proceedings and Field Guide, NASA Ground-water Sapping Conference, Flagstaff, Arizona*, 1985, National Aeronautics and Space Administration.

——(1986b) 'Groundwater sapping experiments and modelling at the University of Virginia', in A.D. Howard, R.C. Kochel and H.E. Holt (eds) *Proceedings and Field Guide, NASA Groundwater Sapping Conference, Flagstaff, Arizona*, 1985, National Aeronautics and Space Administration: 75–83.

——(1990) 'Case study: model studies of groundwater sapping', *Geological Society of America Special Paper* 252: 256–264.

Hughes, P.J. (1972) 'Slope aspect and tunnel erosion in the loess of Banks Peninsula, New Zealand', *New Zealand Journal of Hydrology* 11: 94–98.

Hyett, G.A. (1990) 'The effect of accelerated throughflow on the water yield chemistry under polluted rainfall', unpublished Ph.D. thesis, University of Wales, Aberystwyth.

Jacobberger, P.A. (1988) 'Drought-related changes to geomorphic processes in central Mali', *Geological Society of America Bulletin* 100: 351–361.

Jahn, A. (1968) 'Denudational balance of slopes', *Geographia Polonica* 13: 9–30.

Jones, J.A.A. (1971) 'Soil piping and stream channel initiation', *Water Resources Research* 7: 602–610.

——(1978a) 'Soil pipe networks – distribution and discharge', *Cambria* 5: 1–21.

——(1978b) 'The spacing of streams in a random-walk model', *Area* 10: 190–197.

——(1981) *The Nature of Soil Piping: A Review of Research*, Norwich: GeoBooks.

——(1982) 'Experimental studies of pipe hydrology', in R.B. Bryan and A. Yair (eds) *Badland Geomorphology and Piping*, Norwich: GeoBooks.

——(1986) 'Some limitations to the a/s index for predicting basin-wide patterns of soil water drainage', *Zeitschrift für Geomorphology*, N.F. Supplement Band 60: 7–20.

——(1987a) 'The effects of soil piping on contributing areas and erosion patterns', *Earth Surface Processes and Landforms* 12: 229–248.

——(1987b) 'The initiation of natural drainage networks', *Progress in Physical Geography* 11: 207–245.

——(1988) 'Modelling pipeflow contributions to stream runoff', *Hydrological Processes* 2: 1–17.

——(1989) 'Bank erosion: a review of British research', in M.A. Ports (ed.) *Hydraulic Engineering*, New York: American Society of Civil Engineers.

——(1990) 'Piping effects in humid lands', *Geological Society of America Special Paper* 252: 111–138.

Jones, J.A.A. and Crane, F.G. (1984) 'Pipeflow and pipe erosion in the Maesnant experimental catchment', in T.P. Burt and D.E. Walling (eds) *Catchment Experiments in Fluvial Geomorphology*, Norwich: GeoBooks.

Jones, J.A.A. and Hyett, G.A. (1987) 'The effect of natural pipeflow solutes on the quality of upland streamwater in Wales', *Abstracts, 19th General Assembly of the International Union of Geodesy and Geophysics*, Vancouver, 3: 998.

Jones, J.A.A., Wathern, P., Connolly, L.J. and Richardson, J.M. (1991) 'Modelling flow in natural soil pipes and its impact on plant ecology in mountain wetlands', *International Association of Hydrological Sciences Publication* 202: 131–142.

Jones, N.O. (1968) 'The development of piping erosion', unpublished Ph.D. thesis, University of Arizona.

Kirkby, M.J. (1978) 'Implications for sediment transport', in M.J. Kirkby (ed.) *Hillslope Hydrology*, Chichester: John Wiley.

——(1985) 'Hillslope hydrology', in M.G. Anderson and T.P. Burt (eds) *Hydrological Forecasting*, Chichester: John Wiley.

Kirkby, M.J. and Chorley, R.J. (1967) 'Throughflow, overland flow and erosion',

118

Bulletin of International Association of Scientific Hydrology 12: 5–21.

Knapp, B.J. (1974) 'Hillslope throughflow observations and the problem of modelling', *Institute of British Geographers Special Publication* 6: 23–31.

Laity, J.E. and Malin, M.C. (1985) 'Sapping processes and the development of theater-head valley networks in the Colorado Plateau', *Geological Society of America Bulletin* 96: 203–217.

Lamb, H.H. (1977) *Climate: Present, Past and Future*, vol. 2, London: Methuen.

Leopold, L.B. and Maddock, T., Jr (1953) 'The hydraulic geometry of stream channels and some physiographic implications', *U.S. Geological Survey Professional Paper* 252.

Leopold, L.B. and Miller, J.P. (1956) 'Ephemeral streams: hydraulic factors and their relation to the drainage net', *U.S. Geological Survey Professional Paper* 282-A: 1–37.

McCaig, M. (1983) 'Contributions to storm quickflow in a small headwater catchment: the role of natural pipes and soil macropores', *Earth Surface Processes and Landforms* 8: 239–252.

——(1984) 'The pattern of wash erosion around an upland streamhead', in T.P. Burt and D.E. Walling (eds) *Catchment Experiments in Fluvial Geomorphology*, Norwich: GeoBooks.

Masannat, Y.M. (1980) 'Development of piping erosion conditions in the Benson area, Arizona, U.S.A', *Quarterly Journal of Engineering Geology* 13: 53–61.

Morgan, A.L. (1977) 'An investigation of the location, geometry and hydraulics of ephemeral soil pipes on Plynlimon, mid-Wales', unpublished B.Sc. dissertation, University of Manchester.

Mosley, M.P. (1979) 'Streamflow generation in a forested catchment, New Zealand', *Water Resources Research* 15: 795–806.

Natural Environment Research Council (1975) *Flood Studies Report*, London: Natural Environment Research Council.

Neal, C. and Rosier, P.T.W. (1990) 'Chemical studies of chloride and stable oxygen isotopes in two conifer afforested and moorland sites in the British uplands', *Journal of Hydrology* 115: 269–283.

Neboit, R. (1971) 'Morphogénèse récente des formations tendres en Lucanie (Italie du Sud)', *Méditerranée* 2: 701–719.

Newman, J.C. and Phillips, J.R.H. (1957) 'Tunnel erosion in the Riverina', *Journal of Soil Conservation Service New South Wales* 13: 159–169.

Newson, M.D. (1980) 'The geomorphological effectiveness of floods: a contribution stimulated by two recent events in mid-Wales', *Earth Surface Processes* 5: 1–16.

Parker, G.G., Sr (1963) 'Piping, a geomorphic agent in landform development of the drylands', *International Association of Scientific Hydrology Publication* 65: 103–113.

Parker, G.G., Sr and Higgins, C.G. (1990) 'Piping and pseudokarst in drylands', *Geological Society of America Special Paper* 252: 77–110.

Rasmussen, E.J. (1972) 'Hugskot um air sum renna undir heilum', *Fródsksparrit* 20: 89–98.

Reynolds, B., Neal, C., Hornung, M. and Stevens, P.A. (1986) 'Baseflow buffering of streamflow acidity in five mid-Wales catchments', *Journal of Hydrology* 87: 167–185.

Richardson, J.M. (1992) 'Catchment characteristics and the distribution of natural soil piping', unpublished M.Phil. thesis, University of Wales, Aberystwyth.

Roberts, G., Hudson, J.A. and Blackie, J.R. (1984) 'Nutrient inputs and outputs in a forested and grassland catchment at Plynlimon, Mid Wales', *Agricultural Water Management* 9: 177–191.

Rooyani, F. (1985) 'A note on soil properties influencing piping at the contact zone between albic and argillic horizons of certain duplex soils (aqualfs) in Lesotho, southern Africa', *Soil Science* 139: 517–522.

Savat, J. and De Ploey, J. (1982) 'Sheetflow and rill development by surface flow', in R.B. Bryan and A. Yair (eds) *Badland Geomorphology and Piping*, Norwich: GeoBooks.

Sherard, J.L. and Decker, R.S. (eds) (1977) 'Dispersive clays, related piping, and erosion in geotechnical projects', *American Society for Testing Materials Special Technical Publication* 623.

Sklash, M.G., Beven, K.J., Gilman, K. and Darling, W.G. (in press, a) 'Isotope studies of pipeflow at Plynlimon, Wales, U.K., part 1. Low flow', *Hydrological Processes*.

——(in press, b) 'Isotope studies of pipeflow at Plynlimon, Wales, U.K., part 2. Storm flow', *Hydrological Processes*.

Stagg, M.J. (1974) 'Storm runoff in a small catchment in the Mendip Hills', unpublished M.Sc. thesis, University of Bristol.

Stocking, M.M. (1981) 'Model of piping in soils', *Transactions of Japan Geomorphological Union* 2: 263–278.

Swanson, M.L. (1983) 'Soil piping and gully erosion along coastal San Mateo County, California', unpublished MS thesis, University of California, Santa Cruz.

Taylor, J.A. (1980) 'Man–environment relationships', in J.A. Taylor (ed.) *Culture and Environment in Prehistoric Wales*, Oxford: British Archaeological Reports.

Tsukamoto, Y., Ohta, T. and Nogushi, H. (1982) 'Hydrological and geomorphological studies of debris slides on forested hillslopes in Japan', *International Association of Hydrological Sciences Publication* 137: 89–98.

Weyman, D.R. (1970) 'Throughflow on hillslopes and its relation to the stream hydrograph', *Bulletin of International Association of Scientific Hydrology* 15: 25–33.

——(1974) 'Runoff processes, contributing area and streamflow in a small upland catchment', *Institute of British Geographers Special Publication* 6: 33–43.

Whipkey, R.Z. and Kirkby, M.J. (1978) 'Flow within the soil', in M.J. Kirkby (ed.) *Hillslope Hydrology*, Chichester: John Wiley.

Wolman, M.G. (1955) 'The natural channel of Brandywine Creek, Pennsylvania', *U.S. Geological Survey Professional Paper* 271: 1–56.

6

TECTONICS IN GEOMORPHOLOGICAL MODELS

Michael J. Kirkby

INTRODUCTION

Although plate tectonics has been largely adopted as the dominant paradigm in earth sciences for thirty years, it has had relatively little impact on much of geomorphology. Geomorphologists have spent much of this period concentrating on small-scale process and landform studies for which tectonics has minimum relevance, and have perhaps therefore found difficulty in applying their work at the regional to continental scales at which plate tectonics are most relevant.

Although the detailed conceptual models for slope evolution of W.M. Davis (1899) and Walther Penck (1924) may now appear to be largely irrelevant in the light of subsequent process-based research, their work also contained implicit and explicit tectonic assumptions, and was applied by them and their followers at many landscape scales. One of the most influential commentaries on their tectonic views is Schumm's 1963 paper on rates of denudation and orogeny. He concluded that rates of active tectonic uplift are generally well in excess of regional denudation rates, so that Davis's assumption, of rapid uplift and long periods of stability, was a better first approximation than Penck's assumption of a dynamic balance between uplift and erosion. More recently, there has been an increasing preference for a Penckian dynamic balance between uplift and downcutting (e.g. Adams 1985). It has been suggested that rates might be in balance at the extremes of 25 mm a^{-1} found in the New Zealand Alps, and for relatively stable areas, such as south-eastern England, at rates as low as 0.01 mm a^{-1}.

Working at large scales, geophysicists have frequently adopted a diffusion model for landscape evolution, which is related both to some sediment transport laws and to the influential generalisation of Ahnert (1970), in which denudation rates are directly linked to elevation. It is argued below that a diffusion model, suitably applied, is an adequate asymptotic model for landscape behaviour. Although less suitable for the highly transient situations which arise in response to individual tectonic events, it may be broadly

121

applicable where tectonic activity continues for time spans which are long compared to landscape relaxation times, so that a near equilibrium can be achieved.

In this chapter it is argued from a modelling standpoint that some of the process experience acquired from local-scale studies may still be effectively applied at large scales. There is then scope to improve the geomorphological content of geophysical models, and give geomorphologists back the freedom to explore a new understanding of regional to continental landscapes.

DIFFUSIVE AND NON-DIFFUSIVE PROCESSES

Simulation models for hillslope evolution are commonly based on a continuity equation, combined with a process law for sediment transport in terms of area drained (related to flow discharges) and gradient (Kirkby 1971). These formulations are appropriate, in their simplest form, where sediment transport is at the transporting capacity of the process (transport or flux-limited models), and are usually associated with processes such as soil creep, solifluction, rainsplash and inter-rill or rillwash. Other processes are constrained by the supply of material, commonly through weathering (weathering or supply-limited models). This approach is more appropriate for denudation by rapid mass movements (even where they are approximated as a continuous process) and by solution. It is possible (Kirkby 1992) to combine these two approaches into a single continuum, in which mean travel distance is large (compared to the slope length of interest) for supply-limited processes, and small for flux-limited processes.

Strictly diffusive models are appropriate only for flux-limited processes in which sediment transport is directly proportional to gradient. These are thought to be soil creep, solifluction and rainsplash, and are generally measured as having the rather low transport rates of 10^{-3} to 10^{-2} m^2 a^{-1} per unit gradient.

For the one-dimensional slope profile, and assuming no aerial sediment input, continuity dictates:

$$\frac{\partial z}{\partial t} = U - \frac{\partial S}{\partial x} \tag{1}$$

where z is elevation,
x is horizontal distance from divide,
U is the rate of tectonic uplift,
S is sediment transport and
t is elapsed time.
Using the erosion-limited generalisation, the transporting capacity, C, is defined as the product of a detachment rate, D, and a travel distance, h. A sedimentation balance then gives:

122

$$\frac{dS}{dx} = D - \frac{S}{h} \tag{2}$$

where the difference between dS/dx and $\partial S/\partial x$ may be ignored except for transient cases.

With reasonable generality, the detachment rate and travel distance may be expressed as functions of distance and gradient, L, as in the following simple examples:

(i)   Creep or splash:
      D constant
      $h \propto \Lambda$

(ii) Inter-rill wash:
      D constant
      $h \propto x$

(iii) Rillwash, and by extension fluvial transport:
      $D (x\Lambda - \Theta_c)$ above threshold Θ_c
      $h \propto x$

(iv) Landslides:
      $D \propto (\Lambda - \Lambda_0)/L$ where $S > = 0$
      $h \propto \Lambda/(\Lambda_T - \Lambda)$ where positive, ∞ elsewhere

(v)   Solution:
      D constant
      $h \rightarrow \infty$

Together with initial and boundary conditions, such sets of equations form a basis for generating numerical simulations over time. Figure 6.1 (a)–(c) shows three examples, which differ only in the (constant) rate of down-cutting applied at the slope base. The assumed processes follow the forms above, and include components of all five processes. For the reasonable values used, inter-rill wash is nowhere dominant and there is a zero threshold for rillwash. The initial form consists of a gently sloping plateau, with a deep incision to the slope base.

In Figure 6.1(a), with a fixed basal elevation, the initial evolution for about 50 ka is dominated by landslides, while the steepest part of the profile is above the threshold gradient (22° or 40%). For longer times, the plateau is steadily lowered by solution, while splash and rillwash processes generate a convexo-concave profile which gradually encroaches on the divide and lowers the whole slope, though at much lower rates than for landslides. From about 500 ka, when 80% of the initial relief still remains, the slope appears to be undergoing a general Davisian decline in relief and gradients without appreciable change in form. The extent of the concavity is slight here, but is accentuated if greater emphasis is given to downslope grainsize sorting.

In Figures 6.1 (b) and (c), with 0.05 and 0.10 mm a^{-1} respectively, the

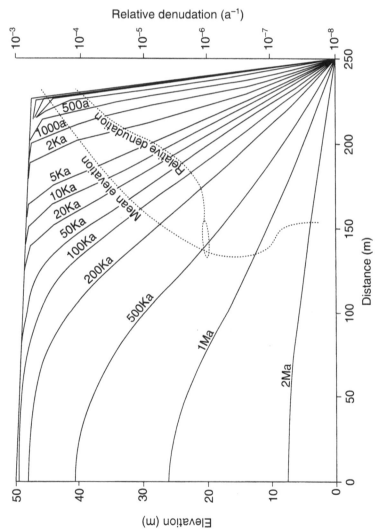

Figure 6.1 Slope profiles generated for the process rates and types shown in Table 6.1. In all cases the initial form is a gently sloping plateau, bounded by a cliff which falls to a fixed basal position. Solid curves show evolution over time. Dotted curves show progressive change over time, with x-axis plotted as the ratio of mean elevation to summit elevation (100% at far right). For mean elevation curve, y-axis is absolute mean elevation. For relative denudation curve, y-axis is the mean rate of denudation divided by the mean elevation, plotted to the logarithmic scale at the right. The three examples differ only in the (constant) rate of downcutting applied at the slope base

    (a) with fixed basal elevation
    (b) constant lowering at 50 mm a^{-1}
    (c) constant lowering at 100 mm a^{-1}

In (b) and (c), the broken line shows the equilibrium profile form

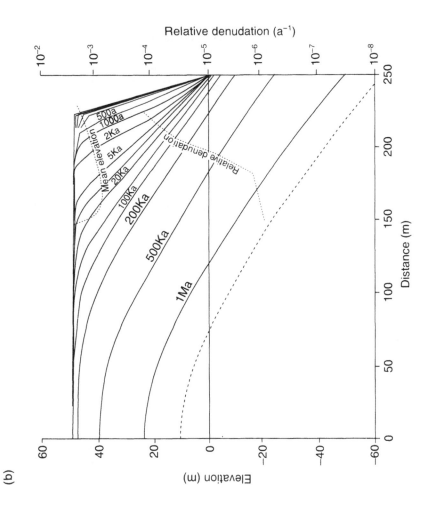

(b)

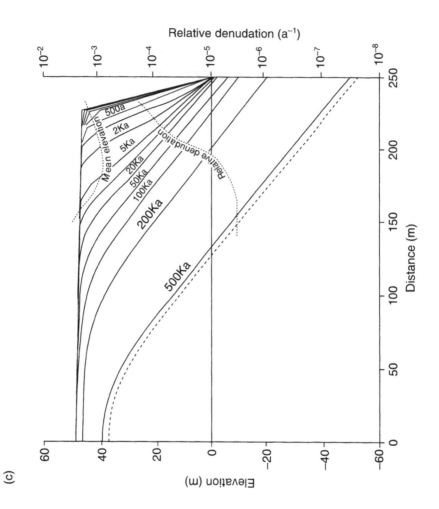

(c)

Table 6.1 File of slope process rates used for simulations shown in Figure 6.1 (a)–(c)

Slope dimensions	
250	Slope length (m)
50	Summit height (m)
20	Number of points downslope
Process rates parameters	
(a)	Soil creep
10	Rate (cm^2/a)
1	Creep travel distance on unit gradient (m)
(b)	Solution (Rate falls in inverse proportion to a/s above critical value given below)
20	Rate of solutional lowering (mm/a)
100	Soil-saturated hydraulic conductivity (m/day)
100	Time to reach chemical equilibrium (hours)
100	Soil residence time in unsaturated phase (hours)
1000	Critical value of a/L (unit area, gradient in m)
(c)	Landslides
10	Rate of free degradation (mm/y)
22	Threshold angle (°)
35	Talus gradient (°)
20	Travel distance (m)
(d)	Splash
1	Rate of detachment (mm/y)
50	Distance (m) for attenuation of detachment by water film per unit gradient
1	Travel distance per unit gradient (m)
(e)	Rillwash
0.01	Rate of detachment at unit × Ü L (mm/my)
0	Detachment threshold (m) per unit gradient
0.01	Travel distance per unit distance in flow

initial evolution is very similar, since total basal downcutting amounts to only 2.5–5 m in 50 ka, but for long time spans the profile shows elements of parallel retreat until the summit plateau has been consumed, and finally approaches a steady state form of constant lowering, in equilibrium with the rate of basal downcutting. At the lower rate of downcutting, the profile remains convex throughout, but at the higher rate the steady state form has a substantial straight section, slightly steeper than the 40% landslide threshold.

Although not shown in Figure 6.1, the stability of the hillslope to small perturbations can be calculated for one-dimensional profiles, using Smith and Bretherton's (1972) criterion for stability: $\partial S/\partial x < S/x$, where the differentiation is performed keeping gradient constant. Instability is likely to be exhibited through enlargement of small rills to permanent tributary valleys, a process which determines the actual length of hillslopes and the drainage density of the landscape (Beven and Kirkby 1993).

CHARACTERISTIC FORMS AND STEADY-STATE PROFILES

It has been shown (Kirkby 1971) that hillslopes dominated by supply-limited processes tend to asymptotic forms, provided that the slope base elevation remains constant, or declines exponentially at an appropriate rate. The 'characteristic form' is one in which denudation at every point is proportional to its elevation above some base level ($- dz/dt \propto z$). Where the slope base itself is held at a fixed elevation, then this elevation is the base level. Characteristic forms also apply where the slope base itself also declines towards some base level elevation. These behaviours are closely analogous to a Davisian view of peneplanation.

Strict parallel retreat ($- dz/dt \propto \Lambda$) and constant downcutting ($- dz/dt$ constant) can occur for any of these processes. Substitution of these conditions into the equations above provides first-order differential equations which normally lead to physically meaningful solutions. Constant downcutting forms, and their variations, seem particularly relevant to interpreting landforms in relation to tectonics, since plate tectonic activity commonly occurs for periods which are long compared to geomorphological evolution. Every point on the steady-state profile is downcutting at the same uniform rate, corresponding closely to Hack's (1960) concept of dynamic equilibrium. There may also be conditions which show at least partial conformity with conditions of parallel retreat associated with Penck, where the slope form is retreating laterally at a uniform rate as a kinematic wave. This equilibrium state is limited by the meeting of divides, and by the extension of flux-limited debris aprons on the lower parts of the hillsides, unless the slope retreat rate is matched by equivalent stream or coastal retreat.

Strictly Davisian decline ($- dz/dt \propto z$) is reached for conditions where h is negligibly small, and the transport capacity, $C = Dh$, is directly proportional to gradient and some well-behaved function of distance from the divide (x). This is a fair approximation for creep, solifluction, splash, and wash processes where there is a negligible threshold (Θ_c) for grain movement. Where travel distances and/or thresholds are significant, no characteristic form solution exists, so that it is not appropriate where rillwash with a significant threshold, landslides or solution play a major part in the landscape evolution.

Figure 6.1 illustrates the extent to which these examples of slope profile evolution approximate to characteristic forms. The two auxiliary curves, labelled as 'Mean elevation' and 'Relative denudation', both have a horizontal scale which is the ratio of current profile mean relief (above the current slope base) to current profile summit relief. This ratio is scaled so that 100% corresponds to the total slope length for the slope profiles. For 'Mean elevation', the vertical scale is the current mean relief for the profile, to the same vertical scale as the slope profiles. Initially the curve for mean elevation descends along the diagonal of the figure, since summit elevation is initially almost totally unaffected by erosion. As the summit begins to be eroded, the curve falls below the diagonal line. If the profile evolves towards a characteristic form, then eventually the elevation ratio on the x-axis tends to a constant value, and the curve then drops vertically. This may be seen to occur over long periods in Figure 6.1(a), where a close approximation to a characteristic form develops (limited only by the presence of a low rate of solution). In (b) and (c), convergence on a steady state requires that the curve converges on a single point, corresponding to the geometry of the steady state form.

For the 'Relative denudation' curve in Figure 6.1, the vertical axis is the ratio of mean denudation rate to mean profile relief, in units of a^{-1}, which is shown as a logarithmic scale on the right of the figure. For a characteristic form, this ratio should tend towards a fixed value, so that the curve converges on a single point. For any steady state form, it is clear that this curve must also converge on a fixed point corresponding to the fixed slope geometry, as may be seen in (b) and (c).

In the early stages of slope evolution, it is clear that there are considerable departures from characteristic forms, as shown by the profiles themselves or the summary curves. Thus the characteristic form is a very poor approximation for landscape response to transient conditions. For the reasonable example process rates used in Figure 6.1, it may be seen that such transient conditions persist for periods of 10^5–10^6 years. At longer time periods, both true characteristic forms (Figure 6.1a), or steady state forms (b and c) can be treated as characteristic forms in the sense that relative denudation tends to a constant value, for the given conditions. It must be noted, however, that the constant value depends on the conditions. Figure 6.2 shows relative denudation expressed in terms of steady state basal downcutting rates. It might equally be expressed in terms of mean or maximum slope gradient, but shows a very great sensitivity to small changes in gradient when the maximum gradients exceed the relevant landslide threshold. This approach has the advantage of combining steady states and true characteristic forms using a single-index parameter.

For steady downcutting in balance with tectonic uplift the simplest appropriate model is of uniform downcutting, again after a transient period of 10^5–10^6 a. The appropriate form can be obtained, either numerically or analytically by solving equations (1) globally, and (2) separately for each applicable process. Figure 6.3 shows some examples, for a combination of creep/splash and landslides. The profiles are expressed as logarithmic plots

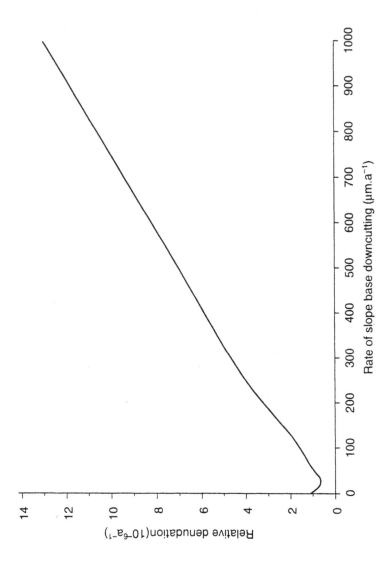

Figure 6.2 Final equilibrium relative denudation rates (mean denudation rate divided by mean profile relief) expressed in terms of steady-state basal downcutting rates, for a series of profile sequences including those shown in Figure 6.1

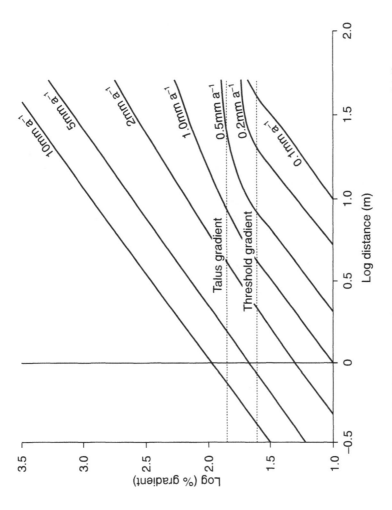

Figure 6.3 Profiles in equilibrium with differing rates of constant downcutting, for a combination of diffusive processes (splash, creep, solifluction) and mass movements. Profiles are shown as logarithmic plots of gradient against distance from the divide. Rates are shown in Table 6.1

of gradient against horizontal distance, for a range of steady downcutting rates.

SCALING UP TO LARGER AREAS

It is possible to use detailed two-dimensional process models for areas ranging from small catchments to continental areas, by extending the one-dimensional models described above, and including stream transport processes. This is a valuable approach, particularly in the context of determining drainage densities, and there are a number of examples of successful models for small areas, for example those of Wilgoose et al. (1991). In the context of relationships with tectonics, however, the amount of computer time required is generally disproportionate to other aspects of the problem. Instead we initially require methods for averaging along one-dimensional sections across the landscape.

In any analysis of larger areas, it is of first importance to add stream processes to the slope processes described above. This is generally done by extending the equations for rillwash, usually with higher exponents of distance (or area) and, in many cases, gradient as well. Along major stream courses, it may be argued that landslides and solution are relatively less important, so that sediment transport can be approximated as flux-limited, at a rate proportional to $(x/u)^m$, for a suitable exponent m.

Two methods are proposed to apply essentially one-dimensional models to larger areas. The first of these (the envelope profile) is to average across the entire landscape, generally referred to average elevations, or to an envelope of the range. The second method (the stream/slope profile) is to combine the branches of the stream network together into an averaged mainstream profile, dynamically linked to a sequence of one-dimensional hillslope profiles which supply sediment to the stream, and are influenced at their base by rates of stream downcutting or aggradation.

The envelope profile method provides some suggestions for the use of characteristic forms, as the simplest possible diffusion approximation. The method is also applied to an area of extensional tectonics in the Basin and Range area of Nevada. The stream/slope approach is applied at a continental scale, first without any tectonic response, and then to an area with isostatic response and elastic flexure, on a similar scale to southern Africa.

ENVELOPE PROFILES

Using the first method, the envelope profile curve near the range divide will be dominated by interfluve behaviour. Downstream, interfluves tend to follow the profiles of their adjacent streams, generally with diminishing relative relief, so that the envelope profile is increasingly dominated by the stream profiles. To give proper weight to the interfluves, it is proposed that an adequate overall profile can be modelled with a composite process law of the form:

$$S = k\Lambda\left[1 + \left(\frac{x}{u}\right)^m\right] \tag{3}$$

where k, u, m are empirical constants.

The effect of flow convergence in the catchment may be to increase the exponent m, whereas it should decrease through flow divergence in alluvial fans. Downstream fining of bed material should also increase m. In semi-arid areas, loss of discharge, through percolation into the stream bed outside the storm area, will substantially decrease m. The appropriate value for the exponent m will therefore generally be obtained empirically, but may be expected to lie between 1.0 and 3.0. Alternatively, some of these effects might be incorporated within a more explicit model.

Under constant downcutting in balance with uplift at rate T, the steady state form of the profile is given by:

$$\Lambda = T/k \, x \, /[1 + (x/u)^m],$$

which is a convexo-concave profile with its maximum gradient at

$$x = (m - 1)^{1/m}u.$$

Characteristic form profiles show a similar convexo-concave form, though with a slightly narrower convexity. In both cases the overall profile is dominated by a stream profile, which shows an overall downstream reduction in gradient as $x^{(1-m)}$. The appropriate choice of m is set by this concavity, while u is associated with the breadth of the summit convexities, and k is the primary rate constant.

CHARACTERISTIC FORMS AS A FIRST-ORDER DIFFUSION MODEL

It will be noted that an expression of the form of equation (3) leads to characteristic forms for the profile as a whole where base level remains fixed over time. As a result, the conclusions described above, for slope profiles, are valid for these larger regional profiles. Approximately, the constant β, in the characteristic form expression $- \, dz/dt = \beta z$, is given by:

$$S \propto k \, \Lambda \, (x/u)^m \propto \beta h \, x$$
$$\Lambda \propto h/x$$

where h is the mean basin relief. Rearranging, we have approximately:

$$\beta \propto k/x^2(x/u)^m$$

It may be seen that if m is close to 2, then the value of β is insensitive to the size of the catchment area, but strongly dependent on the value of u in particular. This is linked to average hillslope length and drainage density by arguments about the stability of hillslope length, and appears to lead to the conclusion that values of β should be most sensitive to climatic and lithological factors, with highest values for semi-arid areas.

It is relevant to compare these conclusions with Ahnert's (1970) relationship, in which rates of denudation calculated from basin sediment yields are found to be proportional either to total elevation or to local relief. This may be viewed as a basin-scale generalisation of the characteristic form. It may be argued that, if individual slope profiles approach characteristic forms, then the entire drainage basin, which is an aggregate of profiles with common highest and lowest points, must also tend towards a characteristic form. In this asymptotic characteristic form, the rate of denudation at every point is proportional to its elevation, internal drainage lines are fixed in plan, and contour lines are also fixed in plan, though changing in elevation value over time.

If it is assumed that characteristic forms have been attained, rates of lowering are directly proportional to local mean elevations. This approach has been widely applied by geophysicists and others for its computational simplicity, but it should be recognised that the implicit assumptions are frequently not met. The characteristic form assumes first that the asymptotic approximation is valid, or in other words that the landscape is 'mature' in relation to a stable base level, and second that the appropriate hillslope and channel processes are dominant. In a tectonically active area, neither of these conditions is likely to be met. Instead, landforms are transient or reacting to constant rates of uplift, and the steep slopes generated are undergoing rapid mass movements which are likely to be the dominant processes, at least in headwater areas. Nevertheless, it may still be possible to use the characteristic form model to represent steady state conditions, provided that it is recognised that the relative denudation rate, β, responds to hillslope gradients, and is therefore ultimately dependent on uplift rate, as was shown above for hillslopes in Figure 6.2.

If the characteristic form method is adopted, the evolution of a landscape, or any point on it, is readily seen. In the simplest case, for denudation at rate $-dz/dt = \beta z$, and uplift at constant rate T, we have:

$$\frac{dz}{dt} = T - \beta z$$

$$z = \frac{T}{\beta}[1 - \exp(-\beta t)] \tag{4}$$

where the solution assumes that $z = 0$ at $t = 0$.

Thus, for these constant uplift conditions, the landscape equilibrates at elevation T/β, with a response time scaled to $1/\beta$. For Ahnert's data set, the value of $1/\beta$ is approximately 10^7 years, corresponding to geological estimates of 10–50 Ma as the period required for peneplanation. Using the same value of β, Himalayan elevations (6000 m) would be in balance with uplift rates of 0.6 mm a^{-1}, while the English Weald (200 m) requires uplift of only 0.02 mm a^{-1}.

Comparison with actual uplift rates suggests that higher β values, and therefore shorter response times, may be appropriate for higher, and therefore generally steeper, mountains. Figure 6.2 suggests that, at high uplift rates, the value of b increases almost linearly with the uplift rate (for a given value of u and, therefore, implicitly climate). Thus if we approximate $\beta = T/L_0$, for $l_0 = 5000$ m, it follows that all areas of steady uplift will ultimately reach an elevation of $L_0 = 5000$ m, and that the uplift rate will only influence the time taken to reach this height. Thus at $T = 0.1$ mm a^{-1}, the time is scaled to $L_0/T = 50$ Ma and for $T = 10$ mm a^{-1}, the time required falls to 0.5 Ma. For a maritime climate, like that of the southern Alps, it may be argued that β is also very strongly influenced directly by elevation, due to strong orographic rainfall effects.

There is thus some scope for using a diffusion model based on the characteristic form assumption to represent the responses to long continued steady uplift. It is argued, however, that there is scope to refine the values of the characteristic constant, β, to allow for a number of effects. The most important of these is that β increases strongly with uplift rates, though with a lower limit. β also appears to respond to climate, with higher values for semi-arid than for temperate climates. These effects are sketched in Figure 6.4. Finally β responds directly to elevation, through orographic rainfall effects which are strongest for maritime climates (i.e. west coasts in the northern hemisphere and east coasts in the southern), and for the margins of other major mountain masses which preferentially catch precipitation.

BASIN AND RANGE TECTONICS

In an area of crustal extension, the movement is commonly taken up in a series of sub-parallel normal faults, which typically dip at about 60° (Jackson et al. 1988). To maintain overall geometry, blocks rotate backwards as indicated in the upper part of Figure 6.5. This geometry is applied here to the Basin and Range area of Nevada, which is a continental area of largely internal drainage. The uplifted footwall blocks undergo erosion, providing material which partly covers the hanging wall block with alluvial fan deposits. In the Central Nevada Seismic Belt, alluvial fan deposits lie on basalts of 10–15 Ma age, which also outcrop locally on uplifted footwalls. Extensional faulting has been accompanied by erosion over this period, and may now be approaching a dynamic equilibrium in which uplift or

135

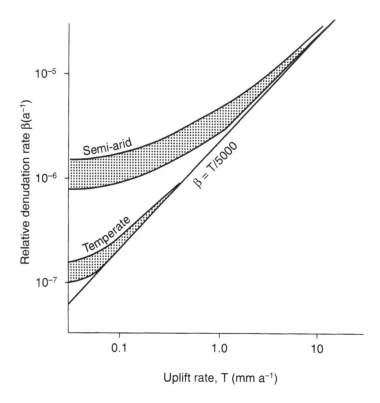

Figure 6.4 Schematic relationship between long-term rate of relative denudation (β) and tectonic uplift rate. Some mass movement activity is assumed in addition to diffusive processes (splash, creep and wash). Relative denudation is expressed in terms of steady uplift rate and climate. Lithology influences the relationship primarily through its impact on mass movement rates

downwarping exactly balance erosion or deposition. Over the basin widths of 20–40 km, isostatic responses are assumed uniform, and therefore ignored for this closed basin context. The centre curve in Figure 6.6 sketches the equilibrium form, and the lower part of the diagram indicates the spatially varying constant rates of tectonics and erosion/deposition.

Using parameter values obtained by back-analysis of the landforms of the Clan Alpine Range, an approximate dynamic equilibrium form may be derived, and is shown in Figure 6.7. The profile is plotted as slope gradient against distance from the divide. The required boundary conditions are (i) that elevations are equal to left and right of the divide (i.e. the areas under the left and right parts of the gradient curve are equal), (ii) that the rate of change of gradient is the same on both sides of the basin axis (i.e. at the zero gradient points at far left and far right) and (iii) that gradient is continuous, though not necessarily changing smoothly, at the fault line. In the analysis shown, the

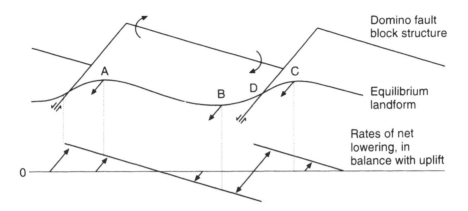

Figure 6.5 Cartoon showing equilibrium landforms associated with domino block extensional faulting. Upper lines show block fault structures. Centre curve shows equilibrium profile for a closed sedimentary basin in balance with faulting. Note modest basin asymmetry, and summits substantially displaced from the fault lines. Lower lines show resulting equilibrium rates of uplift or lowering

effect of stretching has been ignored, as this prevents the achievement of a true equilibrium.

The resulting form shows some asymmetry about the divide, with shorter and steeper slopes on the faulted side of the range. Although the range divide is initially at the hanging wall side of the fault line, slope retreat forces some lateral migration, so that the final range asymmetry is relatively slight. The fault line is not expressed as a surface discontinuity, but appears to follow the locus of steepest slopes on one side of the range. This will not necessarily be the range-front contact between bedrock and alluvial fans, because progressive erosion and deposition continuously increases the volume of sediments, which gradually bury the bedrock more and more. After long periods of time, therefore, the fault line will outcrop well within the fan deposits.

Figure 6.7 shows a time-bound simulation for the same area, taking full account of basin stretching as well as faulting and rotation effects as above. The landform profile is assumed to be indefinitely repeated at left and right, so that the extreme left and right of the profiles may be joined together. The curves here show isochronous sedimentary layers within the fan deposits. The effect of back rotation can be seen in the strong asymmetry of the basin fill, with the lowest point off successive layers closely hugging the footwall slope. As for the equilibrium profile in Figure 6.6, the time-bound model indicates that the highly asymmetric faulting process leads to only modest asymmetry in the form of the bedrock mountain ranges.

The surface and subsurface forms are in broad agreement with the present topography and alluvial stratigraphy (Kirkby, Leeder and White, in press).

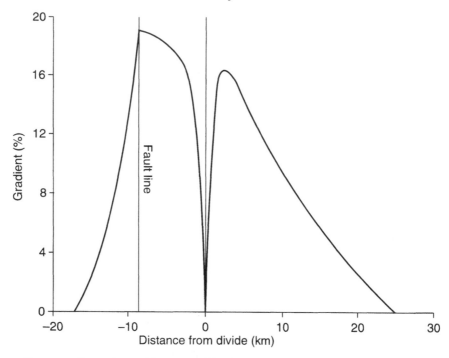

Figure 6.6 Example equilibrium profile obtained for a closed continental basin, in balance with block faulting as outlined in Figure 6.5. Profiles are expressed as gradients. Note that the fault line is only expressed as a discontinuity in rate of change of gradient, here at the point of maximum gradient where the slope changes from convex to concave. Parameters selected with general reference to Basin and Range extension in Nevada

Although isostatic flexure has been ignored here, due to the short basin wavelength, it is recognised that the elastic lithosphere is very thin in this area, and there may be significant small effects, perhaps associated with the generation of range-front faulting and folding on both sides of the valleys, and associated with current seismic activity.

STREAM/SLOPE PROFILES

The proposed alternative to the envelope curve is the use of a single stream profile which follows the lowest flow line in the landscape, representing the major river(s). At successive points along the river course, its profile is linked to a hillslope profile. The two profiles are dynamically linked through the base of the hillslope. Changes in river elevation influence slope base gradient and therefore sediment delivery. Hillslope erosion provides sediment which the river must carry, with consequences for its incision or aggradation. The hillslope divides, which are assumed symmetrical, are lowered by erosion to

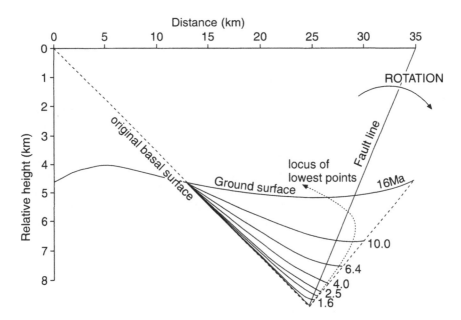

Figure 6.7 Example simulation of closed continental basin evolution over time in the presence of extensional domino block faulting. Parameters are chosen for the Dixie valley/Clan Alpine Range, Nevada. Solid lines show successive isochrons for basin fill over basal basalt flows. Note the position of the main fault line well within the basin sediments, and the modest topographic asymmetry of the basin

Source: Kirkby, Leeder and White in preparation

give summit elevations and typical slope gradients. In this approach it is thus possible to simulate separately the behaviour of stream and summit elevations. At a fault line, for example, the response to uplift may be seen in a rapid adjustment of stream elevations, but a much slower response by summits, so that local relief and local gradients are initially increased. As we will see below, flexure along passive continental margins can similarly lead to streams which cut down in response to uplift, leaving fringing mountain ranges and steep escarpments.

Stream profiles are viewed as strongly dominated by flow processes, so that an equation similar to (3) above is appropriate, with appropriate corrections to allow for the stream network and valley floor geometry. For the slope processes, however, there is no difficulty in applying the full range of both flux-limited and supply-limited processes, as illustrated in Table 6.1.

Linkage between slope and stream profiles is formally obtained through extension of the one-dimensional continuity equation (1 above) for a valley bottom of width w, with allowance for sediment inflows from hillslopes. If x is taken as the downslope direction and y as the downstream direction, we have:

$$\frac{\partial z}{\partial t} = U - \frac{1}{w}\frac{\partial(wS_y)}{\partial y} + \frac{2S_x}{w} \tag{5}$$

where S_x and S_y indicate sediment flows in the downslope and downstream directions.

Equation (1), with S_x replacing S, remains valid for each hillslope profile, with the basal boundary condition set by the value of $\partial z/\partial t$ for the corresponding point on the stream profile.

Figure 6.8 illustrates this approach. The stream initially flows for 1000 km on a level plateau at 1000 m elevation, falling over a steep continental rim to a fixed sea level. Initially identical slope profiles have 100 m of relief, in a 200 m long hillslope. The figure shows the profiles at 50 ka. The continental divide shows little lowering, but the stream profile shows the development

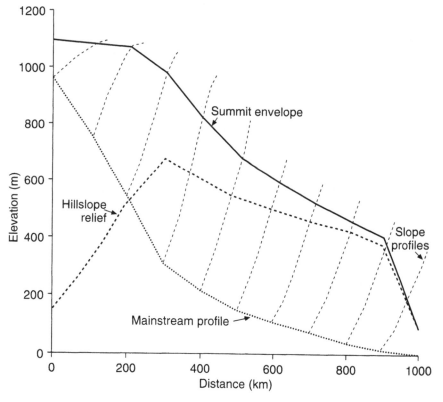

Figure 6.8 Illustration of stream/slope profiles, for erosion of a 1100 m plateau, initially with 100 m of relief, with no tectonic effects. Curves falling from left to right show the mainstream and summit envelope profiles, and the hillslope relief. Curves falling from right to left show 10 constituent slope profiles

of a broadly concave profile. The divides show the remains of a summit plateau, which is increasingly incised downstream. Beyond 300 km, hillslope relief begins to decline progressively, and the summit envelope mirrors the stream convexity. In the centre section (300–600 km), where relief is greatest, slope profiles are dominated by mass movements. This effect has been widely observed, at a variety of scales (e.g. Carter and Chorley 1961; Arnett 1971).

PASSIVE CONTINENTAL MARGINS

For large continental areas, isostatic effects cannot generally be ignored, even if there are no externally imposed tectonic stresses. The section in Figure 6.8 may therefore be seen as representing a passive continental margin, like that around southern Africa, with offshore sedimentation of eroded material. If it is assumed that the initial uplifted configuration is in equilibrium, and that geothermal gradients are neutral in the sense that loading or unloading will not produce thermal stresses, then erosion and deposition produce a pattern of loading on the elastic lithosphere, which bends in response.

The general equation for two-dimensional deflection of a thin elastic plate (Turcotte and Schubert 1982, Chapter 3) is:

$$D\frac{d^4 w}{dx^4} = q(x) - P\frac{d^2 w}{dx} \tag{6}$$

where w is the deflection at distance x along the plate,
D is the flexural rigidity,
P is the horizontal force per unit length
and $q(x)$ is the plate loading at x.

Solving for an unbroken continental plate with a point load at $x = 0$, the deflection is given by:

$$w = \frac{V_0\alpha^3}{8D} - e^{-x/a}\left(\cos\frac{x}{\alpha} + \sin\frac{x}{\alpha}\right)$$

$$\alpha = \left[-\frac{4D}{\rho_m g}\right]^{1/4} \tag{7}$$

where V_0 is the point load at $x = 0$,
α is the flexural parameter defined above
and ρ_m is the mantle density.

The equation is the same for the offshore area, but usually with a different flexural rigidity, D, and with the density term replaced by the difference between mantle and water densities. The solution is symmetrical about $x = 0$.

For the continuous change in loading associated with erosion, the complete solution is obtained as the sum of a series of terms like equation (7), displaced with respect to the point of application of the erosional unloading. In the absence of an adequate model for the distribution of offshore deposition, the marine contribution has been approximated as being anti-symmetric with respect to the coastline.

The concentration of both erosion and deposition close to the shoreline lead to massive flexure at the shoreline, with maximum flexural uplift within 200 km of the coast, and maximum downwarping a similar distance offshore. The example shown in Figure 6.9 is for a mantle density of 3300 kg m^{-3} and a flexural rigidity modulus of 1.1×10^{24} Nm, which corresponds to a 56 km

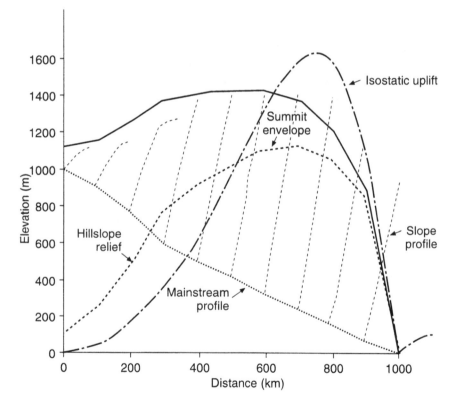

Figure 6.9 Illustration of stream/slope profiles for same process rates and notation as in Figure 6.5. Profiles here represent a passive continental margin, with no external tectonics, but with isostatic uplift, and flexure using average parameters (Turcotte and Schubert 1982, Chapter 3) for continental elastic response. For simplicity, offshore response to sediment loading and flexure is assumed anti-symmetric about the shoreline. Note the evolution of a coastal mountain range with very steep side-slopes, and the folding of initially horizontal strata with the isostatic uplift to generate shoreline-facing escarpments

elastic lithosphere thickness and a flexural parameter of 108 km. Isostatic uplift raises the summits, while incision of stream courses more or less keeps pace with the uplift. Slopes therefore become very steep, generating a coastal range of rugged mountains, enclosing a plateau of more moderate relief. The curve for isostatic uplift also shows the warping undergone by originally horizontal strata, which are deformed into outward-facing escarpments, with a dip which increases with time. There is scope to measure the cumulative amounts and rates of exhumation associated with this process using methods which include apatite fission-track dating (Brown *et al.*, in press) and examination of metamorphic grade.

At many continental margins, the assumption of a neutral geothermal gradient is inappropriate. The continental margin may well be associated with crustal thinning which has strong thermal implications. In addition, the accumulation of sediment offshore may induce thermal stresses which encourage continental growth through uplift and accretion of accumulated offshore sediment.

CONCLUSIONS

Although there is scope for much further development of these themes, we have demonstrated the feasibility of extending relatively simple geomorphological process models up to basin or continental scales. It is possible to retain a link with our understanding of small-scale process, and to link areas together, essentially using a simplified fluvial transport process at the large scale.

Stream processes appear to dominate at large scales, and many aspects of their behaviour can be approximated as flux-limited processes. This approach therefore has some validity, both for the simple one-dimensional envelope profile method and as the linking stream process within a slope/stream scheme. Its greatest weakness may be for very large rivers, where a high proportion of the alluvial sediment is in silt and finer grain sizes, for which there is no well-defined transporting capacity. In such cases it may be necessary to set up an explicitly grainsize selective model.

Where a flux-limited approach is valid, it is reasonable to use characteristic forms as the first-order model for whole landscape evolution. This approach breaks down, however, where there are anomalous variations in local relief, such as those that occur in the example shown in Figure 6.9 for passive continental margins. For such cases it is important to include a distinct hillslope component, and to recognise that it may be dominated by supply-limited mass movement processes which do not lead towards diffusive characteristic forms.

With adequate geomorphological models, we can, for the first time, investigate the links between geophysical and geological processes and landforms at regional to continental scales. It has become clear that these links are strongly interactive, with impacts of erosion and sedimentation on tectonics which significantly influence the rates of geophysical processes. It is vital that geomorphologists contribute their skills to this new understanding, and ensure that measurements of landscape processes are applied in a wider context.

REFERENCES

Adams, J. (1985) 'Large scale tectonic geomorphology of the Southern Alps, New Zealand', in M. Morisawa and J.T. Hack (eds) *Tectonic Geomorphology*, Boston: Allen & Unwin.

Ahnert, F. (1970) 'Functional relationships between denudation, relief and uplift in large, mid-latitude drainage basins', *American Journal of Science*, 268: 243–263.

Arnett, R.R. (1971) 'Slope form and geomorphological process: an Australian example', in D. Brunsden (ed.) *Slopes: Form and Process, Institute of British Geographers, Special Publication* 3: 81–92.

Beven, K.J. and Kirkby, M.J. (1993) *Channel Network Hydrology*, Chichester: John Wiley.

Brown, R.W., Summerfield, M.A. and Gleadow, J.W. (in press) 'The potential of thermochronologic measures of denudation rates in assessing models of long-term landscape development', in M.J. Kirkby (ed.) *Process Models and Theoretical Geomorphology*, Chichester: John Wiley.

Carter, C.A. and Chorley, R.J. (1961) 'Early slope development in an expanding stream system', *Geological Magazine* 98: 117–130.

Davis, W.M. (1899) 'The geographical cycle', *Geographical Journal* 14: 481–504.

Hack, J.T. (1960) 'Interpretation of erosional topography in humid temperate regions', *American Journal of Science* 258A: 80–97.

Jackson, J.A., White, N.J., Garfunkel, Z. and Anderson, H. (1988) 'Relations between normal-fault geometry, tilting and vertical motions in extensional terrains: an example from the southern Gulf of Suez', *Journal of Structural Geology* 10: 155–170.

Kirkby, M.J. (1971) 'Hillslope process–response models based on the continuity equation', in D. Brunsden (ed.) *Slopes: Form and Process, Institute of British Geographers, Special Publication* 3: 15–30.

——(1992) 'An erosion-limited hillslope evolution model', *Catena, Supplement* 23: 157–187.

Kirkby, M.J., Leeder, M.R. and White, N.J. (in press) 'The erosion of actively extending tilt-blocks: a coupled model for topography and sediment budgets', *Tectonics*.

Penck, W. (1924) *Die morphologische Analyse: ein Kapital der physikalischen Geologie*, Stuttgart: J. Engelhorn's Nachschriften. Translated by H. Czech and K.C. Boswell as *Morphological Analysis of Landforms: A Contribution to Physical Geology*, London: Macmillan, 1953.

Schumm, S.A. (1963) 'The disparity between present rates of denudation and orogeny', *U.S. Geological Survey, Professional Paper* 454-H: 1–13.

Smith, T.R. and Bretherton, F.P. (1972) 'Stability and the conservation of mass in drainage basin evolution', *Water Resources Research* 8: 1506–1529.

Turcotte, D.L. and Schubert, G. (1982) *Geodynamics: Applications of Continuum Physics to Geological Problems*, New York: John Wiley.

Wilgoose, G.R., Bras, R.L. and Rodriguez-Iturbe, I. (1991) 'Results from a new model of river basin evolution', *Earth Surface Processes and Landforms* 16: 237–254.

7

PROCESS AND FORM IN THE EROSION OF GLACIATED MOUNTAINS

Ian S. Evans

INTRODUCTION

'Geomorphology ... has for its *objects* of study the geometrical features of the earth's terrain ... within clearly definable ... spatial and temporal scales and in terms of the processes ...' (Chorley 1978: 1). Richard Chorley has led geomorphology to focus on the interactions between process and form, and on the functional study of dynamic environments. One of the most dynamic environments, in terms of erosion rates, is provided by glaciated mountains. Their high relief and the concentration of ice flow along valleys lead to rapid basal sliding, and thus provide excellent conditions for glacial erosion. In rock basins and breached divides, glaciated mountains show the most convincing demonstration of deep erosion by glaciers. Even where glaciers do not survive, mass movements, snow avalanches and energetic streams provide a continuing dynamism which is a heritage of the slopes produced by glaciation. Nevertheless, it has not been easy to relate particular forms to particular processes, or to provide quantitative explanations for variations in landform size and shape. Our excuse for this may be the longer timescales required to study variations of glaciers, together with the greater difficulty of observing the glacier bed compared with the river bed.

Here I will consider the broader-scale erosional forms of glaciated mountains, and their spatial patterns. Older ideas will be reassessed in the light of recent publications. Support will be provided for the propositions that:

1   glacial and closely related (glaciofluvial, snow avalanche and rapid mass movement) processes dominate the geomorphology of glaciated mountains;
2   troughs are glacial palaeochannels calibrated to the discharge of ice and the erodibility of bedrock;
3   fjords are normal glacial troughs, well developed because of high ice velocities;

4   rock basins are most common in zones of upward flow, such as ablation zones;

5   pressure release helps the unloading and collapse of trough sides and cirque headwalls, but not the deepening of troughs: stress concentration is more important there;

6   cirques are developed essentially by glacial processes, rather than by ice-marginal frost-shattering;

7   downward erosion is important in cirque development, but headward erosion is more so;

8   some variations in cirque form can be related to varying glacial and relief environments.

The effects of entirely cold-based glaciers will not be considered here, nor will the processes and forms of deposition. Familiarity with the basic terminology of glaciation will be assumed.

RATES OF EROSION

The most direct evidence of the erosive work of glaciers is the debris load which they release. The suspended sediment load of glacier-fed streams is much greater than that of adjacent non-glacial streams; sediment plumes from glacier-fed streams occur in lakes and in the ocean, for example in the Gulf of Alaska. To measure erosion rates, thorough sampling schemes are needed; otherwise short-lived peaks of sediment discharge such as the first major flood of the ablation season may be missed. Excellent data of this type are now available from Norway; these are summarised in Table 7.1, together with some other relevant measurements.

The Norwegian data relate to glaciers selected for hydro-electric power interest. Mean rates of glacier bed lowering, allowing for bedload as 40% of total proglacial stream load, are between 0.1 and 1.0 mm a^{-1}, both for these glaciers (with records of 5 to 15 years) and for others measured only for 1 to 4 years. For most glaciers, variations between years cover a two- or three-fold range. In some years, subglacial drainage is more active in flushing out subglacial sediment, but five-year average rates probably reflect the true activity of the present-day glacial systems. These glaciers are on resistant, crystalline rocks; they are diminished from their nineteenth-century extent and activity.

Many Soviet papers show sediment yield increasing with the proportion of basin glacier-covered. The results of Chernova (1981) for Central Asian basins mainly covered by glacier ice are included in Table 7.1, despite a lack of methodological detail, because they show an interesting contrast between three small valley glaciers with erosion rates of 0.7 to 0.9 mm a^{-1} and much larger glaciers with four times those rates. This supports the intuitive notion that larger valley glaciers with faster rates of flow should be more erosive.

Table 7.1 Mountain glacier erosion rates from sediment load of proglacial streams
(rates in mm per year)

(a) *Norway*: crystalline rocks (Kjeldsen 1985 and Bogen 1989)
Suspended load sampled 4 times daily; hourly in floods. Where estimates of bedload
are not available, suspended load is taken to be 60% of total load. Glaciers measured
for less than 5 years are excluded but fall within this range of erosion rates.

Years of observation	Glacier	Glacier area (km^2)	Mean erosion rate $(mm\ a^{-1})$	Inter-annual range (-fold)
15	Nigardsbreen	48	0.16	3.0
13	Engabreen	38	0.22	2.3
7	Erdalsbreen	11	0.61	2.9
5	Bondhusbreen	11	0.36	2.1
6	Austre Memurubreen	9	0.31	2.2
6	Vesledalsbreen	4	0.07	2.2
5	Trollbergsdalsbreen (soft bed?)	2	1.02	5.3

(b) *Central Asia* (Chernova 1981; or see Drewry 1986: 87, 91)
Assumptions, and intensity of sampling, not stated.

Duration of study (years)	Glacier	Glacier area (km^2)	Mean erosion rate $(mm\ a^{-1})$
35	Fedchenko	662	2.9
83	Zaravshanskiy	134	3.8
25	R.G.O.	109	2.5
15	Karabatkak	4.7	0.9
6	I.M.A.T.	3.8	0.7
36	AJUTOR-3	3.4	0.7

(c) *Valais Alps, Switzerland* (Small 1987)
Tsidjiore Nouve Glacier, 3.6 km².
Suspended sediment: 0.9–1.0 mm a⁻¹.
Stream load + 20% for moraine dumping: 1.4–1.7 mm a⁻¹.
Whole catchment (4.15 km²), stream load and moraine accumulation: 1.5–2.3 mm a⁻¹
Purging of sediment traps, and suspended load calibrated over 2 years. (Bezinge *et al.*
1989)
1.0–1.1 mm a⁻¹ for whole catchment, excluding moraine accumulation.

(d) *Iceland* (Tómasson, 1987)
Regular monitoring since 1963 but sampling frequency unclear. Suspended load only.
No major outburst floods, but rates are inflated by ash from eruptions, never part of
bed rock.
Vatnajökull 3, Myrdalsjökull 5, Hofsjökull 1, Langjökull 0.2 mm a⁻¹
(Lawler 1991)
1973–1988 Solheimajökull (part of Myrdalsjökull) 5.4 mm a⁻¹

(e) *Washington State, USA* (Metcalf 1979)
Two years' frequent sampling, suspended load only:
Nisqually Glacier 2.5, Blue Glacier 0.05 mm a⁻¹

Although variations in gradient and mass turnover are also involved, we may expect that large, well-fed valley glaciers will erode more than 1 mm a year from their basins. In Switzerland, even a small glacier, the Tsidjiore Nouve, achieves more than this.

Further measurements (Lawler 1991) are available for Iceland, with estimates of 5 mm a^{-1} from weak volcanic rocks. Metcalf's (1979) results are higher for Mount Rainier, on volcanic rocks, than for Blue Glacier, on crystallines.

None of these glaciers surged during the time of observation. Tómasson (1987) pointed out that surges and glacier outburst floods greatly increase sediment yields. Sharp (1988) found sediment concentrations during the 1983 surge of Variegated Glacier, Alaska, which were 550 times inter-surge averages. The sediment yield in late June 1983 was 40 times that three years earlier; plucking and abrasion rates increased with sliding velocity. Hence surging glaciers are likely to be much more effective geomorphologically, and the same probably applies to other glaciers in fast flow mode (Clarke 1987) such as ice streams and many outlet glaciers. Even in the Antarctic, these generate enough heat from deformation that their bases are at pressure melting point and high rates of basal sliding or sediment deformation are maintained. In East Antarctica, ice streams start at subglacial steps – trough heads – where ice accelerates as basal melting conditions are established. Conversely, ice caps and glaciers which are frozen to their beds throughout are not expected to achieve much glacial erosion (England 1986).

To obtain glacial erosion rates over longer periods, a number of studies of moraine, delta and lake infill volumes have been made (Small 1987). There are usually difficulties about some of the assumptions, concerning for example the basal topography of moraines and especially the dates between which they accumulated. One of the more convincingly dated cases is a 0.21 km^2 cirque at 62°N in western Norway which was almost filled by a glacier for some 700 years during the Younger Dryas (Larsen and Mangerud 1981). A lake completed a closed system in which 118,000 m^3 of rock equivalent collected, giving a vertical erosion of 0.56 m at an average rate of 0.8 mm a^{-1}. This is consistent with the present-day results, and would permit formation of this cirque in resistant mica-rich gneiss by erosion of 80 m in 100 ka. Rockfall from the cirque headwall is included here as a component of glacier-related erosion.

It is no longer possible, therefore, to deny the significance of erosion by wet-based cirque and valley glaciers. Compilations of average denudation rates (Saunders and Young 1983) show that only landslides and badland erosion can compare with such glaciers in geomorphic effectiveness. In British Columbia, the highest sediment yields from small and medium-sized basins are confined to those with present-day glaciers. Most of the sediment in other rivers has been eroded from along their banks, and is remobilised from glacier-related Quaternary deposits; hence sediment yield increases with basin area, up to 30,000 km² (Church *et al.* 1989: Fig. 7; Church and Slaymaker 1989). Much of Canada is still in a paraglacial phase of recovery from the Last Glaciation.

This is not to deny that ice sheets are very selective in erosion, and in areas of low relief or cold base (with no meltwater from upstream) they may be ineffective (Sugden 1978; Hall and Sugden 1987). On the other hand, the Cordilleran, British and Scandinavian Ice Sheets have in general been effective in moving sediment out of mountain areas. Despite problems in sampling, and in extrapolating present-day spatial comparisons to former glacial conditions (Harbor and Warburton 1993), we may conclude that the presence of temperate or 'mixed' glaciers increases erosion rates, and that erosion rates were greater in mountains during times of temperate glaciation. The most conspicuous effects of this are the troughs and rock basins eroded.

NIVATION

The efficacy of glacial action may be contrasted with that of 'nivation'. Thorn (1988) found that the concept of nivation is 'so complex that it is operationally unmanageable', but it could be defined simply as weathering and erosion related to the presence of a snowpatch. Traditionally an acceleration of activity was implied, but this is no longer universally accepted. Gelifluction is fastest where water is concentrated (Washburn 1979). Thorn (1976) found that the concentration of water in space (lee slopes and concavities) and time (the melt season) accelerated gelifluction downslope of a snowpatch and wash erosion from unvegetated areas: chemical weathering was enhanced but frost action was more problematic, with water being available at different times to deep frosts. Aeolian erosion was reduced. Total erosion in a Colorado 'nivation hollow' was 0.0074 mm a^{-1}, much faster than the 0.0001 from a nearby snowfree (dry) site, but much less than the 0.14 now, or 2.00 in the past, eroded from bedrock by the nearby Arapahoe cirque glacier.

Measurements in Scandinavia by Hall (1980), Rapp (1986) and Nyberg (1991) support these results; the main effect of snowpatches is the acceleration of solution and of transport immediately downslope, and erosion rates are much less than those reported above for glaciers. More broadly, concerning the efficiency of non-glacial alpine processes such as 'nivation', Caine summarised several studies showing 'no sharp geomorphic discontinuity at timberline' (1978: 124); the onset of glaciation provides the main threshold accelerating activity in mid-latitude mountains.

The morphological diversity of snowpatches suggests that they occupy hollows of diverse origins which have not been greatly modified (Thorn 1988). Indeed, the hollow form around the well-studied Martinelli snowpatch in Colorado is tending to be removed under present conditions (Caine 1992). In Scotland, present snowpatch erosion is confined to redistribution of cohesionless fine sediments by wash, important only where vegetation is sparse (Ballantyne 1985).

Nivation hollows are best developed on weak materials, especially glacial till, and are an order of magnitude smaller than glacial cirques (Evans 1969; Thorn 1988). Given our increasing awareness of the speed and frequency of climatic change during the Quaternary, the traditional model of nivation being the initial stage in development of glacial cirques (Hobbs 1911; Embleton and King 1975: 218) seems more and more far-fetched: a glacier is likely to form before a snowpatch has had much effect. On the same materials, the 'nivation' process-complex is much less effective than glacial action. Snowpatches accumulate in hollows, but not all hollows with snowpatches are nivation hollows.

In maritime climates, melting and refreezing may produce an icy, high-density snowpatch in one winter. This may slide over a wet bed, transport stones and even produce downslope striations (usually shallow) on bedrock (Jennings and Costin 1978; Costin et al. 1964). To a limited extent, these studies suggest a transition between nivation and glaciation. So long as basal sliding is possible, glaciation is much the more effective because of greater stresses due to greater thickness and density.

SNOW AVALANCHE EFFECTS

Rapid flow and high stresses are also achieved in snow avalanches (Perla 1980), mainly on slopes over 25° except for slush avalanches. The importance of avalanches in mountain geomorphology has now been realised (Rapp 1960; Luckman 1977), but there is no space here for a full review of their effects. The geomorphic effect of snow avalanches is limited unless they are full-depth, travelling over the ground: this is most common in spring. They can then erode gullies, and deposit fans or boulder tongues below: the deposits are unsorted and angular. Avalanching pushes talus lower down the slope (Rapp 1960) and makes its profile more concave, its upper margin sharper. Avalanche fans and boulder tongues are produced in the runout zone. Below steep or convex slopes, avalanches may exert considerable impact on alluvial valley-floors, producing small lake basins (avalanche tarns) and concentric mounds of debris (Fitzharris and Owens 1984). In Scotland, present-day avalanche activity is modifying talus slopes, interacting with debris flows (Luckman 1992) or even throwing up a rampart (Ballantyne 1989).

Systems of steep, parallel gullies, clearest on slopes of 35–45°, are commonly due to avalanche activity. These 'chutes' are straighter and more rounded than fluvial gullies, and less angular than rockfall gullies (Matthes 1938). Often the different processes alternate and mountain-side gullies are transitional between these characteristics. Steep 'fluvial' gullies are eroded largely by debris flows. Some avalanche chutes are found above glacial troughs and, more rarely, on cirque headwalls. In many areas of asymmetric glaciation, avalanche chutes and gullies occupy the slopes which have not developed glacial cirques. These slopes have been steepened by valley deepening, largely by glacial erosion. Erosion of bedrock by avalanches is not a rapid process, and debris volumes below account for only small proportions of chute volume: most chute erosion dates back to the last glaciation or earlier.

GLACIAL TROUGHS

In cross profile

Glaciated mountains can be understood only if their 'U-shaped valleys' are regarded as the channels of former glaciers (Davis 1900; Penck 1905; Blache 1952). These troughs are palaeochannels, and should be compared with the channels of rivers, not with river valleys. The slower flow of glaciers means that channels must be some hundred times broader and deeper, and thus occupy a large proportion of the landscape. Like rivers, glaciers sometimes flood beyond their usual banks, but shallower spreads achieve less erosion than the channelled flow. In both, when high flows recede, the bed is alluviated and the banks collapse. The greater scale of glaciers means that these bank collapses may take the form of spectacular rock avalanches (see later section) millennia later.

Both river and glacier channel cross-profiles are roughly parabolic (Gerrard 1990). Graf (1970) fitted a power function to sixty bedrock profiles in the Beartooth Mountains of Wyoming and Montana, with the trough centre as origin, and found average exponents between 1.5 and 2.0, the latter being the value for a parabola as previously found for a Swedish trough by Svensson (1959). Graf explained lower exponents as due to less intense erosion. The broader range of exponents obtained by Aniya and Welch (1981) for the Victoria valley region of Antarctica may not be reliable, and indeed Graf's logarithmic standard errors averaging 0.42 show poor fits. In the only published comparison of different profile models, Doornkamp and King (1971) found that the power function gave a considerably poorer fit than linear, squared or exponential models, for three profiles in the American Rockies. Harbor and Wheeler (1992) discussed the possible pitfalls due to errors in points low on the profile, which have great influence when logarithmic transformations are used, or due to erroneously including deposits as part of the 'bedrock' profile.

Augustinus (1992) fitted quadratics to numerous troughs in southern New Zealand and gave average results for five geologically contrasting regions, each based on at least fifteen curves. He interpreted the quadratic coefficient as a measure of trough steepness and narrowness, with higher values for the plutonic and well-metamorphosed rocks of the Fjordland region than for areas of weaker rocks. Individual coefficients and measures of rock mass strength and intact strength showed great scatter, so Augustinus gave correlations only between averages for the five regions; unfortunately this means that only the strongest correlations can be significant (0.88 at the 0.05 level). On this basis the quadratic and linear coefficients, width/depth ratio and shape factor (area/(depth × perimeter)) intercorrelated significantly, and all but the latter correlated with trough density, which is also greater in Fjordland.

Correlations with averaged measures of intact rock strength (Schmidt Hammer rebound, Los Angeles abrasion value) were not merely insignificant but negligible. Trough density and width/depth ratio did, however, correlate with rock mass strength, reflecting the importance of jointing, weathering and groundwater. Given the rapidity of erosion in the schist and greywacke mountains, Augustinus related these variations in trough form and density to structural influences on glacial erosion during the Quaternary; in Fjordland, where rocks are more resistant to trough widening, older fluvial valleys were deepened by glaciers.

If a glacier flows in a valley with a V-shaped cross-profile, greater drag near the central constriction will reduce basal velocity there. It is easily shown that if erosion increases with basal velocity, it will be greater either side of the centre of the V and will thus transform it into a U shape. In Harbor's (1992) simulations, after considerable erosion, the occupied glacial channel reaches steady state with a near-parabolic profile (b = 2.2 to 2.3) and a depth–width ratio which increases with the power of proportionality between erosion and velocity: raising velocity to the power 2 gives depth = 0.41 times width, while the power 4 gives 0.54, for an initial ratio of 0.5. This steady state is probably achieved after several glaciations. Harbor suggested that his results support a power between 1 and 2, but the support is weak given the starting assumptions. Realistic variations in discharge produce somewhat narrower troughs in Harbor's simulations because for much of the time only the central part is eroded; but the effect of high ice discharges is dominant.

Harbor (1992: 1370) pointed out that lowering of the glacial channel should eventually reduce ice discharge. Such a feedback cannot be incorporated in a 2-D model such as his, since the main control of discharge is the altitude of ice source areas. If these are not lowered significantly, a trunk glacier may continue to erode and thicken until basal ice becomes sluggish, or uncoupled from its bed; investigation of this effect requires a long-profile model, or preferably a 3-D model. If the trunk incises faster than source areas, steepening of glacier surface gradient will maintain basal velocity, perhaps until a deep basin is excavated. In humid or coastal areas, calving into lakes or the sea will keep glacier surface altitude above sea level; such a constraint does not apply to on-land glacier termini in cold, arid areas, but these are likely to be less erosive.

152

For glacial troughs Hirano and Aniya (1988, corrected 1989) derived a catenary profile, which can be approximated by a parabola, as that producing maximum friction and thus most efficient erosion for a given glacial contact perimeter. One might expect such a form to be unstable, and Harbor (1990) showed that the derivation ignored drag past bed irregularities, assumed a water-pressure surface mimicking the bedrock surface, and was inappropriately constrained. He also suggested that existing data sets were inappropriate for tests of developmental hypotheses.

As a general principle, the notion that characteristic forms can be produced by maximising or minimising erosion is curious. Whether transformation is rapid or slow, forms continue to change until erosion is *uniformly distributed*. Until more detailed data with better sampling bases are analysed, we know little more than that trough sides tend to concavity. For example, troughs may tend to semi-circularity in cross-section, but the sides are modified because of the general inability of jointed rocks to maintain vertical cliffs at hectometric scales.

In long profile: steps and basins

As Linton (1963) pointed out, glacial troughs are more deeply concave in long profile than are river valleys. Wheeler (1979: 250) has confirmed that glaciated valleys are steeper in their upper reaches, and flatter in the lower reaches. Clearly glaciers are more effective at valley deepening than rivers: their unlimited competence and high capacity to transport debris must help, but in addition they break up rock where basal ice velocity is high. Boulton (1974) demonstrated how this can result from differential pressure around a bump. Bumps do not seem to be eliminated; perhaps this is because of the different processes up-ice and down-ice. If new fractures are produced low on the down-ice side, the whole form may erode downwards as well as up-ice.

Nye and Martin (1968) provided an explanation for the way long profiles consist of smooth concavities separated by much sharper convexities. 'Slip lines' in glaciers are two orthogonal sets of curves following directions of maximum shear stress (they are not lines of slip, or shear planes; these are not common in glaciers). There is a limit to their concavity, of the order of 100 m radius on gentle slopes. Hence ice in sharp concavities will remain inactive until the adjacent rock is worn down and the whole bed follows either slip lines, or envelopes of slip lines. Glacial erosion is thus a co-operative process, controlled not by local factors (geology, ice velocity) alone, but also by erosion in the surrounding area. This is one type of form–process feedback. Nye and Martin excluded lee cavity formation, a limitation which now seems more serious than in 1968, and which explains the persistence of sharp concavities at the foot of steps.

In long profile, troughs are invariably stepped or irregular. Major steps or

153

valley obstructions are known as 'verrous' in French (Blache 1952) or 'riegels' in German. Hooke (1991) explained the exaggeration of steps in glaciated valleys in terms of localised variations in water pressure. He found a diurnal twofold variation in water pressure in boreholes on a riegel in Störglaciaren, northern Sweden. Pressure variation lags air temperature by only four hours, suggesting the effect of meltwater input nearby in the moulin field on the riegel. This surface water reaches the bed within a few hundred metres horizontally.

Water pressure fluctuations in cavities in the lee of bumps help both fracture and entrainment of bedrock blocks. An abrupt fall in pressure promotes fracture by the growth of cracks where pressure lags behind that in the cavity, i.e. where rock permeability is low (Iversen 1991). The extra weight of ice above a step increases differences between principal compressive stresses, giving tensile stress at the tips of favourably oriented cracks. Modelling a 1 m vertical step, Iversen suggested that near-vertical fractures (parallel to the most compressive principal stress) were likely to form. When water pressure increases again, it takes part of the glacier weight and increases sliding velocity, permitting freezing on to bumps where pressure is reduced (Iversen 1991). Together with reduced effective normal pressure, which reduces frictional resistance to dislodgement, this helps entrainment of blocks; hence the quarrying process is accelerated on the lee slopes of bumps (Röthlisburger and Iken 1981). As these become headwalls of steps, steepening increases crevassing and there is a positive feedback as water input and pressure variations increase. This provides an attractive model for glacial valley step development even in massive rocks: more permeable, jointed rocks are quarried even more easily.

Hooke and Iversen's water pressure variations are added to glacier thickness or velocity variations causing opening and closing of lee cavities (Boulton 1974) in making lee slopes favoured sites for glacial quarrying. This contradicts and replaces some early ideas (Machatschek 1969), that reduced glacier loading on steeper slopes reduced erosion, and preglacial steps could maintain their position. Sugden et al. (1992) applied Hooke and Iversen's concepts to the plucking and up-glacier retreat of the lee slopes of transverse ridges in a broad Scottish mountain valley, which produced large-scale roches moutonnées intimately associated with ice scouring and meltwater channels. The last phase of plucking was beneath thin ice, at the very end of the last ice-sheet glaciation.

Reversed slopes make it more difficult for water conduits to remain open, and Hooke (1991) suggested that this encourages till deposition rather than erosion. Subglacial basins are likely to be deeper at the up-ice end.

Matthes (1930) showed how new bumps or steps may be produced by plucking working on rock alternately well-jointed and massively jointed. Major steps are commonly related to lithology: Monjuvent (1974) found this in all his examples in the Drac and Romanche valleys. Other riegels are related to position in the valley system, but some have retreated up-valley. This makes it difficult to test whether steps in the preglacial profile were needed to initiate the glacial positive feedback.

In glacially transformed valleys, rock basins are complementary to riegels. Basins generally relate to weaker or more broken rock, but in some of the larger examples such as those at the edge of the Alps, position has played a role. Long 'ribbon' lakes are commonly divided into several basins by underwater sills. The existence of deep basins just up-ice from terminal moraines suggests that flow upward towards the ice surface (as required in the ablation zone of an equilibrium model) increases towards the snout in large glaciers and greatly facilitates the excavation of basins. Hence the deepest basins do not seem to form at the equilibrium line, but further downstream, with reversed slopes where discharge is diminishing rapidly. This is an obvious consequence of *calibration* of trough size to ice discharge, with erosional deepening from a previous river profile increasing rapidly downstream to the upper ablation area, then declining.

Fjords and ribbon lakes

Holtedahl (1967) showed that the rock thresholds of fjords (e.g. Sogne) are where they emerge from the mountains, permitting lateral spreading of ice and hence upward flow. He related the more complex long profile of Hardangerfjord to several diffluences, where ice was lost down side-channels. This too demonstrates the broad applicability of the calibration principle.

Are fjords a special type of glacial trough, or are they simply well-developed troughs which happen to have been invaded by the postglacial sea? Blache (1952, 1960) inclined to the latter opinion, arguing that erosion of deep fjords on the west coasts of southern Chile, South Island New Zealand, Norway and British Columbia to southern Alaska came from high ice velocity due to heavy snowfall as these coastal mountains intercept the westerlies. Rapid calving of ice into the ocean also produces fast flow (Clarke 1987) and the on-land ablation zone is shortened: high discharge and fast flow are thus found immediately up-ice of the terminus. Flow of the order of a kilometre a year is likely to have profound effects on the glacier bed unless it is buffered by thick deposits. The instability of calving glaciers may aid the excavation of basins in fjords and large lakes.

In each of the fjord regions just mentioned, deep ribbon lakes are found on the landward side of the mountains. Like fjords, they are flat-bottomed because of thick postglacial deposits reducing their apparent depth. There are transitional cases; in British Columbia between Vancouver and Hope, some ribbon lakes north of the Fraser River have in the past functioned as fjords. If ribbon lakes are not quite so deep as fjords, they are often set in less dramatic relief, where glacier or ice-sheet gradients were less, for example in

Sweden compared with Norway (Blache 1960). Also in the more continental climate there was less snowfall, and reduced altitudinal gradients of both accumulation and ablation; the reduced ice discharge per unit area reduced average velocities. Ablation at the lower ends of these long glaciers or ice streams produced upward flow strong enough to mimic the effects of flotation in the ocean. In fjords, calving, high glacier activity and high gradient favoured production of somewhat more spectacular forms, but there is no need for further special effects as proposed by Crary (1966).

East of the Scandinavian mountains, dozens of ribbon lakes are 20 to 40 km long (Rudberg 1992): the longest is 117 km, and the maximum water depth is 460 m. In Scandinavia (Porter 1989), in Scotland and in British Columbia, many ribbon lakes could have developed under average Quaternary conditions, rather than during the ice-sheet maxima favoured by Blache (1960). In the Alps, however, the major lake basins on both northern and southern margins seem to relate rather to maximum glacial conditions. Other regions with ribbon lakes, unrelated to fjords, include the western slope of the Putora Mountains in north-central Siberia and the eastern slope of the Kilbuck Mountains in south-west Alaska.

In three dimensions: calibration

Playfair's 'law of accordant junctions' applies only to river and glacier *surfaces*, at equilibrium. Viewing troughs as glacier palaeochannels, Penck (1905) formulated a 'law of adjusted cross sections' in which trough cross-sectional area is proportional to (former) ice discharge. This was developed as the principle of 'calibration' by Blache (1952), who provided the best explanation of 'hanging valleys' (tributary palaeochannels), each eroded in proportion to the discharge of its glacier. Like rivers, glaciers obey the necessary geometric relationship that

discharge = cross-sectional area × mean velocity.

Both Penck and Blache assumed that mean velocity does not vary so much as to accommodate all change in discharge. As a first approximation this seems justifiable and almost unavoidable; adjustment to increase in discharge is likely (again as in rivers) to be by increases in both area and velocity. Studies such as that of Meier (1960) confirm that, within a valley glacier, mean velocity does vary in sympathy with cross-sectional area. But velocity varies with gradient as well as with ice depth, and this may explain some deviations from calibration.

A quantitative test of calibration was applied by Haynes (1972) to nineteen troughs in granodioritic gneisses on the northwest side of Sukkertoppen Ice Cap, all opening on to Søndre Strømfjord in West Greenland. Cross-sectional areas measured near the mouths of these troughs correlated well with ice catchment areas; a logarithmic regression accounted for 51% of variance. Moreover, the addition of 'distance from the sea' (a surrogate for precipitation at a given altitude) in multiple regression increased this to 87%. Hence the troughs are well adjusted to potential ice discharge from this local ice cap, assuming that present ice divides bear some relation to those when the glacial troughs were eroded.

Roberts and Rood (1984) considered the calibration of thirty-three fjords in southern British Columbia. Average depth along fjord centre-line was predicted from present-day fluvial drainage area. The logarithmic regression gave an R^2 of 61% despite exclusion of much of the trough depth (that part above present sea level). Some residuals can be explained by transfluence within the Cordilleran Ice Sheet, greatly increasing the ice catchment of, for example, Howe Sound, where 81% of variation in fjord length is accounted for by present drainage area, despite the rather arbitrary position of fjord heads, beyond which alluvial plains have varying extents. The strength of these relationships, and those of Haynes, is as high as can be expected given the number of other factors involved.

In a simple linear glacier, the greatest ice discharge and thus cross-section is around the equilibrium line (as it crosses the glacier surface). In dendritic glacier systems, however, maximum discharge may come from the confluence of major branches, even some way below the equilibrium line. The greatest glacial troughs are therefore expected in ablation zones. This applies not only to land-terminating glaciers, but especially to tidewater glaciers where the ablation zone is truncated and a deep trough may extend to the terminus. Calibration is clearly evident in the pattern of north Icelandic troughs illustrated on the cover of Sugden and John (1976): as troughs join, discharge increases and troughs become broader at least to the present coast. The troughs are calibrated to an extensive glaciation, covering the continental shelf. Their simplicity may reflect erosion in the fairly homogeneous lithology of Late Cenozoic basalts. On an even grander scale, Linton (1963) suggested elimination of divides between the broadening troughs of well-supplied glaciers above snowline in the Antarctic Peninsula.

Former glacier size and form are clearly engraved in the landscape in Figure 7.1, which shows the headwaters of Taseko River in British Columbia around 123.3°W, 51.1°N. Between the broad trough of the Taseko (right) and Granite Creek (centre) six hanging tributary valleys curve northward to join the Taseko at small angles; these are eroded in quartz-diorite or granodiorite of the Coast Plutonic Complex. Larger tributaries joining the Taseko from the east are separated by narrow, tapered rock ridges, several of which are off this photograph. This pattern can be explained only by adjustment to ice flow down the Taseko valley. Since this is an area of low gradient, near an ice divide and away from major ice discharge routes, the pattern of calibrated forms survived inundation by the last Cordilleran Ice Sheet. The down-valley flow relates to average Quaternary conditions (Porter 1989) and was probably more prolonged than the ice sheet

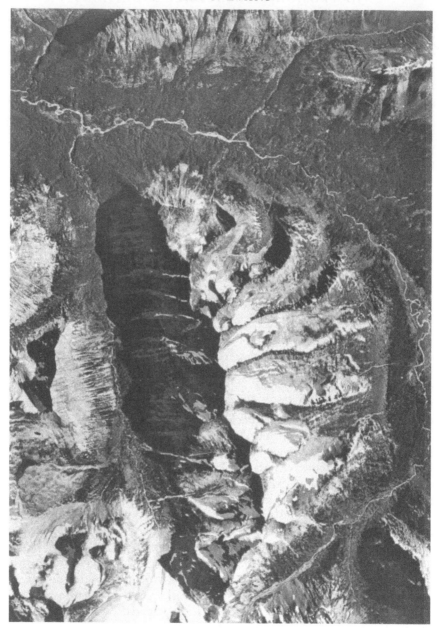

Figure 7.1 Calibration of cirques and troughs in the British Columbia Coast
Mountains: Gibson (Granite) Creek (left) and Taseko River (top and right)

Source: Aerial photograph A13248–10 © 1951 Her Majesty the Queen in Right of Canada;
reproduced from the collection of the National Air Photo Library with permission of
Natural Resources Canada. North at top; extent approximately 8.5 × 11.8 km, on topographic
map sheet 92 o 3.

maximum: the forms it produced dominate the present landscape.

Diffluence of ice to a valley without a glacier, or transfluence to one with a lower glacier surface, modifies the pattern of valleys by producing intersecting troughs (Clayton 1974; Haynes 1977). Creation of a glacial trough where formerly there was only a col provides clear evidence of deep glacial erosion (Linton 1963). Faster erosion on the down-ice side of the col, due to the steeper gradient, may displace the postglacial drainage divide up-ice initially by a small amount, but by a large jump if the col is eroded below the level of the up-ice trough. Linton (1951) and Dury (1955) gave examples of such displacement. Haynes (1977) used the topological α index to measure the increase in valley connectivity due to watershed breaching by ice sheets. In Scotland, this is a function of glacial accumulation patterns, land altitude and dissection, and geological variety; connectivity is highest in the west, where rapid basal sliding of the ice sheet was widespread. Varying degrees of ice sheet modification are superimposed on the troughs and cirques of alpine glaciation. Purely alpine glaciation would retain the roughly dendritic valley patterns inherited from fluvial dissection.

CIRQUES

Erosion by rotational flow

Clear cirque forms are accepted as one of the best proofs of glaciation in mountains. The cirque form consists essentially of a steep headwall slope arcing around a gentler floor (Evans and Cox 1974): it is concave both in profile and in plan. Compared with most fluvial valley-heads, cirque floors have been broadened and headwalls steepened. The 'problem of cirque erosion' resides in finding a process to achieve this, for example by undercutting headwalls. Resolution of the problem should answer the question whether cirques are essentially glacial or largely periglacial features.

The older literature on this problem was thoroughly reviewed by Embleton and King (1975, Chapter 7), and by Charlesworth (1957). To account for the enhanced concavity, maximum erosion needs to be at the heart of the cirque, just below the headwall; the lower part of the slope which is being steepened to become a headwall must be eroded faster than the upper part. It is difficult to achieve this without rotational flow of the glacier, giving rapid basal slip in this central zone, and thus more plucking and abrasion. Rotational flow is most readily achieved with a cirque glacier, where both accumulation and ablation zones are within the cirque (Lewis 1948; McCall 1960; Grove 1960). Rotation is encouraged by net ablation in the lower part requiring upward movement (relative to the ice surface) to maintain equilibrium; and by net accumulation high on the glacier, especially where enhanced by avalanched or wind-drifted snow, giving a wedge-shaped annual increment. This combination is analogous to encouragement of rotational

159

landslides by loading the upper part and removing support lower down. Given this pattern of glacier balance, rotation is likely; erosion from the related basal sliding will develop concavity of the glacier bed, giving a positive feedback.

Field mapping shows that many classic 'armchair'-shaped cirques, with gently sloping floors or even rock basins, are in the source areas of former valley glaciers, which may or may not have been preceded by discrete cirque glaciers. This implies either that the cirques developed before the valley glacier stage, or that they could develop when ice flowed out even to the extent of the whole cirque being in the accumulation zone. The latter is supported by suggestions of multiple 'rotational cells' in valley glaciers (Lewis 1948). Development of such a cell at the very head of a glacier is encouraged by 'wedge-shaped accumulation' caused by wind-drifting and, once a steep headwall is developed, by avalanching. If this is important in the development of classic armchair cirques, we may hypothesise that leeward glaciers will be more effective eroders of cirques than 'shadeward' glaciers, and that cirque aspect may differ from glacier aspect more than was found by Evans (1977). More evidence on this would be welcome: we may tentatively conclude that a cirque glacier stage is helpful but not indispensable to cirque development.

Wearing down versus wearing back

Emphasis on glacial erosion (as in Charlesworth 1957: 299) supports downward as well as lateral erosion as essential factors in cirque development. White (1970) went too far in emphasising downward erosion, but he was reacting to a long-standing bias towards lateral erosion which goes back to Johnson's (1904) hypothesis of 'frost-shattering in the bergschrund' and continued through the work of Lewis (1940, 1960) and of McCall (1960). The limitations of the bergschrund hypothesis are well rehearsed: bergschrunds – large crevasses at the sources of mountain glaciers – are mostly less than 30 m deep and rarely penetrate to the base of large headwalls. Bergschrunds are often absent from glaciers in cirques, especially in polar and tropical mountains. If bergschrunds caused headwall retreat, 'headwall-foot zones' would extend upwards and headwall height would be reduced. Yet Haynes (1968) has argued convincingly that where they occur, headwall-foot zones and the 'schrundlines' at their upper limits are structurally controlled. Likewise, if periglacial processes such as frost-shattering on the exposed headwall were all-important, the headwall would retreat and be destroyed rather than enhanced. In general, subglacial erosion at the base of the headwall is required for cirque development, while periglacial processes degrade cirque headwalls and fill in floors. Subglacial erosion by external ice overriding headwalls will also damage cirque form.

The morphometric evidence suggests that cirque length and width

160

increase with size more rapidly than does depth (Gordon 1977; Olyphant 1981). Large cirques are not necessarily older than small ones, but cirque erosion is a one-way process; cirque area will not decline unless adjacent troughs and cirques are eroded more vigorously. Larger cirques are better-developed, with larger, flatter floors and more enclosing headwalls, biting more deeply into the mountains (Evans and Cox 1995); they are more likely to have cols and lakes.

Headward erosion of 1 or 2 km is evident in the asymmetry of valley heads in parts of British Columbia (Evans 1972). Location of the highest summits on cirque rims is further evidence of headward erosion (Rudberg 1992). Because local glaciation is often concentrated on one side of a mountain range (Evans 1977), considerable headward erosion is possible without deep cols being produced by the intersection of opposing cirques. The valley diversion attributed to cirque headward erosion by Waitt (1975) is an unusual case. Since headwall gradient is maintained during retreat, the importance of back-wearing does not imply that ice-marginal erosion dominates, but that erosion is concentrated at the headwall foot, where ice is thick. Maintenance of gradient is facilitated by active quarrying at the headwall foot, as well as by abrasion.

There is no apparent limit to the change of shape with size; the development of glacial cirques must be regarded as allometric. In principle cirque area, and even cirque depth, could go on increasing, but almost all cirques are within a limited size range, 200 to 2500 m in width and length, 75 to 1000 m in depth. This variation over only one order of magnitude is a clear deviation from 'self-similar' fractal models of the land surface (Evans and Cox 1995).

Cirques are highly varied in form; weak relationships between floor gradient, headwall gradient and closure in plan deny the existence of a single path of development and suggest the influence of different situations and different initial sites (gullies, convergent gullies, landslide scars and other concavities). Cirques develop in different ways in response to climatic, topographic and geological controls. The sequence of development proposed by Hobbs (1911), and widely quoted, illustrates some of these spatial contrasts but is not acceptable as a general model. Blache (1960) related these contrasts to snow abundance and the spacing of glaciers, not to a temporal progression. Glacier spacing relates also to initial topography and thus to tectonic environment.

High-alpine cirques

Classic, concave 'armchair' cirques do not dominate all mountain ridges. Blache (1952, 1960) noted the concentration of 'cirques en van' in the highest alpine massifs. These high-alpine forms cover the whole landscape at high altitudes and are relatively shallow, with steeply sloping, abraded floors (aprons) and low but continuous headwalls. They are also found in coastal British Columbia (Derbyshire and Evans 1976; Ryder 1981), where headwalls are sometimes broken or missing. Here and in Oisans, France, median apron gradients are around 26°.

Figure 7.2 portrays a model of the contrast between the two types. I propose that high-alpine, 'van' cirques form entirely above the snowline, and since they cover most of the land surface they gain little snow from wind drifting and avalanching. Hence there is no 'accumulation wedge' to encourage rotation; thin glaciers with many crevasses – like repeated bergschrunds – slide rapidly over their beds. The contrast is essentially glaciological. Often, as near Mount Tzoonie, British Columbia, high-alpine cirques cluster around a deep trough head, containing an oval rock basin. More complex forms may relate to changing glaciological conditions over time.

Frost shattering

Frost shattering does contribute to cirque enlargement. The essential requirements are abundant liquid water and temperatures sometimes low enough to freeze this in joints and other partings in the rock. At high altitudes, meltwater will freeze in contact with rock at sub-zero temperatures; the rock surface will be glazed with ice, inhibiting further frost action. At lower altitudes, meltwater may be abundant but the problem is how to freeze it beneath a glacier: this probably requires changes in pressure.

Battle (1960) established that temperatures deep in bergschrunds varied only slowly, usually at rates below 0.2°C hour^{-1}, and ranged between 0 and –2°C in summer. Marked diurnal cycles were lacking. He concluded that, so long as water is available, frost action should be greater on the headwall, where temperature changes are faster and more extreme: shallow bergschrunds and open randklufts at the margin may provide both the water and the temperature regime required. Gardner (1987) found plenty of meltwater and active frost shattering at several such randkluft sites in the Canadian Rockies. He supported Battle's conclusion, and pointed out that the randkluft lip (glacier edge) was lowered by 1.5 to 3.5 m each month during the ablation season, broadening the zone favourable to weathering: temperature range was greater when the sites were exposed.

There has been little support for Battle's emphasis on rate of freezing (McGreevy and Whalley 1982). Speed of temperature fall is important in 'closed-system' frost shattering, where volume increase on freezing is the operative mechanism; but models of 'open-system' frost shattering, with slow ice lens growth as in soil frost heave, are becoming more popular for rock weathering (McGreevy 1981; Walder and Hallet 1985, 1986). Here slow temperature change helps ice lens growth, permitting water to migrate. Much of the damage to rock is done in the temperature range –5 to –15°C, and

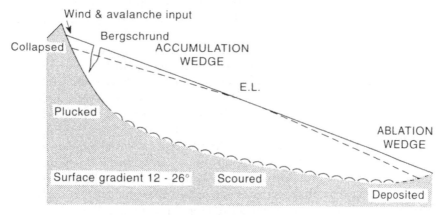

'Armchair' Cirque
marginal (usually asymmetric) glaciation

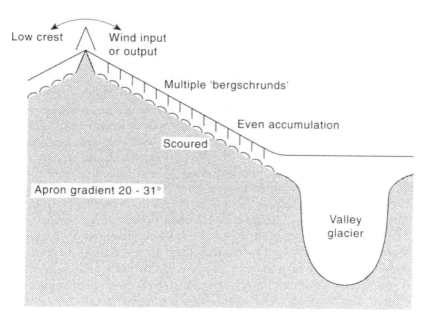

'Van' Cirque
high-alpine glaciation, usually symmetric
Well above snowline

Figure 7.2 A model of the contrast between classical armchair cirques and high-alpine 'cirques en van' (Tzoonie-Phantom or Tantalus-type cirques in British Columbia)

annual or even longer-term temperature cycles are effective. In Lautridou's (1988) experiments, damage to rocks with medium-sized pores increased down to –8 to –10°C.

The best conditions for 'open-system' frost shattering are very cold rock adjacent to a supply of liquid water. Ice lensing could be important on exposed headwalls or beneath 'cold' ice, but not beneath ice at the pressure-melting point. Very low-frequency changes, as the 0°C isotherm migrates in response to change of climate or ice thickness over centuries, may produce frost shattering if water is available in the rock (Fisher 1955, 1963). While also influenced by glacier balance and geothermal heat flow, this process would relate frost action more closely to temperature conditions: if it is important in cirque erosion, it would produce uniformity in cirque altitudes, which would vary locally with aspect and regionally rather gradually. Evidence of steep regional gradients in cirque altitude in Snowdonia (Unwin 1973), in Washington (Porter 1964) and in British Columbia suggests that variations in snowfall are more important than such variations in temperature. Some cirques are in the alpine permafrost zone and contain rock glaciers, while others (in snowier, maritime climates) have floors well below permafrost level. Cirque floors (and headwall feet) seem to relate to past snowlines rather than to past zero isotherms.

In Lewis's (1940) modification of Johnson's bergschrund hypothesis, meltwater comes from the headwall above the bergschrund and penetrates below it; frost action on the lower headwall facilitates glacial plucking. Meltwater has been observed in several tunnels through alpine glacers. Lewis (1940: 81) was not convincing on how bulk meltwater can refreeze under a glacier; if the ice is isothermal and below the winter 'cold wave', only rock at sub-zero temperatures can absorb the latent heat liberated. Local falls of water pressure might help, as in Hooke's (1991; see above) update of the meltwater hypothesis, but Hooke proposed a direct effect on fracture growth without the intervention of frost wedging. If Hooke's model for valley steps can be extended to cirques, it supports the importance of subglacial quarrying rather than ice-marginal freeze–thaw in their development.

The 100,000 years required for cirque erosion, proposed by Larsen and Mangerud (1981; see above), seems reasonable as an order of magnitude. A shallow cirque on massive rock, or a large one on weak rock, might develop during a single, prolonged glaciation. Most well-developed cirques have developed over successive glaciations. This is not problematic, since even if muted by interglacial erosion the cirque form itself encourages formation of a glacier (Graf 1976). There may instead be a problem as to why cirques are not much larger, given the length of the Quaternary and the number of glaciations. Four possible explanations are:

(i)   the most favourable conditions, with ice gradients encouraging rota-
      tion, occupy only part of each glaciation;

(ii)   where conditions are most favourable, cirques are competing against adjacent cirques, side by side if not back to back;

(iii) valleys had to be eroded first, to provide gradients sufficient for cirque development; and

(iv)   Quaternary uplift brought some mountain ranges above snowline only for the last few glaciations.

Of these, the first seems most widely applicable.

STRESS RELEASE AND ROCK AVALANCHES

Glacial troughs and cirques are not formed by continual small-scale events alone; large, sudden mass movements are also important, the more so as slopes become higher and steeper. These are controlled by the state of rock stress, which in mountain areas is highly variable and may owe more to local topography than to regional tectonic stress fields (Kohlbeck et al. 1979). The load of mountains is spread by deformation at depth, and stress in valleys is much greater than predicted from local loading.

The concept of stress concentration in valleys provides a more viable explanation of trough deepening than does traditional pressure release by erosional unloading (Lewis 1954; Linton 1963). The latter model suggests that as glaciers erode deeper, rock removal reduces total load and new joints in the rock below open up parallel to the surface. The weakness of this model should be readily apparent: as Addison (1981) stated, 'pressure release joints' have often been misidentified. First, the spacing of pressure release joints increases rapidly with depth. Under thick ice, the considerable hydrostatic load will prevent opening of new joints of this type. On deglaciation, the ice load in deep rock basins such as Sognefjord or Lake Chelan is replaced by a considerable water load; again the hydrostatic pressure is always high and there is little scope for pressure release jointing. Second, formation of a glacier increases loading on the land surface. Only when it has eroded one-third of its own thickness is there any reduction in loading. The thickness of a glacier capable of eroding so much means that any pressure release joints will be widely spaced.

As erosion progresses, increased ice thickness (especially in basins) gives a negative feedback to pressure release jointing. By contrast, the concentration of stress in a valley floor increases as the valley is deepened. In ductile rocks, the phenomenon of 'valley bulging' is a response to stress concentration due to the load of ridges or plateaus on either side. In brittle rocks, fracturing is likely below valley floors and at the feet of steep slopes. Higher, steeper slopes are more effective in increasing the stresses, which are further augmented by the residual horizontal stresses found in many parts of the world (Hast 1967; Zoback 1992). Unlike most rivers, glaciers have the competence to remove large blocks of fractured rock. As glacial troughs and

165

cirques are deepened, levels of greater stress are reached; stress is concentrated at low points, causing fracturing and providing positive feedback to erosion. This process is limited only when basal ice decelerates as the trough or cirque basin cross-section becomes 'too large' for the available ice discharge.

Pressure release is important for cirque headwalls and trough sides, but in a more general way than the creation of new joints. Having been steepened by glacial erosion at the foot, trough and cirque walls often exceed the critical gradient-height combination for the jointed rock, and so collapse (Whalley 1974, 1984). Mountain glaciation pushes slope stability to its limits and provides a field test of rock mass strength. An early stage of slope collapse is 'sagging' ('Sackung'), producing slope trenches and uphill-facing scarps principally on high, linear ridges between glaciated valleys, or behind cirque headwalls (Radbruch-Hall 1978; Varnes et al. 1989). Bovis (1990) measured ongoing movements in 30M m^3 of monzonite and young volcanics on a trough-side in British Columbia. Downslope extension over a deep-seated shear zone produced antislope scarps by near-surface flexural toppling. The stress field reflected debuttressing from recent glacial retreat, rather than isostatic rebound.

At the least, such slope deformation breaks up rock and greatly eases erosion during the next glacier advance. Often, however, slope collapse goes further and produces spectacular rock avalanches (Nicoletti and Sorriso-Valvo 1991). Many of these are on glacially oversteepened trough sides, for example in the Alps (Eisbacher and Clague 1984). In Kärkevagge, Sweden, Rapp (1960: 122) estimated that rockwall breakdown on deglaciation produced cliff retreat of 10 to 20 m, at least ten times that during postglacial time. Yet the response to deglaciation need not be rapid; especially on massive rocks, joint opening through stress release may continue for millennia before collapse occurs. For a 10,000 km^2 area of high mountains in southern New Zealand, over the last 1700 years, Whitehouse and Griffiths (1983) estimated a frequency of one large rock avalanche (over 1M m^3 in volume) per 94 years; the triggers for most are probably major earthquakes. In the Karakoram, rock avalanches provide a major component of material in lateral moraines and 'ablation valleys' (Hewitt 1989): the sediment yields of proglacial streams are high.

Glaciated mountains are littered with the debris of rock avalanches and smaller rockfalls. These have been produced in the short interval since the last glaciation and will easily be swept away in the next major glaciation. Viewed over the long timescale of numerous mountain glaciations in the Quaternary, slope collapse must have made a major contribution to trough and cirque widening. Removal of the shattered material creates a new geometry; with or without further glacial erosion this may facilitate the next round of slope collapse. Gerber and Scheidegger (1969) showed that many features of mountain cliffs – wedge-shaped breakouts, triangular peaks, pyramids and tapering pillars – relate to conjugate fractures. 'Generally, cliff-surfaces recede by stress-induced joint surfaces' (Gerber 1980: 104).

Most rock avalanche deposits are undated, and it is difficult to estimate either current or Late Glacial frequencies of major events. If we concentrate on historic events, it is striking how many of these are from cirque headwalls or other ice-source cliffs; for example, the 1975 event in South Georgia witnessed by Gordon *et al.* (1978). All of the events collated in Table 7.2 travelled beyond their source cirques; horizontal travel was 1.5 to 5.0 times vertical drop. All but two were on to glaciers, which increased their mobility (S.G. Evans *et al.* 1989), and increased their volume by adding ice, snow and moraine.

The Alaska earthquake of 27 March 1964 produced many large rock avalanches from cliffs 'currently undergoing glacial erosion' (Post 1967: 31). Without giving precise heights of fall, Post tabulated 50 rock avalanches exceeding 1 km in length and 0.5 km^2 in area which were first observed shortly after the earthquake, along with 4 first observed the following year (probably lagged effects of the earthquake), and 11 from 1945–1963. Most of those he illustrated came from cirque headwalls. Further details for the much-discussed slide on to Sherman Glacier were given by McSaveney (1978), while Hoyer (1971) provided the details for a smaller rock avalanche.

Eight of the rock avalanches in Table 7.2 occurred in the short interval 1962–1975. Even during this period, only the more celebrated events are included. Clearly during earthquakes, and on young volcanic rocks, collapse of cirque headwalls is a common event. Five of the events (two on Huascaran and two on the Mont Blanc range) affected granitic or gneissic rocks, generally considered to be among the most resistant rocks, with few joints and great mass strength. Although these properties increase resistance to smaller collapses and rockfall events, they did not prevent the cirque headwalls from succumbing to major collapses.

Collapse is most likely as glacier wastage reduces 'buttressing' support of the headwall, but the ultimate cause is the removal of rock from the toe by glacial erosion. The implication is that headwall retreat may be a very discontinuous process. Most deglaciated cirques contain no major rock avalanche deposit, but abundant talus from smaller events. Some cirques may never have been affected by a major event, but many of those in areas of high relief have been.

CONCLUSION

Progress has been made both in the analysis of landforms and in the observation or inference of glacial processes. We are at an early stage in the development of models to link these two. There is now much less room to doubt the reality of glacial erosion by a variety of mechanisms. The existence of highly distinctive landforms in glaciated mountains is thus less puzzling

Table 7.2 Rock avalanches from cirque headwalls

Place	Rock type	Date	Rock vol. (M m³)	Velocity (km hr⁻¹)	Fall (km)	Travel (km)	Source
Lyell Glacier, S Georgia	Greywackes & slates	6 Sept 1975	2–3	60	1.6	4.0	Gordon et al. 1978
Huascaran, C Blanca, Peru	Granodiorite	10 Jan 1962	10	170	4.1	15.9	Plafker and Ericksen 1978
		31 May 1970	45–95	280	4.1	16.0	Plafker and Ericksen 1978
Triolet Glacier, Mt Blanc	Granite	12 Sept 1717	16–20	125	1.86	7.2	Porter and Orombelli 1980
Pandemonium Creek, W-C British Columbia	Gneiss	1959	5	290	2.0	8.6	S.G. Evans et al. 1989
Brenva Glacier, Mt Blanc	Granite	14–19 Nov 1920	6–7	100	2.89	5.0	Valbusa 1921, in Porter and Orombelli 1980
Sherman Glacier, S Alaska	Sandstone & argillite	27 March 1964	10.1	240	1.1	6.0	McSaveney 1978
Puget Peak, S Alaska	Argillite	27 March 1964	0.5	80	1.2	2.13	Hoyer 1971
Fairweather Glacier, S Alaska	–	pre–22 Aug 1965	–	–	3.35	10.5	Post 1967
Little Tahoma Peak, Mt Rainier	Andesite	14 Dec 1963	11	105–150	1.89	6.9	Crandell and Fahnestock 1965
Devastation Glacier, SW British Columbia	Altered volcanics	22 Jul 1975	27	–	1.15	6.5	Mokievsky-Zubok 1978
Mt Col. Foster, Vancouver Island, BC	Volcaniclastic	23 Jun 1946	1.5	–	1.07	1.6	S.G. Evans 1989

than it was a century ago. Stress fields are modified by topography and by the flow of ice, leading to distinctive types of fracture and slope failure. The results are strongly influenced by rock structure and lithology. A variety of processes which are literally periglacial (in space) or paraglacial (in time) play their parts, but powerful glacial erosion is the basic motivator.

Currently identifiable cirques have developed mainly during the last few glaciations. In ice sheet source mountains (British Columbia, Scotland, Norway), and in source massifs of the Alps (Blache 1960), cirques may relate largely to average Quaternary conditions. Downvalley, and in marginal areas of purely local glaciation, they relate to glacial maxima. Present conditions are geomorphologically most effective for steep, well-fed glaciers such as those on Mt Rainier, and the Bossons, Bionassay, Miage and Brenva on Mont Blanc.

ACKNOWLEDGEMENTS

Comments from Nick Cox and Jeff Blackford led to improvements in the text.

REFERENCES

Addison, K. (1981) 'The contribution of discontinuous rock-mass failure to glacier erosion', *Annals of Glaciology* 2: 3–10.

Aniya, M. and Welch, R. (1981) 'Morphological analyses of glacial valleys and estimates of sediment thickness on the valley floor – Victoria Valley system, Antarctica', *The Antarctic Record* 71: 76–95.

Augustinus, P.C. (1992) 'The influence of rock mass strength on glacial valley cross-profile morphometry: a case study from the Southern Alps, New Zealand', *Earth Surface Processes and Landforms* 17: 39–51.

Ballantyne, C.K. (1985) 'Nivation landforms and snowpatch erosion on two massifs in the Northern Highlands of Scotland', *Scottish Geographical Magazine* 101: 40–49.

——(1989) 'Avalanche impact landforms on Ben Nevis, Scotland', *Scottish Geographical Magazine* 105: 38–42.

Battle, W.R.B. (1960) 'Temperature observations in bergschrunds and their relationship to frost shattering', in W.V. Lewis (ed.) 'Norwegian cirque glaciers', *Royal Geographical Society Research Series* 4: 83–95.

Bezinge, A., Clark, M.J., Gurnell, A.M. and Warburton, J. (1989) 'The management of sediment transported by glacial meltwater streams and its significance for the estimation of sediment yield', *Annals of Glaciology* 13: 1–5.

Blache, J. (1952) 'La sculpture glaciaire', *Revue de Géographie Alpine* 40: 31–123.

——(1960) 'Les résultats de l'érosion glaciaire', *Méditerranée* 1: 5–31.

Bogen, J. (1989) 'Glacial sediment production and development of hydro-electric power in glacierized areas', *Annals of Glaciology* 13: 6–11.

Boulton, G.S. (1974) 'Processes and patterns of glacial erosion', in D.R. Coates (ed.) *Glacial Geology*, Binghamton: State University of New York.

Bovis, M.J. (1990) 'Rock-slope deformation at Affliction Creek, southern Coast Mountains, British Columbia', *Canadian Journal of Earth Sciences* 27: 243–254.

Caine, N. (1978) 'Climatic geomorphology in mid-latitude mountains', in J.L. Davies and M.A.J. Williams (eds), *Landform Evolution in Australasia*, Canberra: Australian National University Press.

——(1992) 'Sediment transfer on the floor of the Martinelli snowpatch, Colorado Front Range, U.S.A.', *Geografiska Annaler* 74A: 133–144.

Charlesworth, J.K. (1957) *The Quaternary Era, with Special Reference to its Glaciation*, London: E. Arnold, 2 volumes.

Chernova, L.P. (1981) 'Influence of mass balance and run-off on relief-forming activity of mountain glaciers', *Annals of Glaciology* 2: 69–70.

Chorley, R.J. (1978) 'Bases for theory in geomorphology', in C. Embleton, D. Brunsden and D.K.C. Jones (eds) *Geomorphology: Present Problems and Future Prospects*, Oxford: Oxford University Press.

Church, M., Kellerhals, R. & Day, T.J. (1989) 'Regional clastic sediment yield in British Columbia', *Canadian Journal of Earth Sciences* 26: 31–45.

Church, M. and Slaymaker, O. (1989) 'Disequilibrium of Holocene sediment yield in glaciated British Columbia', *Nature* 337: 452–454.

Clarke, G.K.C. (1987) 'Fast glacier flow: ice streams, surging and tidewater glaciers', *Journal of Geophysical Research* B92: 8835–8841.

Clayton, K.M. (1974) 'Zones of glacial erosion', *Institute of British Geographers Special Publications* 7: 163–176.

Costin, A.B., Jennings, J.N., Black, H.P. and Thom, B.G. (1964) 'Snow action on Mount Twynam, Snowy Mountains, Australia', *Journal of Glaciology* 5: 219–228; and (1973) *Arctic and Alpine Research* 5: 121–126.

Crandell, D.R. and Fahnestock, R.K. (1965) 'Rockfalls and avalanches from Little Tahoma Peak on Mt. Rainier, Washington', *United States Geological Survey Bulletin* 1221-A: 1–30.

Crary, A.P. (1966) 'Mechanism for fjord formation indicated by studies of an ice-covered inlet', *Geological Society of America Bulletin* 77: 911–930.

Davis, W.M. (1900) 'Glacial erosion in France, Switzerland and Norway', *Proceedings of the Boston Society for Natural History* 29: 273–322. (Reprinted in D.W. Johnson (ed.) *Geographical Essays*, Boston (1909): Ginn, and in C. Embleton (ed.) 1972: 38–69.)

Derbyshire, E. and Evans, I.S. (1976) 'The climatic factor in cirque variation', in E. Derbyshire (ed.) *Geomorphology and Climate*, London: Wiley.

Doornkamp, J.C. and King, C.A.M. (1971) *Numerical Analysis in Geomorphology*, London: Edward Arnold.

Drewry, D.J. (1986) *Glacial Geologic Processes*, London: Edward Arnold.

Dury, G.H. (1955) 'Diversion of drainage by ice', *Science News* 38: 48–71.

Eisbacher, G.H. and Clague, J.J. (1984) 'Destructive mass movements in high mountains', *Geological Survey of Canada Paper* 84-16: 1–230.

Embleton, C. (ed.) (1972) *Glaciers and Glacial Erosion*, London: Macmillan.

Embleton, C. and King, C.A.M. (1975) *Glacial Geomorphology*, 2nd edition, London: Edward Arnold.

England, J. (1986) 'Glacial erosion of a High Arctic valley', *Journal of Glaciology* 32: 60–64.

Evans, I.S. (1969) 'The geomorphology and morphometry of glacial and nival areas', in R.J. Chorley (ed.) *Water, Earth and Man*, London: Methuen.

——(1972) 'Inferring process from form: the asymmetry of glaciated mountains', in W.P. Adams and F.M. Helleiner (eds) *International Geography 1972* 1, Toronto: University of Toronto Press.

——(1977) 'World-wide variations in the direction and concentration of cirque and glacier aspects', *Geografiska Annaler* 59A: 151–175.

Evans, I.S. and Cox, N.J. (1974) 'Geomorphometry and the operational definition of cirques', *Area* 6: 150–153.

————(1995) 'The form of glacial cirques in the English Lake District, Cumbria', *Zeitschrift für Geomorphologie* N.F. 39: 175–202.

Evans, S.G. (1989) 'The 1946 Mount Colonel Foster rock avalanche and associated displacement wave, Vancouver Island, British Columbia', *Canadian Geotechnical Journal* 26: 447–452.

Evans, S.G., Clague, J.J., Woodsworth, G.J. and Hungr, O. (1989) 'The Pandemonium Creek rock avalanche, British Columbia', *Canadian Geotechnical Journal* 26: 427–446.

Fisher, J.E. (1955) 'Internal temperatures of a cold glacier and conclusions therefrom', *Journal of Glaciology* 2: 582–591.

————(1963) 'Two tunnels in cold ice at 4000 m on the Breithorn', *Journal of Glaciology* 4: 513–520.

Fitzharris, B.B. and Owens, I.F. (1984) 'Avalanche tarns', *Journal of Glaciology* 30: 308–312.

Gardner, J.S. (1987) 'Evidence for headwall weathering zones, Boundary Glacier, Canadian Rocky Mountains', *Journal of Glaciology* 33: 60–67.

Gerber, E. (1980) 'Geomorphological problems in the Alps', *Rock Mechanics* Suppl. 9: 93–107.

Gerber, E. and Scheidegger, A.E. (1969) 'Stress-induced weathering of rock masses', *Eclogae Geologicae Helvetiae* 62: 401–415.

Gerrard, A.J. (1990) *Mountain Environments*, London: Belhaven.

Gordon, J.E. (1977) 'Morphometry of cirques in the Kintail-Affric-Cannich area of northwest Scotland', *Geografiska Annaler* 59A: 177–194.

Gordon, J.E., Birnie, R.V. and Timmis, R. (1978) 'A major rockfall and debris slide on the Lyell Glacier, South Georgia', *Arctic and Alpine Research* 10: 49–60.

Graf, W.L. (1970) 'The geomorphology of the glacial valley cross section', *Arctic and Alpine Research* 2: 303–312.

————(1976) 'Cirques as glacier locations', *Arctic and Alpine Research* 8: 79–90.

Grove, J.M. (1960) 'A study of Veslgjuv-breen', in W.V. Lewis, (ed.) 'Norwegian cirque glaciers', *Royal Geographical Society Research Series* 4: 69–82.

Hall, A.M. and Sugden, D.E. (1987) 'Limited modification of mid-latitude landscapes by ice sheets: the case of northeast Scotland', *Earth Surface Processes and Landforms* 12: 531–542.

Hall, K. (1980) 'Freeze–thaw activity at a nivation site in northern Norway', *Arctic and Alpine Research* 12: 183–194.

Harbor, J.M. (1990) 'A discussion of Hirano and Aniya's (1988, 1989) explanation of glacial valley cross profile development', *Earth Surface Processes and Landforms* 15: 369–377.

————(1992) 'Numerical modelling of the development of U-shaped valleys by glacial erosion', *Geological Society of America Bulletin* 104: 1364–1375.

Harbor, J.M. and Warburton, J. (1993) 'Relative rates of glacial and non-glacial erosion in alpine environments', *Arctic and Alpine Research* 25: 1–7.

Harbor, J.M. and Wheeler, D.A. (1992) 'On the mathematical description of glaciated valley cross sections', *Earth Surface Processes and Landforms* 17: 477–485.

Hast, N. (1967) 'The state of stresses in the upper part of the earth's crust', *Engineering Geology* 2: 5–17.

Haynes, V.M. (1968) 'The influence of glacial erosion and rock structure on corries in Scotland', *Geografiska Annaler* 50A: 221–234.

————(1972) 'The relationship between the drainage areas and sizes of outlet troughs of the Sukkertoppen Ice Cap, West Greenland', *Geografiska Annaler* 54A: 66–75.

————(1977) 'The modification of valley patterns by ice-sheet activity', *Geografiska Annaler* 59A: 195–207.

Hewitt, K. (1989) 'The altitudinal organisation of Karakoram geomorphic processes and depositional environments', *Zeitschrift für Geomorphologie* N.F. 76: 9–32.

Hirano, M. and Aniya, M. (1988) 'A rational explanation of cross-profile morphology for glacial valleys and of glacial valley development', *Earth Surface Processes and Landforms* 13: 707–716. Correction (1989) 14: 173–174.

Hobbs, W.H. (1911) *Characteristics of Existing Glaciers*, New York: Macmillan.

Holtedahl, H. (1967) 'Notes on the formation of fjords and fjord-valleys',*Geografiska Annaler* 49A: 188–203.

Hooke, R. Le B. (1991) 'Positive feedbacks associated with erosion of glacial cirques and overdeepenings', *Geological Society of America Bulletin* 103: 1104–1108.

Hoyer, M.C. (1971) 'Puget Peak avalanche, Alaska', *Geological Society of America Bulletin* 82: 1267–1284.

Iversen, N.R. (1991) 'Potential effects of subglacial water pressure fluctuations on quarrying', *Journal of Glaciology* 37: 27–36.

Jennings, J.N. and Costin, A.B. (1978) 'Stone movement through snow creep, 1963–75, Mount Twynam, Snowy Mountains, Australia', *Earth Surface Processes* 3: 3–22.

Johnson, W.D. (1904) 'The profile of maturity in alpine glacial erosion', *Journal of Geology* 12: 569–578 (reprinted in C. Embleton (ed.) 1972: 70–78).

Kjeldsen, O. (1985) 'Materialtransportundersøkelser ved Nigardsbreen', in E. Roland and N. Haakensen (eds) *Glasiologiske Undersøkelser i Norge 1982*, Norges Vassdragsdirektoratet, Hydrologisk Avdeling, Rapport nr. 1–85: 74–86.

Kohlbeck, F., Scheidegger, A.E. and Sturgl, J.R. (1979) 'Geo-mechanical model of an alpine valley', *Rock Mechanics* 12: 1–14.

Larson, E. and Mangerud, J. (1981) 'Erosion rate of a Younger Dryas cirque glacier at Kråkenes, western Norway', *Annals of Glaciology* 2: 153–158.

Lautridou, J.-P. (1988) 'Recent advances in cryogenic weathering', in M.J. Clark (ed.) *Advances in Periglacial Geomorphology*, Chichester: J. Wiley.

Lawler, D. (1991) 'Sediment and solute yield from the Jökulsá Sólheimasandi glacierized river basin, southern Iceland', in J. Maizels and C. Caseldine (eds) *Environmental Change in Iceland*, Dordrecht: Kluwer.

Lewis, W.V. (1940) 'The function of meltwater in cirque formation', *Geographical Review* 30: 64–83.

——(1948) 'Valley steps and glacial erosion', *Transactions of the Institute of British Geographers* 13: 19–44.

——(1954) 'Pressure release and glacial erosion', *Journal of Glaciology* 2: 417–422.

——(1960) 'The problem of cirque erosion', in W.V. Lewis (ed.) 'Norwegian cirque glaciers', *Royal Geographical Society Research Series* 4: 97–100.

Linton, D.L. (1951) 'Watershed breaching by ice in Scotland', *Transactions of the Institute of British Geographers* 15: 1–16.

——(1963) 'The forms of glacial erosion', *Transactions of the Institute of British Geographers* 33: 1–28 (reprinted in C. Embleton (ed.) 1972: 149–172).

Luckman, B.H. (1977) 'The geomorphic activity of snow avalanches', *Geografiska Annaler* 59A: 31–48.

——(1992) 'Debris flow and snow avalanche landforms in the Lairig Ghru, Cairngorm Mountains, Scotland', *Geografiska Annaler* 74A: 109–121.

McCall, J.G. (1960) 'The flow characteristics of a cirque glacier and their effect on glacial structure and cirque formation', in W.V. Lewis (ed.) 'Norwegian cirque glaciers', *Royal Geographical Society Research Series* 4: 39–62 (reprinted in C. Embleton (ed.) 1972: 205–228).

McGreevy, J.P. (1981) 'Some perspectives on frost shattering', *Progress in Physical Geography* 5: 56–75.

McGreevy, J.P. and Whalley, W.B. (1982) 'The geomorphic significance of rock temperature variations in cold environments: a discussion', *Arctic and Alpine Research* 14: 157–162.

McSaveney, M.J. (1978) 'Sherman Glacier Rock Avalanche, Alaska, U.S.A.', in B. Voight (ed.) *Rockslides and Avalanches* vol. 1, Amsterdam: Elsevier.

Machatschek, F. (1969) *Geomorphology*, 9th edn, translated by D.J. Davis, Edinburgh: Oliver & Boyd.

Matthes, F.E. (1930) 'Geologic history of the Yosemite valley', *United States Geological Survey Professional Paper* 160: 1–137 (reprinted, abridged, in C. Embleton (ed.) 1972: 92–118).

——(1938) 'Avalanche sculpture in the Sierra Nevada of California', *Bulletin of the International Association of Scientific Hydrology* 23: 631–637.

Meier, M.F. (1960) 'Mode of flow of Saskatchewan Glacier, Alberta, Canada', *United States Geological Survey Professional Paper* 351: 1–70.

Metcalf, R.C. (1979) 'Energy dissipation during subglacial abrasion', *Journal of Glaciology* 23: 233–246.

Mokievsky-Zubok, O. (1978) 'A slide of glacier ice and rocks in western Canada', *Journal of Glaciology* 20: 215–217.

Monjuvent, G. (1974) 'Considérations sur le relief glaciaire à propos des Alpes du Dauphiné', *Géographie Physique et Géologie Dynamique* 2nd series, 16: 465–502.

Nicoletti, P.G. and Sorriso-Valvo, M. (1991) 'Geomorphic controls of the shape and mobility of rock avalanches', *Geological Society of America Bulletin* 103: 1365–1373.

Nyberg, R. (1991) 'Geomorphic processes at snowpatch sites in the Abisko mountains, northern Sweden', *Zeitschrift für Geomorphologie* N.F. 35: 321–343.

Nye, J.F. and Martin, P.C.S. (1968) 'Glacial erosion', *International Union of Scientific Hydrology Publication* 79: 78–86.

Olyphant, G.A. (1981) 'Allometry and cirque evolution', *Geological Society of America Bulletin* 92: 679–685.

Penck, A. (1905) 'Glacial features in the surface of the Alps', *Journal of Geology* 13: 1–19.

Perla, R.I. (1980) 'Avalanche release, motion and impact', in S.C. Colbeck (ed.) *Dynamics of Snow and Ice Masses*, New York: Academic Press.

Plafker, G. and Ericksen, G.E. (1978) 'Nevados Huascarán avalanches, Peru', in B. Voight (ed.) *Rockslides and Avalanches* vol. 1, Amsterdam: Elsevier.

Porter, S.C. (1964) 'Composite Pleistocene snowline of Olympic Mts. and Cascade Range, Washington', *Geological Society of America Bulletin* 75: 477–482.

——(1989) 'Some geological implications of average Quaternary glacial conditions', *Quaternary Research* 32: 245–261.

Porter, S.C. and Orombelli, G. (1980) 'Catastrophic rockfall of 12 September 1719 on the Italian flank of the Mont Blanc massif', *Zeitschrift für Geomorphologie* N.F. 24: 200–218.

Post, A. (1967) 'Effects of the March 1964 Alaska earthquake on glaciers', *United States Geological Survey Professional Paper* 544-D: 1–42.

Radruch-Hall, D.H. (1978) 'Gravitational creep of rock masses on slopes', in B. Voight (ed.) *Rockslides and Avalanches* vol. 1, Amsterdam: Elsevier.

Rapp, A. (1960) 'Recent development of mountain slopes in Kärkevagge and surroundings, northern Scandinavia', *Geografiska Annaler* 42: 65–200.

——(1986) 'Comparative studies of actual and fossil nivation in north and south Sweden', *Zeitschrift für Geomorphologie*, N.F. *Supplement Band* 60: 251–263.

Roberts, M.C. and Rood, K.M. (1984) 'The role of the ice-contributing area in the morphology of transverse fjords, British Columbia', *Geografiska Annaler* 66A: 381–393.

Röthlisberger, H. and Iken, A. (1981) 'Plucking as an effect of water-pressure variations at the glacier bed', *Annals of Glaciology* 2: 57–62.

Rudberg, S. (1992) 'Multiple glaciation in Scandinavia – seen in gross morphology

173

or not?' *Geografiska Annaler* 74A: 231–243.

Ryder, J.M. (1981) 'Geomorphology of the southern part of the Coast Mountains of British Columbia', *Zeitschrift für Geomorphologie* N.F., *SupplementBand* 37: 120–147.

Saunders, I. and Young, A. (1983) 'Rates of surface processes on slopes, slope retreat and denudation', *Earth Surface Processes and Landforms* 8: 473–501.

Sharp, M. (1988) 'Surging glaciers: geomorphic effects', *Progress in Physical Geography* 12: 533–559.

Small, R.J. (1987) 'Moraine sediment budgets', in A.M. Gurnell and M.J. Clark (eds) *Glaciofluvial Sediment Transfer: An Alpine Perspective*, Chichester: J. Wiley.

Sugden, D.E. (1978) 'Glacial erosion by the Laurentide ice sheet', *Journal of Glaciology* 20: 367–391.

Sugden, D.E., Glasser, N. and Clapperton, C.M. (1992) 'Evolution of large roches moutonnées', *Geografiska Annaler* 74A: 253–264.

Sugden, D.E. and John, B. (1976) *Glaciers and Landscape*, London: Edward Arnold.

Svensson, H. (1959) 'Is the cross-section of a glacial valley a parabola?', *Journal of Glaciology* 3: 362–363.

Thorn, C.E. (1976) 'Quantitative evaluation of nivation in the Colorado Front Range', *Geological Society of America Bulletin* 87: 1169–1178.

——(1988) 'Nivation: A geomorphic chimera', in M.J. Clark (ed.) *Advances in Periglacial Geomorphology*, Chichester: John Wiley.

Tómasson, H. (1987) 'Glacial and volcanic shore interaction, Part 1: On land', in G. Sigbjarnarson (ed.) *Iceland Coastal and River Symposium Proceedings*, Reykjavík: National Energy Authority.

Unwin, D.J. (1973) 'The distribution and orientation of corries in northern Snowdonia, Wales', *Transactions of the Institute of British Geographers* 58: 85–97.

Varnes, D.J., Radbruch-Hall, D.H. and Savage, W.Z. (1989) 'Topographic and structural conditions in areas of gravitational spreading of ridges in the western United States', *United States Geological Survey Professional Paper* 1496: 1–28.

Waitt, R.B. (1975) 'Late Pleistocene alpine glaciers and the Cordilleran Ice Sheet at Washington Pass, North Cascade Range, Washington', *Arctic and Alpine Research* 7: 25–32.

Walder, J.S. and Hallet, B. (1985) 'A theoretical model of the fracture of rock during freezing', *Geological Society of America Bulletin* 96: 336–346.

——(1986) 'The physical basis of frost weathering: towards a more fundamental and unified perspective', *Arctic and Alpine Research* 18: 28–32.

Washburn, A.L. (1979) *Geocryology*, London: Edward Arnold.

Whalley, W.B. (1974) 'The mechanics of high-magnitude, low-frequency rock failure and its importance in a mountainous area', *University of Reading Department of Geography Geographical Papers* 27: 1–48.

——(1984) 'Rockfalls', in D. Brunsden and D.B. Prior (eds) *Slope Instability*, Chichester: J. Wiley.

Wheeler, D.A. (1979) 'The overall shape of longitudinal profiles of streams', in A.F. Pitty (ed.) *Geographical Approaches to Fluvial Processes*, Norwich: Geo Abstracts.

White, W.A. (1970) 'Erosion of cirques', *Journal of Geology* 78: 123–126.

Whitehouse, I.E. and Griffiths, G.A. (1983) 'Frequency and hazard of large rock avalanches in the central Southern Alps, New Zealand', *Geology* 11: 331–334.

Zoback, M. (1992) 'First- and second-order patterns of stress in the lithosphere: the world stress map project', *Journal of Geophysical Research* B97: 11703–11728.

8

LAND-USE CHANGES AND TROPICAL STREAM HYDROLOGY

Some observations from the upper Mahaweli Basin of Sri Lanka

C.M. Madduma Bandara

INTRODUCTION

It was nearly two and half decades ago that the writer as a student from the tropics had the privilege of coming within the intellectual aura of Chorley, who proved to be a teacher with enduring impacts on his students. His teachings on drainage basins as fundamental geomorphic units at that time created not only an intense academic stimulus for the writer, but also fresh insights on their potential use for political and regional demarcations. His rationalism and deep respect for objectivity permeated not only one's academic pursuits but also many matters in one's personal life. It was a time when Chorley was riding high on the waves of the 'quantitative revolution' in geography, and when there were deliberate moves towards the development of a modern geographical theory with the popularisation of systems, models and paradigms. However, one thought which seemed to linger occasionally in Chorley's mind was how empirical realities, particularly in the tropics, might fit into those beautiful models and other theoretical constructions which were so close to his heart. The writer reminisces, with some nostalgia, how Chorley referred often to the 'vigour and dynamism' of tropical landscape processes and restrained himself from going beyond that, in some of those infrequent encounters with him as an academic supervisor.

Since those days in Cambridge in the late 1960s, in the distant tropical island of Sri Lanka, as it has been often said, much water had been flowing under the stupendous bridges of the Mahaweli Ganga constructed by the British during a bygone colonial era, into the great natural harbour of Trincomallee through its crocodile-infested delta and finally into the Indian Ocean along the less well-known submarine canyons off the eastern coast. The long-term trend of flow of this magnificent river, which is now heavily

175

dammed to produce hydro-power and water for irrigation, had, however, been changing significantly over the decades in response to changes in land use in its upper catchment areas which form the focus of this chapter.

THE MAHAWELI GANGA

The Mahaweli Ganga is the largest river basin of Sri Lanka, covering nearly one-fifth of the total area of the island (Figure 8.1). It has attracted some

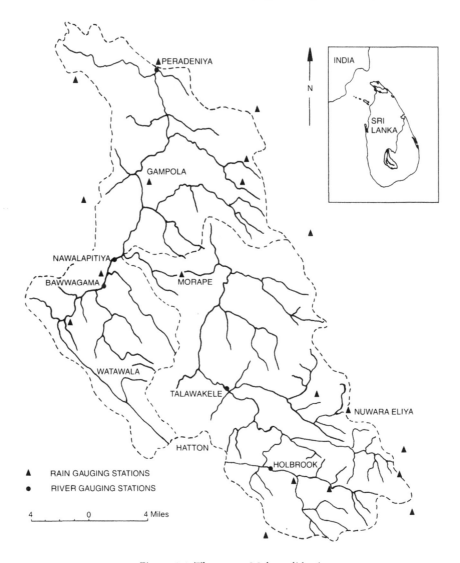

Figure 8.1 The upper Mahaweli basin

176

international attention in recent decades due to the massive, foreign-aided development projects based on its land and water resources. The upper Mahaweli basin above Peradeniya, which covers about 10% of the total basin area, contributes nearly 25% of the total volume of water carried by the river. Before the advent of plantation agriculture during the colonial era, a large proportion of the basin was under tropical montane rainforests. Since the middle of the last century, these forests were cleared to pave the way, first for coffee and then for tea. This has resulted in serious disturbances to the natural forest ecosystems, while leading to increased water and sediment discharges. In one of the early estimates the sediment yield from the basin above Peradeniya was computed to be in the region of 130,000 to 820,000 tons per annum. Although more recent estimates tend to place this around 486,000 tons per annum, available records are inadequate to identify any significant temporal trends.

SOURCES OF DATA

Streamflow data for the Mahaweli basin above Peradeniya were recorded at six gauging stations since the early 1940s (Table 8.1). The longest and perhaps the most reliable record was for Peradeniya, where observations were commenced in 1944. Until major engineering and related development activities were undertaken in the late 1970s, Peradeniya recorded virtually the virgin flows of the Mahaweli Ganga without serious disturbances except in catchment land use. On the other hand, rainfall records were available for much longer periods extending over 100 years. Both rainfall and streamflow data were analysed through conventional techniques of temporal analysis such as moving averages and simple linear regression models. Land-use

Table 8.1 Hydrological data for stream gauging stations in the upper Mahaweli basin

Stations and streams	Basin area (km²)	Mean annual basin rainfall (mm)	Mean annual discharge (mm)	Runoff–rainfall ratio (%)
Peradeniya (Mahaweli)	1108	3004	1953	65.51
Bawwagama (Mahaweli)	169	4274	2649	61.90
Watawala (Hatton Ganga)	65	3777	2282	60.42
Horape (Kotmala Oya)	554	2736	1935	70.71
Talawakele (Kotmala Oya)	290	2372	1361	57.31
Holbrook (Agra Oya)	120	2349	1304	55.48

Source: Hydrology Division, Department of Irrigation, Colombo

information was derived mainly from sequential aerial photographs and from available land-use maps. Thus land use as depicted on 1956 aerial photographs was compared with that of 1979/81.

RESULTS OF TEMPORAL ANALYSIS

The analysis of rainfall data was based on both point rainfall values for individual stations as well as areal values derived through Thiessen Polygon averaging methods. Thus the annual rainfall at Nuwara Eliya, a station located in the upper elevations of the basin, was analysed to detect any time trends for a 112-year period of record (Figure 8.2). The preliminary results indicate a significant decline in annual rainfall amounting to about 500 mm or 20% of the total during the last hundred years. This trend is visible even in the shorter time span of 40 years from 1940 to 1980 for which streamflow records were available. As could be seen from Figure 8.2, this decline was characterised by some cyclic fluctuations, although their periodicity is not regular enough to make any definitive predictions.

The results of the analysis of annual discharge data for Peradeniya gauging station are given in Figure 8.3, which shows the time trends of discharge and the runoff–rainfall ratios. This indicates that annual discharge, as well as runoff–rainfall ratios, had witnessed a significant increase during the four decades of available record. This was despite the general decline of annual rainfall over the basin as a whole, as referred to earlier in relation to Nuwara Eliya. In order to examine this trend further, the monthly values for the same period were plotted separately for wet and dry seasons (Figure 8.4). Here it becomes clear that, while wet season flows were increasing rapidly over the years, the dry season flows were declining slightly. Therefore, it may be surmised that the wet season flows were primarily responsible for the observed increases in annual discharge.

The change of land use in the Mahaweli basin above Peradeniya is indicated by Table 8.2, which summarises information available for two time periods, namely 1956 and 1979. In the 1950s tea occupied over 60% of the basin area while forests covered nearly 17% of it. It is generally believed that a tea plantation when well-managed acts very much like natural forest cover through protection of the ground by its thick canopy foliage. By 1979, the extent of land under tea dropped below 40%, while forests recorded a 2% decline. Much of the land lost from tea and forests was taken up by homesteads and crops other than tea. Only about 2% of such lands was brought under forest plantations. The built-up areas under both rural and urban settlements have witnessed a significant expansion.

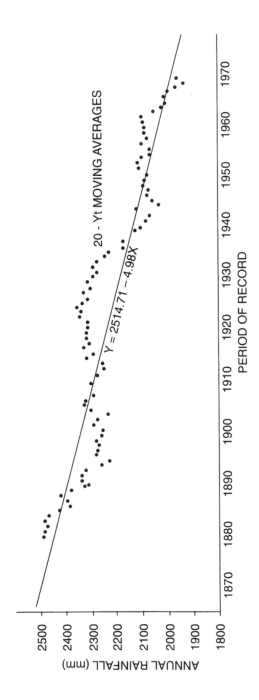

Figure 8.2 Annual rainfall trend at Nuwara Eliya

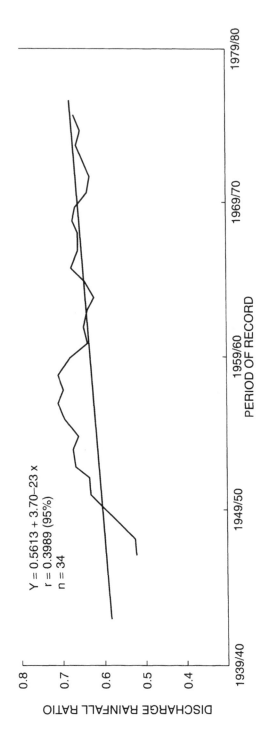

Figure 8.3 Changing runoff–rainfall ratios of the upper Mahaweli above Peradeniya

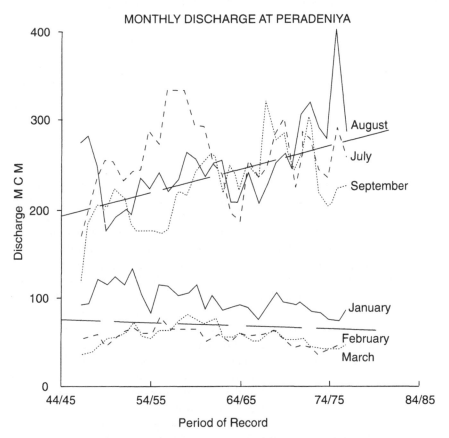

Figure 8.4 Seasonal trends in streamflow at Peradeniya

Table 8.2 Land-use change in the Mahaweli basin above Peradeniya

Land use	% Extent under each type		
	1956	1979/81	% Change
Tea	61.10	38.81	−22.29
Homesteads	6.56	16.67	+10.11
Grasslands	5.51	7.79	+2.28
Forests	16.94	14.55	−2.39
Forest plantations	2.54	4.15	+1.65
Other croplands	6.68	15.10	+8.32
All other uses	0.66	3.03	+2.37

Sources: Land Use Maps by Hunting Air Surveys (1956); District Land Use Maps by Survey Department Colombo

181

RAINFALL, STREAMFLOW AND CHANGING LAND USE

The significant declining trend of rainfall at high elevations in the central highlands of Sri Lanka since the latter half of the last century, as seen in Nuwara Eliya, has eluded any proper explanation so far. In the late 1960s when such a trend was suspected, some believed that it was related to the expansion of tea plantations, which were opened up in areas of high montane forests during that period (Hamamori 1968). However, a decrease as large as 20% cannot easily be explained by any land-use change alone. It is more likely to have something to do with monsoonal air circulation patterns and global climatic changes than with land-use conversion.

The observed increases in annual streamflow in the upper Mahaweli basin, despite a negative long-term trend of rainfall at higher elevations, may seem somewhat paradoxical at first sight. It means, however, that if the rainfall had been increasing or remained static in the catchment areas, the magnitude of the discharge increase should have been even greater. The annual flow of the Mahaweli Ganga in general had witnessed a 10% increase compared with the period of record prior to 1964, on which Mahaweli Master Plan estimates of water resources were based (Table 8.3). In other words, the large Mahaweli reservoir systems which have been created since the mid-1970s have actually been 'benefiting' from this increase. The low reservoir replenishments that were experienced in recent years and which caused serious problems for hydro-power generation and regulation of irrigation water supplies could have been much worse if not for these increased discharges, although they do not reflect a condition of 'good health' in the catchment. Similar increasing trends of discharge were observed in many other streams in the hill country of Sri Lanka such as those in the Knuckles Range and in the Kirindi Oya basin (Madduma Bandara 1985; Madduma Bandara and Kuruppuarachchi 1989).

The supply of water around the year, particularly during the dry weather, is specially important for watersheds in areas of hydro-power reservoirs. In

Table 8.3 Mean annual discharges (in millions of cubic metres) of the Mahaweli Ganga prior to the development of large reservoir systems

Gauging station (downstream)	Basin area (km²)	1944–1964 (average)	1964–1976 (average)	% Change
Morape	554	974	957	−09.82
Peradeniya	1108	2074	2137	+10.30
Randenigala	2343	3643	3822	+10.49
Weragantota	4046	5192	5594	+10.77
Manampitiya	7371	7851	8289	+10.56

Sources: UNDP/FAO (1968), Hamamori (1968)

such areas, there is a need to develop watershed strategies that would enhance the 'production' of water as much as timber, both valued by the cubic metre, with minimum conflict of interest. The changing patterns of discharge observed in the upper Mahaweli basin, as indicated above, at first sight appear to be beneficial to large reservoir schemes where water can be stored throughout the year. However, the rapid increases in wet season discharges also indicate high rates of overland flow in the catchments and consequent loss of soil and its eventual deposition in river and reservoir beds. From an engineering perspective, it has been argued that, in view of the large capacity of most reservoirs, the observed rates of sedimentation may not impair their productive life within the next 50–100 years (TAMS 1980). This cannot by any means be considered a sustainable approach to water resources management. It has also been contended that reservoir silt can be used more productively for brick and tile industries and river gravel for gemming. But then the loss of soil fertility in agricultural lands in the catchment and the consequent need for heavy use of fertilisers, insecticides and pesticides can lead to other environmental problems. Similarly, the declining water yields observed during the dry season tend to offset any gains in water yield during the wet season.

SOME THEORETICAL ISSUES

There appears to be hardly any consensus among hydrological scientists with regard to the actual impacts of land-use conversions in medium- and large-sized tropical watersheds. The conventional view was that, when tree cover is maintained in a watershed, its impacts on water resources were necessarily positive. While it was believed that the establishment of forest cover would generally improve the quantity as well as the quality of water yields, it was also thought that tree cover would have a 'ponding effect' leading to a steady enhancement of dry weather flows. These ideas were given authenticity by the Gathering Grounds Committee of Britain, which concluded that 'water undertakers should be encouraged to adopt a policy of afforestation in all gathering grounds where soil and climatic conditions are suitable'. The Committee thus disregarded the views of early scientists in the field such as Frank Law (1956), whose empirical studies were almost discredited by the scientific community at that time. There appears to be some reversal of thinking on this subject in recent years and with regard to the uplands of Britain where Frank Law worked, Leeks and Roberts (1987) concluded that 'afforestation of upland catchments is not necessarily beneficial to water resources'.

With regard to the rest of the world, Bosch and Hewlett (1982), who reviewed the results of nearly a hundred paired basin experiments, where land-use–hydrological relationships were investigated, came to the conclusion that 'no experiments in deliberately reducing vegetation cover caused any reductions in water yields, and nor have any deliberate increases in cover caused increases in water yield'. Similarly, Hamilton *et al.* (1985), summarising the state of knowledge on the protective role of tropical forests, arrived at almost parallel conclusions:

> reducing forest cover on a forested catchment increases the water yield from that catchment with the majority of the increase occurring in the base flow component; ... changing forest cover of head water catchments will at most have only a minor impact on downstream flooding particularly on large river basins.

This debate resurfaced again when the recent floods in Bangladesh and those in the Philippines took a heavy toll where this had been quickly attributed to deforestation of uplands (Hamilton 1992). With regard to Bangladesh floods, it was contended that, even if the entire upper region of the River Ganges including Nepal is clothed with forest, Bangladesh will still continue to suffer from floods.

In some recent research studies such as that of Trimble *et al.* (1987) some attempt has been made to quantify the negative effects of afforestation on water resources, with reported reductions as high as 25% of the water yields. Further, the work of Oyebande (1988) in Nigeria, Rahim in Malaysia and Bruijnzeel in Indonesia has demonstrated that tropical stream hydrology can be more sensitive to forest cover changes than that in the temperate regions. However, as Anderson and Spencer (1991) observed, these impacts are likely to be much more short-lived due to rapid regeneration of tropical vegetation.

In general, there appears to be more consensus on the impacts of land-use changes on increasing annual yields than on the dry season flows which are of critical importance for power generation and water supply. Here, the established view among most ecologists appears to be that the establishment of forest cover will result in the improvement of dry weather flows. On the other hand, most hydrologists seem to believe that it can be the reverse due to resultant increases in evapotranspiration. Here, the crucial issue is the nature of trade-off between changes in evapotranspiration and the rates of infiltration following land-use conversion. The results reported from different experiments across the world seem to vary widely. In some cases (Lal 1983), increases in overland flow accompanied by proportionate increases in base flow have been reported from deforested areas. In others (Bruijnzeel 1990), dry season flows recorded progressive decreases. This had been the case of the upper Mahaweli basin, where there had been significant reductions in the tea cover. Such a trend can set in with decreased infiltration resulting from soil degradation and increased rates of overland flow both due to changes in agricultural land-use and also due to expansion of human settlements.

In this context the model presented by Cassels, Hamilton and Saplaco

(1982) is both interesting and controversial. Here, the most controversial arrow paths appear to be those which finally lead to increases in dry weather flows. This is an area where many imaginations have wandered, while the actual research input had been obviously insufficient. In order to understand the trade-offs between evapotranspiration and base flow we need more information from tropical forest catchments, particularly for different tree species at varying rates of growth. There appear to be hardly any precise techniques of measuring evapotranspiration losses from mature trees. Casual observations on the amounts of water oozing out of stumps of certain species of trees cut for timber, such as Jak Wood (*Artocarpus heterophyllus*) and Hawari Nuga (*Alstonia macrophylla*) indicate substantial losses. Similarly, the practice of growing certain eucalyptus species in damp valleys in tea estates as a means of making such places less swampy shows that early planters understood this process. It is obvious that a deeper understanding of the relationship between water and trees, be they forest trees or planted species, is of paramount importance for the development of forest policies in the tropics. In Sri Lanka reservation of riparian lands and stream banks was provided under the Crown Lands Ordinance for many decades, although its enforcement had been far from satisfactory. Even if such hydrological reservations are protected, there is some uncertainty as to what particular types of tree species could be utilised to conserve soil as well as water, in addition to other conservation needs. This is an area where a major research effort directed at policy formulation is needed.

REFERENCES

Abdul Rahim Nik (1988) 'Water yield changes after forest conversion to agricultural land use in peninsular Malaysia', *Journal of Tropical Forest Science* 1: 67–84.

Anderson, J.M. and Spencer, T. (1991) 'Carbon, nutrient and water balances of tropical ecosystems subject to disturbance: management implications and research proposals', *MAB Digest* 7, Paris: UNESCO.

Bosch, J.M. and Hewlett, J.D. (1982) 'A review of catchment experiments to determine the effect of vegetation changes on water yield and evapotranspiration', *Journal of Hydrology* 55.

Bruijnzeel, L.A. (1990) *Hydrology of Moist Tropical Forests and Effects on Conversion. A State of Knowledge Report*, Amsterdam: UNESCO/IHP, Faculty of Earth Sciences, Free University.

FAO (1992) *The Forest Resources of the Tropical Zone by Main Ecological Regions*. UN Conference on Environment and Development, Rio de Janeiro, June 1992. Forest Resources Assessment, 1990 Project, Rome: FAO.

Gilmour, D.A., Cassells, D.S. and Bonell, M. (1982) 'Hydrological research in the tropical rainforests of north Queensland: some implications for landuse management', First National Symposium on Forest Hydrology, Melbourne, Australia.

Hamamori, A. (1968) *Climate and Hydrology: Mahaweli Ganga Irrigation and Hydro-power Survey*, vol. 2, Colombo: FAO and Irrigation Department.

Hamilton, L.S. (1992) 'Philippine storm disaster and logging: the wrong villain?' *Asia-Pacific Uplands – A Newsletter for Scientists*, published by R.D. Hill, Department of Geography and Geology, University of Hong Kong.

Hamilton, L.S. *et al.* (1985) *The Protective Role of Tropical Forests: A State of Knowledge Review*, Honolulu: Environment Policy Institute, East–West Center.

Lal, R. (1983) 'Soil erosion in the humid tropics with particular reference to agricultural land development and soil management', *Publications of the International Association of Scientific Hydrology* 140: 221–240.

Law, Frank (1956) 'The effect of afforestation on the yield of water catchment areas', *Journal of the British Water Works Association* 38: 489–494.

Leeks, G.J.L. and Roberts, G. (1986) 'The effects of forestry on upland streams – with special reference to water quality and sediment transport', in *Environmental Aspects of Plantation Forestry in Wales*, Institute of Terrestrial Ecology, Natural Environmental Research Council.

Madduma Bandara, C.M. (1985) 'Water resources of the Kirindi Oya basin: a study in river basin water balance', in *Some Aspects of the Water Resources in Sri Lanka*, Colombo: Irrigation Department.

———(1985) 'Some aspects of the hydrology of the upper Mahaweli basin', in *Some Aspects of the Water Resources of Sri Lanka*, Colombo: Irrigation Department.

Madduma Bandara, C.M. and Kuruppuarachchi, T.A. (1989) 'Landuse changes and hydrological trends in the upper Mahaweli basin', in 'Hydrology of the natural and man-made forests', Kandy: Sri Lanka–German Upper Mahaweli Watershed Management Project (unpublished).

Newson, M.D. (1979) 'The results of ten years' experimental study of Plynlimon, mid-Wales and their importance for the water industry', *Jour. Inst. Wat. Eng. Soc.* 33: 321–333.

Oyebande, L. (1988) 'Effects of tropical forest on water yield', in E.R.C. Reynolds and F.B. Thompson (eds) *Forests, Climate and Hydrology: Regional Impacts*, Tokyo: United Nations University.

TAMS (1980) *Environmental Assessment: Accelerated Mahaweli Development Programme*, vol. II: *Terrestrial Environment*, New York.

Trimble, S.W. (1990) 'Geomorphic effects of vegetation cover and management: some time and space considerations in prediction of erosion and sediment yield', in J.B. Thornes (ed.) *Vegetation and Erosion*, Chichester: John Wiley.

UNDP/FAO (1969) *Mahaweli Ganga Irrigation and Hydro-power Survey. Final Report*, Rome: FAO/SF:55/Cey-7.

9

PALAEOCLIMATOLOGY, CLIMATE SYSTEM PROCESSES AND THE GEOMORPHIC RECORD

Roger G. Barry

INTRODUCTION

Geomorphologists have a long tradition of interest in climate processes and awareness of their significance for landscape developments. Formal recognition of the role of exogenic processes, particularly climatic conditions, in shaping landscapes can be traced at least to the beginning of this century in the writings of William Morris Davis (1899), Albrecht Penck (1910) and Emmanuel de Martonne (1913). Davis distinguished the 'normal' temperate fluvial cycle of rivers and subsequently the glacial and arid 'climatic accidents', a tripartite division followed by Penck and many subsequent geomorphologists. Penck noted that the moisture budget (precipitation $\lessgtr$ evaporation; snowfall $\leqq$ ablation) accounted for the type of hydrological control, but he provided no link to surface morphology. In an attempt to develop a systematic basis for climatic geomorphology, Büdel (1948) identified eight morphoclimatic zones categorised, for example, by frost debris, temperate mature soils with and without permafrost, desert debris, sheetwash and tropical mature soils. Peltier (1950) elaborated the conceptual foundation for climatic geomorphology. Nine morphogenetic regions, based only on annual temperature and precipitation, were suggested and inferred relative intensities of weathering, frost action, mass movement, wind action and fluvial erosion were graphed in relation to these two climate variables. The limitations of this two-variable approach were readily pointed out by geomorphologists (see Stoddart 1969b) and a revised scheme was subsequently put forward (Peltier 1975).

Although these authors sought to relate weathering and erosion processes, and ultimately landforms, to *present* climatic conditions, the implications of major climatic shifts during the Quaternary were already evident to de Martonne (1913). In a critique of a conference on the morphology of climatic zones that took place in Düsseldorf, Passarge (1926) stated that 'present-day

187

surface landforms are for the most part ... the product of Pleistocene processes'. He concluded that the standard approach to morphological–climatic zones was untenable. The 1950s–1960s witnessed the full emergence of the paradigm of palaeoclimatic variations and their geomorphic significance as represented in relict landforms and polygenetic landscapes (Wilhelmy 1958; Birot 1968; Tricart and Cailleux 1972; also Derbyshire 1973; see Beckinsale and Chorley 1991 for a detailed review). Many erosion surfaces in mid-latitudes were observed to be undergoing progressive destruction and the timescale required for planation surfaces to develop was now extended to the Tertiary. These ideas are most fully developed in a formal pedagogic sense in the writings of Julius Büdel, most notably in his textbook *Climatic Geomorphology* (Büdel 1977, 1982). His geomorphological ideas evolved over some fifty years, from the 1930s to 1981 (see Kiewietdejonge 1984) and were based on field investigations that ranged from Spitsbergen to North Africa and India, with extensive work in Europe. The polygenetic origin of most landscapes came to be recognised, not only in formerly glaciated mid-latitude areas, but also in subtropical and tropical latitudes where humid and arid phases have alternated during the Pleistocene (Street-Perrott and Harrison 1984). The implication of these profound changes in environmental processes is that many landforms, and perhaps most landscapes, are out of equilibrium with current processes (Stoddart 1969a; King 1980). Moreover, similar landforms, such as inselbergs and tors, can result from different processes (the concept of 'convergence' or 'equifinality'), particularly as a result of structural influences.

As the magnitude and global character of past environmental changes became fully appreciated, geomorphologists turned attention from form to process. Large-scale geomorphic processes are determined by a mix of endogenic factors (orogeny, tectonics), climatic factors which operate directly as well as through the hydrologic regime and the vegetation cover, and by processes internal to the geomorphic system. Many attempts have been made to relate weathering to climate including proposed schemes of climatically determined weathering and soil leaching regimes (Huggett 1991, Chapter 4). However, these generalisations ignore the varying age of different land surfaces.

The probable response of fluvial systems, soil systems and depositional/denudational regimes in general to changes in climate forcing have been elaborated by many geomorphologists. Chorley (1957), Melton (1957) and Ahnert (1987), for example, investigate the role of spatial differences in climate, whereas Starkel (1987) and Kirkby (1989) consider the effects of temporal changes of regime. The concept of intrinsic thresholds in fluvial and other geomorphic systems, whereby a change in the system occurs without any shift in external controls (such as base level, climate or vegetation cover), introduced by Schumm (1973), rapidly became broadened in its usage and application. Thus, subsequently, Schumm (1979) defined 'a threshold of landform stability that is exceeded either by intrinsic change of the landform itself, or by a progressive change of an external variable'. Progressive weathering of the wastle mantle that eventually triggers slope failure and mass movement would be an example of an intrinsic threshold. McDowell *et al.* (1990) suggest, however, that threshold response, and oscillatory behaviour in response to a single perturbation ('complex response') are 'concentrated at time scales less than 1,000 years'. Nevertheless, the idea of thresholds and feedback processes serve as a basis for most geomorphological studies of Pleistocene and Holocene features (Chorley *et al.* 1984).

These ideas are parallel to modern concepts of the global climate system as exemplified by the work of Lorenz (1968) on non-deterministic theories of climate change. Lorenz proposes that the climate system may demonstrate *almost intransitive behaviour*, meaning that for constant (time independent) forcing the system may exhibit several quasi-stable states with rapid transitions between them. An intransitive system is one for which more than one solution (a climatic probability distribution) may exist for a specific set of external forcing functions. The possibility exists that interannual variability of climate may be a result of intransivity and chaotic behaviour in the atmosphere (Lorenz 1990). It is worth emphasising that any complex non-linear system with multiple interacting feedback processes, such as are present in the climate system and in geomorphic systems, can demonstrate significant shifts in state that are independent of any external forcing. These ideas have now evolved into the widely applied concepts of *strange attractors* and *limit cycles* (Tsonis and Elsner 1989), although the extent to which they provide a useful practical approach to climatic or geomorphic studies remains uncertain.

The objectives of this discussion are to outline the nature of the climatic system and its spatio-temporal variability, and to illustrate the results of recent modelling studies of palaeoclimatic regimes and climate processes that may be of relevance in geomorphology. My caveat is that this is a climatologist's perspective and it may risk belabouring the obvious. My hope is that this climatological viewpoint may serve to stimulate some new ideas and applications, or help confirm previous speculations and hypotheses about climatic factors and climate–surface interactions.

THE CLIMATE SYSTEM

Components of the system

The components of the complete climate system are: the atmosphere, oceans, cryosphere, the land surface, and the biomass on land and in the oceans. Energy is received by the system through solar radiation and continually redistributed by motion in the atmosphere and oceans. The response of each component varies over a range of timescales. The lower atmosphere has a

189

response time to imposed thermal changes of about a month whereas for the deep ocean it is of the order of centuries. Snow cover and sea ice vary greatly over the annual cycle, whereas glacier and ice-sheet volumes change over decades to millennia. Biomass changes also have seasonal and long-term responses that affect climate through physical processes (energy absorption, evaporation, surface roughness), as well as through exchanges of trace gases and aerosol production. Variations in the land surface due to tectonic movements are extremely slow (10^6–10^8 yr) as regards elevation and location, except for short-lived volcanic releases into the atmosphere, but are highly significant for global climate on geologic timescales.

The major climate components interact through a crucial set of internal feedback mechanisms (Hansen *et al.* 1984). The principal ones are: ice–albedo feedback linking snow and ice extent to absorbed radiation and air temperature (a positive, amplifying feedback); water vapour–radiation feedback, also positive, whereby increased air temperature leads to increased atmospheric water vapour, which traps more infra-red radiation, raising temperatures further; and cloud–radiation feedback. The last process is complex since increased cloud cover may reduce surface temperatures through a decrease in solar radiation receipts, or increase them through more infra-red re-radiation to the surface, depending upon the cloud type, height and surface albedo.

Timescales

The recognition of major changes in global climate by glacial geologists in the mid-nineteenth century provided the foundation for a comprehensive theory of climate change. Richard Foster Flint (1974) identified the elaboration of such a time-based theory as a challenge and opportunity comparable with the nineteenth-century biological achievement in understanding organic evolution and the twentieth-century geophysical theory of lithosphere dynamics.

Long-term mean climatic states, such as have been identified for the early Cenozoic era (Axelrod 1984), can be regarded as providing the setting for the 'slow physics' of geomorphic processes in terms of weathering regimes and the formation of soil horizons. Ollier (1988) notes, for example, that Australia, where some regolith is of Mesozoic–Tertiary age, drifted northward between 55 and 10 Ma with a climate shifting from warm and wet to hot and arid. Mid-latitude glacial cycles and subtropical pluvial conditions, in contrast, represent episodic changes that are firmly linked to astronomical forcings of the climate system with climate responses that are amplified by the feedback processes noted above.

The difficulty in relating process geomorphology to the interpretation of landforms is caused by the immense complexity of causal interactions, the existence of local azonal influences, and the limited knowledge of the *rates* of various processes. Despite the seminal work of Wolman and Miller (1960) on the relative importance of frequent minor events and large-scale events that are infrequent, but not rare, the role of exceptional catastrophic processes is not fully resolved. Ives (1987) summarises rates of denudation and cliff-face retreat estimated in mountain environments of Scandinavia by A. Rapp and in Colorado and Tasmania by N. Caine. However, he emphasises that we know little about the representativeness of such data in space and time. One major difficulty is posed by short hydrologic records. In the Colorado Front Range, for example, Jarrett (1990) considers that the 1976 flash flood in the Big Thompson Canyon, which was triggered by a severe thunderstorm over the headwaters, was the largest flood in 10,000 years based on studies of deposits in the canyon, whereas frequency analysis of floods recorded between 1877 and 1990 suggests a recurrence interval of only 100–300 years. This illustrates the importance of palaeohydrologic research.

Several estimates have been made of overall continental rates of chemical and mechanical denudation (see Huggett 1991, Table 4.3). These tend to differ widely depending on the relative weights accorded to the effects of climate and relief. Saunders and Young (1983) suggest that denudation rates in temperate continental areas range from 10 to 100 mm/10^3a on normal relief, increasing to 100 to 200+ mm/10^3a on steep slopes. In semi-arid climates, and on steep slopes in tropical rainforest, rates are 100 to 1000 mm/10^3a. The highest rates quoted are 1000–5000 mm/10^3a on steep, glaciated terrain and 10^3 to 10^6 mm/10^3a on badlands. However, Phillips (1990) indicates that slope gradient accounts for 70% of the expected variation in erosion rates. Rainfall erosivity and rainfall retention by the surface are the main secondary factors.

Climatic parameters of special relevance to geomorphology

From an earth-process viewpoint, various climatic characteristics can be identified as having particular significance for geomorphological studies. They are:

(a) the freezing threshold which determines the possibility for glaciers and/ or permafrost to develop and the occurrence of periglacial processes (French and Karte 1988) and periglacial hydrologic regimes (Clark 1988);

(b) the occurrence of mean temperatures ≥20°C in all months of the year which influences chemical weathering rates;

(c) the equatorward and altitudinal limits of snowfall which determine nivation processes. The rain/snow threshold is closely correlated with monthly mean temperature (Barry 1992: 240);

(d) the annual regime of the water-budget of the regolith, determined by the potential evapotranspiration (PE) and precipitation (R). Rock

191

decomposition has been related to a simple 'weathering climate index' based on the PE/R ratio by Weinert (1965);

(e) the boundary of arid and semi-arid conditions which delimits aeolian processes and normal fluvial processes;

(f) the typical intensities of precipitation which affect infiltration and runoff characteristics, and therefore erosion potential;

(g) the annual and diurnal amplitudes of the surface–atmosphere exchanges of energy and moisture. Maritime regimes in mid-latitudes and tropical humid regimes have high rates of seasonal moisture exchange and small annual and diurnal temperature ranges, whereas continental regimes in mid-latitudes and dry tropical regimes have high rates of seasonal and/or diurnal energy exchange and correspondingly large temperature ranges; and

(h) the occurrence of persistently strong winds in arid and semi-arid regions that cause deflation through removal of dust.

Generalised world maps of temperature and precipitation data considered to be of importance to the geomorphologist have been published by Common (1966), but there appears to have been no expanded treatment of this topic.

Climate models

Figure 9.1 illustrates the typical range of physical processes represented in global climate models. They include a surface hydrologic cycle and bio-spheric interactions in addition to horizontal and vertical atmospheric transfer of momentum, heat and moisture, as well as radiative and condensation processes (Meehl 1984). Most palaeoclimate reconstructions employ low-resolution atmospheric general circulation models. Typical grid resolutions are 4–5° latitude × 5–8° longitude with 9–12 vertical levels which implies that:

(a) land/sea outlines are simplified and many regional-scale geographical features are only crudely represented;

(b) orography is greatly smoothed;

(c) synoptic-scale atmospheric processes are poorly resolved;

(d) mesoscale convective processes have to be parameterised (i.e. represented in a statistical manner).

In spite of these inherent limitations, global climate models are able to reproduce many of the zonal and continental-scale features of the earth's climate. Comparisons of the ability of three different models to simulate broad patterns of present climate are provided by Gates et al. (1990). Apart from zonal averages of temperature and its annual range, large-scale aridity is also reasonably captured since it is determined by large-scale pressure systems and subsidence. A recent intercomparison of fourteen climate models by Boer et al.

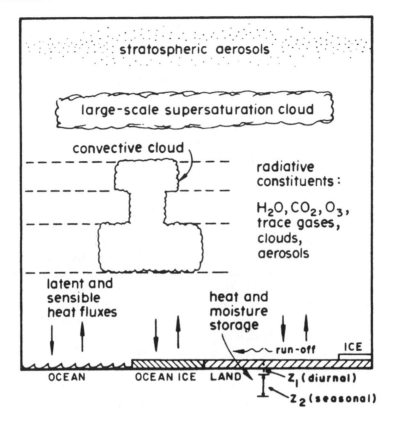

Figure 9.1 Sub-gridscale processes represented in a single grid box of the GISS climate model

Source: Hansen *et al.* 1983; reproduced by courtesy of the American Meteorological Society

(1992) shows that zonally averaged precipitation is poorly simulated, except by high-resolution models in June–July–August (in both hemispheres). Regional variations in moisture balance, especially in mid-latitudes, are not yet well handled by most climate models. The moisture balance depends on precipitation minus evaporation and since the latter is strongly affected by soil moisture and transpiration, which require detailed biosphere–atmosphere process submodels to represent their variability adequately (Henderson-Sellers and Pitman 1992), it is not surprising that calculated moisture budgets differ considerably between models. Surface energy exchanges are also strongly dependent on cloud cover and surface albedo. The radiative components are reasonably well predicted, but the turbulent heat fluxes are strongly modified by vegetation and atmospheric boundary layer processes that are not treated in detail in most global climate models.

Most climate simulations treat equilibrium conditions for the present day (a control case) and for changed boundary conditions that approximate some elements of a past climatic state. Commonly, for ease of interpretation, a series of model sensitivity studies are performed where the effects of changing a single, or limited numbers of, factor(s), can be more readily identified. Such sensitivity studies have included changes in external forcings (solar constant, orbital parameters, volcanic dust and greenhouse gases in the atmosphere) or in the surface boundary conditions (orography, continental location, ice sheets, sea surface temperature and vegetation cover). Models may also be run with or without daily and annual cycles depending on the nature of the problem and the degree of averaging that is appropriate. For an equilibrium experiment for perpetual January (or July) conditions, with prescribed boundary conditions, the model may be run for 400 days to obtain a climatological sample (Sloan and Barron 1992). A climate model with a simple ocean mixed-layer formulation may be run with a seasonal cycle for 15 to 20 model years to achieve equilibrium. Using twice-daily samples, seasonal mean outputs can then be calculated for the last 3 to 5 model years. Coupled ocean–atmosphere models are now being employed on the latest supercomputers. They enable the effects of ocean heat storage and transport in the entire ocean depth to be treated, rather than necessitating prescribed surface temperatures, or a simple ocean mixed layer (Meehl 1990). Experiments with such models require runs of 50–100 model-years, in view of the slow response time of the deep ocean, with monthly statistics being retained for analysis.

There is now a large number of atmospheric general circulation models (GCMs). Table 9.1 lists those that are commonly used in climate sensitivity

Table 9.1 GCMs commonly used in climatic studies

Model and institution	References to model description
Geophysical Fluid Dynamics Laboratory (GFDL), NOAA[1]	Gordon and Stern (1982); Manabe and Hahn (1981)
Goddard Institute for Space Studies (GISS), NASA[2]	Hansen et al. (1983)
Max Planck Institute (MPI) for Meteorology/University of Hamburg[3]	Simmons et al. (1989); Lautenschlager and Herterich (1990)
National Center for Atmospheric Research (NCAR)	Williamson et al. (1987); Washington and Meehl (1989)
Oregon State University (OSU)	Schlesinger and Gates (1980)
UK Meteorological Office (UKMO)	Corby et al. (1977); Slingo and Pearson (1987)

Notes:
[1] National Oceanographic and Atmospheric Administration (USA)
[2] National Aeronautics and Space Administration (USA)
[3] This model is based on that of the European Centre for Medium Range Weather Forecasting (ECMWF)

analyses and palaeoclimatic studies. Increasingly, attention is being given to their intercomparability in terms of their simulation of the modern climate (Gates *et al.* 1990). Differences between models are being investigated particularly in terms of their treatment of such key factors as albedo–temperature feedback (Cess *et al.* 1990) and cloud processes. Up to now, coupled models have been mostly used to analyse the temporal aspects of increasing CO_2 concentrations on climate.

PRE-QUATERNARY PALAEOCLIMATIC STUDIES

Improved knowledge of the Earth's climatic history on long timescales has been made possible by advances in three principal areas: palaeogeographic reconstructions for the Palaeozoic, Mesozoic and Cenozoic eras, based on plate tectonic theory and palaeomagnetic studies; syntheses of palaeoceanographic conditions from palaeobiogeographic interpretations of planktonic and benthic species, oxygen isotope temperature determinations, and distributions of marine sediments rich in carbonates and organic carbon (Lloyd 1984; Hay 1988); and the application of global numerical climate models (Kutzbach 1985). Highlights of these numerous and diverse investigations are all that can be presented.

Relationships between plate tectonic processes and climate are noted by Wold and Hay (1990). The three primary ones involve:

(1) rapid sea-floor spreading in the Cretaceous leading to high concentrations of atmospheric CO_2, an enhanced global greenhouse and an Earth with ice-free poles (Barron and Washington 1985);
(2) changes in continental distribution and relief causing modification to atmospheric circulation patterns and climatic regimes (thermal and hydrologic) (Barron and Washington 1984; Barron *et al.* 1989); and
(3) changes in the area of shallow seas resulting in regional changes in surface albedo and cloudiness and, in high latitudes, affecting the likelihood of sea ice formation and/or the growth of marine-based ice sheets and ice shelves.

On geological timescales, global intervals of rapid uplift and mountain building are inferred by Wold and Hay (1990) from estimated rates of sediment flux. These rates are shown to have been high over the last 10 Ma, and about 90–100 (late Cretaceous), 240 (middle Triassic), 380–400 (Devonian) and 540–550 Ma (Cambrian); rates were particularly low between 160 and 200 Ma (Jurassic). However, estimates for individual geosynclinal basins are necessary for regional interpretations to be developed. Aeolian dust, which makes up 50% of siliceous and red clays in the North Pacific zone of westerly winds, reveals major fluctuations in circulation intensity and continental aridity during the Tertiary (Rea *et al.* 1985).

Key palaeoceanographic findings relate to the evolution of the ocean's thermal structure (Hay 1988). During Cretaceous time, ocean surfaces were as warm as, or warmer than, present in the tropics and probably high in the polar regions. In marked contrast to the late Cenozoic and modern ocean, bottom waters were also warm. Bottom waters cooled sharply in the early Oligocene (38 Ma), apparently linked to episodes of Antarctic glaciation. Since mid-Miocene time, there has existed a 20°–25°C difference between tropical surface water temperatures and those of the ocean deep water (Hay 1988). Under this regime, deep water has a polar or sub-polar origin and its formation is controlled by thermohaline processes that cause density to increase (cooling due to ocean–atmosphere heat flux supplemented by salinity increases through evaporation and sea ice formation). In contrast, during the Cretaceous period and early Tertiary, the ocean circulation appears to have been halothermal with bottom waters formed by salinity increases mainly in shallow seas in subtropical arid zones off the western continental margins.

Understanding of the long-term changes in global climate has been revolutionised over the last decade through the application of general circulation models (GCMs) to palaeoclimatic questions. Improved palaeo-geographic reconstructions of continental configurations and ocean areas have enabled simulations of climate to be made for the Cretaceous period (Barron and Washington 1984, 1985, for example) and earlier times (Kutz-bach and Gallimore 1989). While not resolved completely, the processes that may account for pronounced global warmth during the Cretaceous and a reduced equator–pole temperature gradient, can be roughly quantified.

Geologic evidence for the Cretaceous suggests warmer climatic conditions in high latitudes, although absolute values are uncertain and it is unclear whether temperatures were high year-round (Crowley and North 1991: 155–160). Simulations with the NCAR Community Climate Model (CCM), using annually averaged solar radiation and a 'swamp' ocean (without heat capacity), by Barron and Washington (1984, 1985) indicate that the changed land–sea distribution for mid-Cretaceous time (~100 Ma), with an equatorial Tethyan Seaway, sea level 100–200 m higher than now, and land connections between North America and Europe, South America and Africa, and Australia and Antarctica, resulted in a global warming of 4.8°C. About two-thirds of this is attributable to the changed land–sea configuration and the remainder to the absence of polar ice cover. The analysis, however, suggests that the simulated temperatures in high latitudes are below those indicated by the geological data. Schneider *et al.* (1985) and Barron and Washington (1985) propose that an increase of atmospheric CO_2 concentrations to at least × 4 present values, and possibly as much as × 8 more, would be required to match the geological evidence for Cretaceous warmth (Figure 9.2). This is in line with geochemical estimates of much higher atmospheric CO_2 levels as a result of increased volcanism. Higher CO_2 levels would also have enhanced the intensity of the hydrologic cycle with increased subtropical evaporation,

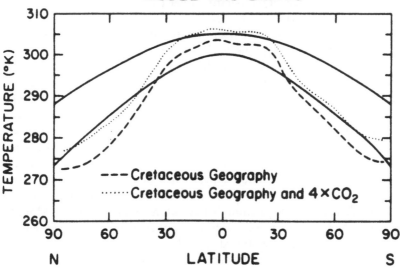

Figure 9.2 Zonally averaged surface temperatures for the Cretaceous (100 Ma) from geological evidence (solid lines show the inferred range) compared with values given by simulations with the NCAR CCM for Cretaceous geography (dashed line) and for geography plus × 4 CO_2 concentration (dotted line)

Source: Barron and Washington 1985; reproduced by courtesy of the American Geophysical Union

higher precipitation rates in equatorial and middle latitudes, and increased runoff (Barron and Washington 1985). The geomorphic implications of high CO_2 levels for weathering rates needs to be explored. An alternative explanation for an ice-free state near the poles involves increased ocean heat transport. Rind and Chandler (1991) find that a 50 to 70% increase in poleward ocean heat transport would maintain ice-free polar oceans and reproduce the warmth of Mesozoic times (230–65 Ma).

Recent studies have focused on basic issues of global climate such as the effects of continental location and elevation, and of topography, on regional- and continental-scale climates. Several findings have important implications for geomorphology and palaeohydrology. Simulation experiments for the Triassic period (*c.* 200 Ma) megacontinent of Pangaea, using the NCAR CCM (Kutzbach and Gallimore 1989), show that net effective precipitation (P–E) was close to zero between 30°N and 30°S, despite strong surface heating, apparently as a result of the long overland path of the air. Effective precipitation, averaged over the Pangaean land mass, was about half that of the present day, implying great aridity. However, it is possible that a more realistic biosphere–atmosphere moisture coupling would increase this ratio.

197

Moisture recycling over Amazonia accounts for about half of present rainfall amounts (Salati 1985) and this process is not properly accounted for in the model. The annual cycle of temperature over Pangaea was also large and the model simulated high-latitude sea ice.

Experiments comparing the global climatic effects of continents located in north and south 'polar' regions (45° latitude to the poles) with a global tropical (17°N–17°S) continent have been performed by Hay et al. (1990a, 1990b). The land/sea proportions and mean elevations were maintained as at present except for Antarctica, which was lower to reflect reduced ice mass. The tropical configuration gave an earth that is 8°C warmer than present with reduced evaporation and precipitation in the tropics. The earth with polar land masses has a mean temperature close to present and a strong hydrological cycle in low latitudes. The continental interiors, however, are dry. A polar continental location appears insufficient, by itself, to promote glaciation but the specification of snow cover on a polar continent sets up a positive feedback for precipitation that could maintain an ice cap. Mountain ranges on either the east or west coasts of a pole-to-pole continent enhance the transport of moisture eastward into the interior, relative to a low flat plateau (Hay et al. 1990a) but in different latitude zones in each case. West coast mountains create an equatorial wet belt, whereas east coast mountains eliminate this and cause considerable aridity, except for narrow belts in subpolar latitudes. Importantly, all continental configurations show zonal patterns of precipitation, soil moisture and runoff, implying that the climatic pattern proposed by Köppen (1931), for an idealised continent featuring a south-west–north-west slope of the arid zone boundary from 20° latitude on the west coast to 40°–60° latitude in the east, may be determined by the present-day orography of Asia and the American Cordilleras. This conclusion is also supported by an experiment using the GFDL model without any topography (Broccoli and Manabe 1992). For this case, the Köppen arid and semi-arid climates are shown to be nearly continuous over land areas between 15° and 35° latitude with scarcely any extensions into mid-latitudes. The present-day role of orography on precipitation regimes in tropical and subtropical latitudes is summarised by Meehl (1992). He shows that the high plateau and mountain areas of both hemispheres strengthen summer monsoon regimes.

Model simulations by Ruddiman and Kutzbach (1989), Manabe and Broccoli (1990) and Broccoli and Manabe (1992) demonstrate that the present-day dry continental interiors of Asia and North America are caused by the large-scale orography of Tibet and the Cordilleras. In the absence of such mountain barriers, the continental interiors would experience moist climatic regimes (Figure 9.3). Rapid Plio-Pleistocene uplift in Tibet and the Himalayas is a candidate causal factor in the inception of glaciation in the Northern Hemisphere around 2.4 Ma according to Ruddiman and Raymo (1988). They propose a modification of the planetary wave structure in

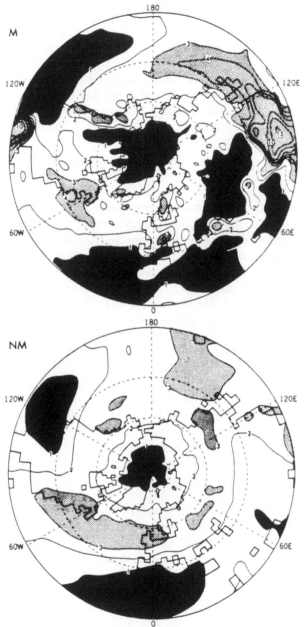

Figure 9.3 Annual mean precipitation (mm day^{-1}) for simulations with the GFDL global climate model for a case with global topography (M) and with no mountains (NM). Stippling indicates precipitation > 3 mm day^{-1}; solid black is precipitation < 1 mm day^{-1}. Contours are at 1, 2, 3, 4, 5, 6, 8, 10, 15, 20 and 30 mm day^{-1}

Source: Broccoli and Manabe 1992; reproduced by courtesy of the American Meteorological Society

199

winter and the transition seasons, due to uplift of the Tibetan Plateau, in such a way as to favour snow accumulation and/or decreased ablation. Rind and Chandler (1991) suggest that topographic uplift during the late Cenozoic may have altered surface winds sufficiently to permit the growth of sea ice as a result of decreased poleward ocean heat transport. However, Molnar and England (1990) critique these views and argue that the increasing aridity in central Asia during the late Tertiary was related to the global effects of Tertiary cooling.

QUATERNARY PALAEOCLIMATIC STUDIES

The climatic character of the last glacial maximum and subsequent time has been extensively investigated and current efforts are now directed especially at the last interglacial and Late Glacial events. Extensive and detailed geological, palynological, ocean sediment and ice core records have also become available for model validation purposes, as well as documenting the timescales of change. A major finding is that interglacial conditions, as warm as now, have occupied only about 10% of the last 750,000 years or so. The initial applications of GCMs to examine the effects of surface boundary conditions at 18,000 BP on global atmospheric circulation and climate (Williams et al. 1974; Gates 1976; Manabe and Hahn 1977) were rapidly followed by sensitivity experiments incorporating the Milankovitch orbital forcings (Kutzbach and Guetter 1986), by studies of Holocene conditions for 9000 BP, and subsequently for the entire Late Glacial and postglacial period at 3000-year intervals (COHMAP Members 1988). Using primarily the surface boundary conditions provided by the CLIMAP project members (1976, 1981), recent modelling studies of the last glacial maximum have addressed the consequences of changes in orbital forcing, land albedo and carbon dioxide concentrations (Hansen et al. 1984; Kutzbach and Guetter 1986; Rind 1987; Broccoli and Manabe 1987; Hyde et al. 1989; Lautenschlager and Herterich 1990). A preliminary attempt has also been made to model changes, between the last glacial maximum and modern conditions, in global patterns of desert dust removal (Joussaumé 1989) although the mobilisation of the aerosols is not treated directly.

There is considerable observational evidence that the major global wind systems were strengthened in both hemispheres under ice-age conditions. Aeolian dust and pollen in deep-sea cores, and indices of oceanic upwelling off Peru, suggest increases in strength of 20% in the North Pacific westerlies, 30% in the North Pacific trades, 30–50% in the South Pacific trades and 50% in the North Atlantic trades (Janecek and Rea 1985; Crowley and North 1991: 57). Chloride (from sea salt) aerosols from ice cores in Greenland and Antarctica have also been used to infer increased North Atlantic westerlies and Southern Ocean westerlies compared with present averages. It is unclear whether there are concomitant latitudinal displacements of the zonal wind systems. Rind (1986) finds no displacement of the location of the subtropical jet stream between 'warm' ($\times 2$ CO_2) and 'cold' (ice-age) climates simulated by the Goddard Institute for Space Studies (GISS) model, whereas earlier experiments with different models reported an equatorward shift (Williams *et al.* 1974; Gates 1976). Shinn and Barron (1989) show that the Northern Hemisphere jet stream shifted 5° *poleward* in winter for a 'maximum' ice-sheet case, based on Hughes (1985), in comparison with a minimum ice-sheet reconstruction. Several models (Kutzbach and Wright 1985; Manabe and Broccoli 1985) suggest that the Laurentide and Fenno-Scandinavian Ice Sheets caused jet stream splitting. Orographic forcing predominates over thermodynamic influences according to Rind (1987) and hence the question of absolute ice-sheet elevation and size is important. Shinn and Barron's results indicate a more distinct split jet stream flow over North America for the maximum ice-sheet case, and a stronger downwind Atlantic jet, whereas no clear split is apparent for either case over Europe. The northern branch of the westerly flow around the Laurentide Ice Sheet is expected to maintain cold outflows over the western North Atlantic and the presence of sea ice. Interstadial fluctuations in ice mass and elevation may have been sufficient to promote jet stream shifts that greatly modified ocean circulation patterns and sea ice extent in the western North Atlantic, causing abrupt alterations of climatic regime.

An extensive comparison of results from the NCAR Community Climate Model (CCM) ice-age simulation and observational evidence for 18,000 yr BP was performed by Kutzbach and Wright (1985). Figure 9.4a shows the jet streams and surface winds simulated by the NCAR CCM. Below the diagram are approximate regional departures of temperature, precipitation and (P–E) from present-day values. Figure 9.4b presents a summary of geological and palynological evidence for changes in the United States at approximately 18,000 yr BP (Barry 1983). There is a generally good degree of agreement. However, Davis (1989) notes contradictions between the Kutzbach and Wright model results for jet stream displacement and Holocene patterns of moisture in the Pacific Northwest and American Southwest. He argues for changes in the strength, but not the position, of circulation features.

A less obvious mechanism for extratropical climatic anomaly patterns differing from those of the present day is suggested by climate model studies of the El Niño–Southern Oscillation (ENSO). From coupled model experiments, Meehl and Branstator (1992) find that changes in external forcing (such as CO_2-doubling) permit ENSO oscillations to continue in the tropics, but may cause substantial shifts in extratropical teleconnections. For example, the circulation in the Northern Hemisphere becomes more zonal with displacements of regional anomalies. Hence, modern tropical–extratropical teleconnections may have been transformed during the times of changed external forcing.

A persistent enigma in model–observation comparisons for 18,000 yr BP is

(a)

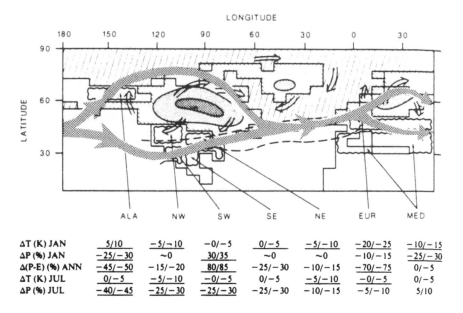

LONGITUDE

ΔT (K) JAN	ALA	NW	SW	SE	NE	EUR	MED
ΔT (K) JAN	5/10	−5/−10	−0/−5	0/−5	−5/−10	−20/−25	−10/−15
ΔP (%) JAN	−25/−30	~0	30/35	~0	~0	−10/−15	−25/−30
Δ(P-E) (%) ANN	−45/−50	−15/−20	80/85	−25/−30	−10/−15	−70/−75	0/−5
ΔT (K) JUL	0/−5	−5/−10	−0/−5	0/−5	−5/−10	−0/−5	0/−5
ΔP (%) JUL	−40/−45	−25/−30	−25/−30	−25/−30	−10/−15	−5/−10	5/10

Figure 9.4(a) Climate simulation for the NCAR CCM for northern mid-latitudes at 18,000 BP and comparison with geological and palynological data. The jet streams are shown, surface winds (double-shafted arrows), the locations of the main ice sheets (hatched lines). A band of increased precipitation along the southern jet stream is enclosed by dashed lines.
  Area-averaged departures (18,000 BP minus present) are tabulated for temperature (K), precipitation (%) and precipitation minus evaporation (%) for the seven areas indicated. Underlined values are significant above the 90% level compared to the model's natural variability
  Source: Kutzbach and Wright 1985; reproduced by courtesy of Pergamon Press

(b) Estimated changes in summer temperature (°C) (annual temperature change in Wyoming in parentheses) and in annual precipitation for 18,000 BP from palaeoenvironmental evidence over the contiguous United States
  Source: Barry 1983; reproduced by permission of the University of Minnesota Press

the discrepancy between observed snowline lowering of the order of 1000 m on the mountains in east Africa, the Andes, Hawaii and New Guinea and changes of only about 2°C in tropical lowland or sea surface temperatures. The snowline depressions imply 5°–6°C cooling above 2000 m elevation (Webster and Streten 1978; Rind and Peteet 1985). Attempts to explain this larger cooling primarily invoke the fact that lower sea surface temperatures reduce the moisture content in the boundary layer and a drier atmosphere has a steeper lapse rate (Kraus 1973). A detailed calculation of tropical boundary layer processes by Betts and Ridgway (1992) takes account of the

(b)

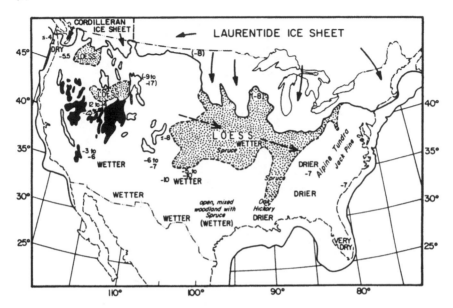

effects of changes in sea surface temperature, pressure (increased 14 mb due to 120 m lowering of sea level), a CO_2 concentration lowered to 195 ppm, increased windspeed and energy export on equivalent potential temperature (θ_e) in the boundary layer. They suggest that the necessary 14°C reduction in θ_e to account for a 950 m lowering of snowline would require a 2°C sea surface cooling, a 14 mb increase in sea-level pressure and a 30% increase in windspeed. However, recent reconstruction of tropical sea-surface temperature from coral isotope data at Barbados show a 5°C cooling ca. 18,000 years ago (Guilderson *et al.* 1994) which is supported by noble gas measurements in ground water from lowland Brazil (Stute *et al.* 1995).

Major revisions of palaeoclimate paradigms have occurred over the last two to three decades. In the 1960s, it was widely thought that tropical pluvial conditions corresponded with mid-latitude glacial intervals but by the 1980s it was recognised that the tropics were largely arid during glacial maxima (Warren 1985). Nevertheless, the degree of precipitation reduction is uncertain. It is possible that a reduction in soil moisture, caused by a modest precipitation decrease or a shift in its timing, could enhance the initial precipitation change by diminishing the continental recycling of moisture (Broccoli and Manabe 1992).

Stages of high and low lake levels in the tropics and subtropics show a complex spatio-temporal pattern. Figure 9.5 indicates that the zones 5°S–15°N and 35°–45°N had distinctly different behaviour. A late Pleistocene to mid-Holocene high lake level phase is absent in the

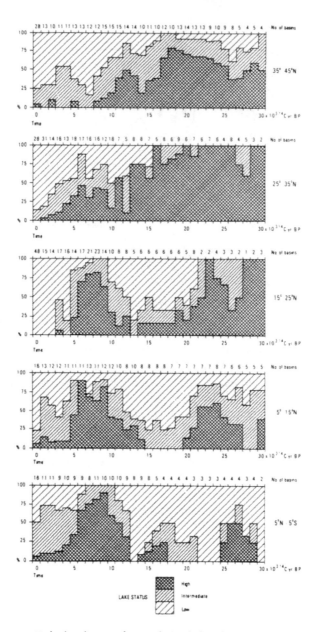

Figure 9.5 Lake level status for 10° latitude bands between 5°S and 45°N. Low = 0–15% of the altitudinal range of fluctuations (including dry lakes), intermediate = 15–70% of the range, and high = 70–100% of the range, including overflowing lakes

Source: Street-Perrott and Harrison 1984; reproduced by courtesy of the American Geophysical Union

extratropics, where levels were high from 30,000 until 16,000 yr BP (Street-Perrott and Harrison 1984). However, calculations of energy and moisture budgets for such lakes suggest that precipitation need not have exceeded present values due to reduced evaporation (caused by seasonal lake ice and lower air temperature) resulting in a higher effective precipitation. The forcing of these changes in moisture regime appears to be driven by the Milankovich orbital effects which augmented incoming solar radiation in summer by 7–8% (above present) in the early Holocene (Kutzbach and Street-Perrott 1985).

The moist phase of the Holocene in North Africa appears to have been greatly surpassed during the last (Eemian) interglacial. Petit-Maire (1989) reports large lakes with a rich fauna and palaeolithic artefacts in western Libya (27°–28°N) and northern Mali (22°N) and cites similar lacustrine episodes in western Egypt, Algeria and Tunisia. She proposes a long-lasting (c. 50 ka) lacustrine phase in contrast to the 4,000–5,000 yr duration of the early Holocene humid regimes. The mechanisms that may account for such 'pluvial' regimes have also been controversial. One argument favours north–south displacement of the subtropical high-pressure belts (Nicholson and Flohn 1980) whereas another interpretation invoked expansion of the belts during cold periods and contraction during warm periods, with associated intensification/weakening of the wind regimes (Sarnthein 1978). Both palaeoclimatic evidence and modelling studies have been used to investigate this problem. Sarnthein et al. (1981) analysed the modern and past deposition of Saharan dust in the North Atlantic. Summertime instability and Easterly Wave disturbances loft dust to about 3000 m over the southern Sahara and Sahel from where this harmattan dust is transported westward by a mid-tropospheric Easterly Jet Stream. Low-level North-Easterly trades over the north-west African coast are a relatively minor contributor at present. Grain-size data, used to estimate palaeo-wind velocities, indicate that the harmattan dust outbursts were in a similar position to now (16°–20°N) during glacial times, although speeds were reduced. However, Sarnthein et al. (1981) conclude that the trades were stronger. At 6000 yr BP the harmattan transport remained active and its position unchanged whereas the trade wind component was much reduced.

A further climatic argument has concerned the relationship between the mean state of global climate and the degree of variability. Several meteorologists have suggested that cold intervals were more variable, but instrumental records and various proxy time series fail to support this argument (Schuurmans and Coops 1984) although the central Greenland ice core records show a clear contrast between the climate stability of the Holocene and the general instability of the preceding 240,000 years (Dansgaard et al. 1993). Climatic warming might also be expected to lead to more extreme temperatures, but examination of changes in frequency distribution suggests that more frequent 'moderate' extremes may be more typical.

DISCUSSION

The preceding literature survey suggests a number of issues that require attention and resolution both in the areas of climate research and geomorphological applications of palaeoclimatic information.

Climatic topics meriting attention include the following:

1 There are suggestions from analyses of moisture balance data in Amazonia (Salati 1985), and from discussions of modelling results (Broccoli and Manabe 1992), that there is significant recycling of moisture over land areas. This question was addressed over forty years ago by Benton *et al.* (1950), who concluded that only perhaps 15–20% of the moisture advected by large-scale atmospheric transport over the central United States is actually precipitated. Nevertheless, there is also evidence that large-scale irrigation in areas of semi-arid conditions can augment summer precipitation (Pielke and Avissar 1990). The related question of the effects of deforestation/afforestation of large watersheds has also received extensive studies by hydrologists. The exhaustive review of catchment experiments by Bosch and Hewlett (1982) concludes that there are increases in water yield from deforested catchments, especially in drier climates where the vegetation recovers more slowly. The general applicability of these results remains to be determined. Evidently, moisture recycling appears to operate under some conditions, but its effectiveness as a net contributor to regional precipitation on a long-term basis – say, with extensive and permanent replacement of tropical rainforest by pasture and agricultural plots – needs to be evaluated by modelling and careful review of all available data.

2 The role of large-scale orography in generating the quasi-stationary waves in the troposphere is relatively well understood (Wallace 1983). However, the contribution of orography and stationary waves to blocking regimes is less clear and this question is important for understanding regional differences in moisture balance, especially in terms of the uplift of mountain ranges and plateaus on geologic timescales.

3 The modelling studies of Rind (1986, 1988) indicate an apparent stability of the jet stream location for widely different climatic states. Rind suggests that increased eddy flux of sensible heat under cold climatic conditions offsets a reduction in the latent heat flux. However, other modelling studies produce different results and these dissimilarities may be a result of differing model grid resolutions and the parameterisation of biospheric processes. Inter-model comparisons are now being actively pursued by the climate modelling community but such discrepancies are not always readily isolated. Using observational data on the seasonal and interannual variation in latitude of the subtropical anticyclone, Flohn and Korff (1969) showed a clear linear

relationship with the meridional temperature gradient in the mid-troposphere; the high shifts equatorward as the gradient intensifies.

4 The characteristics of the global climate associated with a presumed unipolar glaciation in the Southern Hemisphere during most, or all, of the Oligocene, Miocene and early Pliocene are virtually unknown. Flohn (1981) drew attention to the possibility of a hemispheric climate asymmetry during the mid- to late-Cenozoic era, although recent studies suggest fluctuations during this time in Antarctic ice extent (late Miocene – early Pliocene) and in Northern Hemisphere ocean currents (Crowley and North 1991: 199–201).

Palaeoclimatic topics that need resolving are as follows:

1 The loosely defined notion of an 'equable' climate has had long-standing geological usage. Sloan and Barron (1990) attempted to clarify the concept, which is of particular relevance to the question of the 'normal' long-term state of the Earth's climate through the Phanerozoic era (~550 Ma) and to the supposed ice-free state of the Cretaceous. However, Sloan and Barron's postulates are misleading (Axelrod 1992). Since strictly 'equable' climates have a narrow geographical range (Bailey 1964), it is recommended that the term be abandoned and some more suitable ones be developed for the particular intended usages.

2 The paradox of substantial glacial maximum cooling on tropical mountains compared with modest changes in adjacent lowlands and in the oceans has been considerably clarified, although Betts and Ridgway (1992) consider that 'it remains hard to reconcile a sea surface temperature decrease of only 1–2 K and 950 m snowline lowering'. The discussions assume that the apparent differences are closely synchronous, but the dating of the land and ocean records are subject to error bars (one standard deviation) of approximately ±1000 years. The glacier terminus–equilibrium line altitude–freezing-level relationship in the tropics also merits closer scrutiny (Hastenrath 1990), particularly under different external climate forcings, as does the reconstruction of temperature at the actual sea surface from foraminiferal records via transfer functions. Nevertheless, there is evidence from the Little Ice Age and recent variations in the European Alps that changes in snow cover may vary non-linearly with altitudinal changes in temperature (Barry 1992: 371).

There is considerable opportunity for global climate models to provide palaeoclimatic reconstructions of geomorphologically pertinent variables. However, attention needs to be given to the following:

1 Determination of the duration of specific climate regimes that are

considered to be relevant to the imprinting of geomorphic and soil characteristics that survive to the present. Obviously, difficulties arise because the rates of different weathering and denudation processes depend heavily on such factors as rock type and the factor of timescales may make some climatically determined differences irrelevant, apart from the question of thresholds of geomorphic response and of the role of relatively rare (but not catastrophic) large-magnitude events;

2   The modelling of climatic effects on landform processes, along the lines of Kirkby's (1989) model for changes in slope and soils, leading to improved definition of climatically determined landforms and their age;

3   The extension of local and basin-scale process studies to scales commensurate with the outputs of global climate models;

4   Determination of the spatial extent of landscapes of different ages. This has been little addressed, but it would seem that attempts such as those of Brown (1979) and Ollier (1979) could be repeated and improved, given the increasing variety and sophistication of methods of dating materials; and

5   The development of quantitative relationships between global palaeoclimates, landscape evolution and sediment cycles.

It is apparent that while progress is being made in understanding climatic processes and climate history relevant to the geomorphic record, much remains to be learnt about the rates of the processes involved and the quantitative roles of climatic and other variables. The benefits of increased interdisciplinary co-operation are also evident in aiding the interpretation of proxy data using model experiments. Such collaboration needs to be strengthened and expanded in scope.

ACKNOWLEDGEMENT

I wish to thank Dr N. Caine, Institute of Arctic and Alpine Research, University of Colorado, and Dr G. Meehl, National Center for Atmospheric Research, Boulder, for their helpful comments on a first draft.

REFERENCES

Ahnert, F. (1987) 'An approach to the identification of morphoclimates', in V. Gardiner (ed.) *International Geomorphology 1986, Part II*, Proceedings of the First International Conference on Geomorphology, Chichester: John Wiley.

Axelrod, D.I. (1984) 'An interpretation of Cretaceous and Tertiary biota in polar regions', *Palaeogeography, Palaeoclimatology, Palaeoecology* 45: 105–147.

——(1992) 'What is an equable climate?', *Palaeogeography, Palaeoclimatology, Palaeoecology* 91: 1–12.

Bailey, H.P. (1964) 'Toward a unified concept of the temperate climate', *Geographical Review* 54: 526–545.

Barron, E.J., Hay, W.M. and Thompson, S.L. (1989) 'The hydrologic cycle: a major variable during earth history', *Palaeogeography, Palaeoclimatology, Palaeoecology (Global Planet. Change)* 75: 157–174.

Barron, E.J. and Washington, W.M. (1984) 'The role of geographic variables in explaining paleoclimates: results from Cretaceous climate model sensitivity studies', *Journal of Geophysical Research* 89: 1267–1279.

——(1985) 'Warm Cretaceous climates: high atmospheric CO_2 as a plausible mechanism', in E.T. Sundquist, and W.S. Broecker (eds) *The Carbon Cycle and Atmospheric CO_2: Natural Variations Archean to Present, Geophysical Monographs* 32, Washington, D.C.: American Geophysical Union.

Barry, R.G. (1983) 'Late-Pleistocene climatology', in S.C. Porter (ed.) *Late Quaternary Environments of the United States*, vol. 1: *The Late Pleistocene*, Minneapolis: University of Minnesota Press.

——(1992) *Mountain Weather and Climate*, 2nd revised edn, London: Routledge.

Beckinsale, R.P. and Chorley, R.J. (1991) *The History of the Study of Landforms or the Development of Geomorphology*, vol. III, London: Routledge.

Benton, G.S., Blackburn, R.T. and Snead, V.O. (1950) 'The role of the atmosphere in the hydrologic cycle', *Transactions of the American Geophysical Union* 31: 61–73.

Betts, A.K. and Ridgway, W. (1992) 'Tropical boundary layer equilibrium in the Last Ice Age', *Journal of Geophysical Research* 97(D2): 2529–2534.

Birot, P. (1968) *The Cycle of Erosion in Different Climates*, transl. C.I. Jackson and K.M. Clayton, London: B.T. Batsford.

Boer, G.J. and 13 others (1992) 'Some results from an intercomparison of the climates simulated by fourteen atmospheric general circulation models', *Journal of Geophysical Research* 97(D12): 12,771–12,786.

Bosch, J.M. and Hewlett, J.D. (1982) 'A review of catchment experiments to determine the effects of vegetation changes on water yield and evapotranspiration', *Journal of Hydrology* 55: 2–23.

Broccoli, A.J. and Manabe, S. (1987) 'The influence of continental ice, atmospheric CO_2 and land albedo on the climate of the last glacial maximum', *Climate Dynamics* 1: 87–100.

——(1992) 'The effects of orography on midlatitude northern hemisphere dry climates', *Journal of Climate* 5: 1181–1201.

Brown, E.H. (1979) 'The shape of Britain', *Transactions of the Institute of British Geographers* NS 4: 449–460.

Büdel, J. (1948) 'Das System der klimatischen Geomorphologie', *Verhandlungen deutscher Geographie* 27: 65–100 (English translation in E. Derbyshire (ed.) *Climatic Geomorphology*, London: Macmillan).

——(1977) *Klima–Geomorphologie*, Berlin: Borntraeger (English translation in L. Fischer and D. Busche (eds) *Climatic Geomorphology*, Princeton: Princeton University Press).

Cess, R.D. *et al.* (1990) 'Intercomparison and interpretation of climate feedback processes in 19 atmospheric general circulation models', *Journal of Geophysical Research* 95(D10): 601–615.

Chorley, R.J. (1957) 'Climate and morphometry', *Journal of Geology* 65: 628–638.

Chorley, R.J., Schumm, S.A. and Sugden, D.A. (1984) *Geomorphology*, London: Methuen.

Clark, M.J. (1988) 'Periglacial hydrology', in M.J. Clark (ed.) *Advances in Periglacial Geomorphology*, Chichester: John Wiley.

CLIMAP Project Members (1976) 'The surface of the Ice-Age Earth', *Science* 191: 1131–1136.

——(1981) 'Seasonal reconstruction of the Earth's surface at the Last Glacial Maximum', Geological Society of America, Map Chart Series, MC-36.

COHMAP Members (1988) 'Climatic changes of the last 18,000 years: observations and model simulations', *Science* 141: 1043–1052.

Common, R. (1966) 'Slope failure and morphogenetic regions', in G.H. Dury (ed.) *Essays in Geomorphology*, London: Heinemann.

Corby, G.A., Gilchrist, A. and Rowntree, P.R. (1977) 'United Kingdom Meteorological Office five-level general Circulation Model', *Methods in Computational Physics* 17: 67–110.

Crowley, T.J. and North, G.R. (1991) *Paleoclimatology*, New York: Oxford University Press.

Davis, O.K. (1989) 'The regionalization of climatic changes in western North America', in M. Leinen and M. Sarnthein (eds) *Paleoclimatology and Paleometeorology: Modern and Past Patterns of Global Atmospheric Transport*, Dordrecht: Kluwer Academic Publishers.

Dansgaard, W. *et al.* (1993) 'Evidence for general instability of past climate from a 250,000-year ice-core record', *Nature* 364: 218–220.

Davis, W.M. (1899) 'The geographical cycle', *Geographical Journal* 14: 481–504.

Derbyshire, E. (ed.) (1973) *Climatic Geomorphology*, London: Macmillan.

Flint, R.F. (1974) 'Three theories in time', *Quaternary Research* 4: 1–8.

Flohn, H. (1981) 'A hemispheric circulation asymmetry during late Tertiary', *Geologische Rundschau* 70: 725–736.

Flohn, H. and Korff, H.Cl. (1969) 'Zusammenhang zwischen Temperaturgefälle Äquator-Pol und den planetarischen Luftdruckgürteln', *Annalen der Meteorologie* NS 4: 163–164.

French, H.M. and Karte, J. (1988) 'A periglacial overview', in M.J. Clark (ed.) *Advances in Periglacial Geomorphology*, Chichester: John Wiley.

Gates, W.L. (1976) 'Modeling the Ice Age climate', *Science* 191: 1138–1144.

Gates, W.L., Rowntree, P.R. and Zeng, Q.-C. (and contributors) (1990) 'Validation of climate models', in J.T. Houghton, G.J. Jenkins and J.J. Ephraums (eds) *Climate Change: The IPCC Scientific Assessment*, Cambridge: Cambridge University Press.

Gordon, C.T. and Stern, W.P. (1982) 'A description of the GFDL global spectral model', *Monthly Weather Review* 110: 625–644.

Guilderson, T.P., Fairbanks, R.G. and Rubenstone, J.L. (1994) 'Tropical temperature variations since 20,000 years ago: modulating interhemispheric climate change', *Science* 263: 663–665.

Hansen, J., Lacis, A., Rind, D., Russell, G., Stone, P., Fung, I., Ruedy, R. and Lerner, J. (1984) 'Climate sensitivity: analysis of feedback mechanisms', in J.E. Hansen and T. Takahashi (eds) *Climate Processes and Climate Sensitivity*, Geophysical Monographs 29, Washington: American Geophysical Union.

Hansen, J., Russell, G., Rind, D., Stone, P., Lacis, A., Lebedeff, S., Ruedy, R. and Travis, L. (1983) 'Efficient three-dimensional global models for climate studies. Models I and II', *Monthly Weather Review* 3: 609–662.

Hastenrath, S. (1990) *Climate Dynamics of the Tropics*, Dordrecht: Kluwer Academic.

Hay, W.W. (1988) 'Paleoceanography: a review for the GSA Centennial', *Geological Society of America Bulletin* 100: 1934–1956.

Hay, W.W., Barron, E.J. and Thompson, S.L. (1990a) 'Results of global atmospheric circulation experiments on an earth with a meridional pole-to-pole continent', *Journal of the Geological Society of London* 147: 385–392.

——(1990b) 'Global atmospheric circulation experiments on an earth with polar and tropical continents', *Journal of the Geological Society of London* 147: 749–757.

Henderson-Sellers, A. and Pitman, A.J. (1992) 'Land-surface schemes for future climate models; specification, aggregation and heterogeneity', *Journal of Geophysical Research* 97(D3): 2687–2696.

Huggett, R.J. (1991) *Climate, Earth Processes and Earth History*, Berlin: Springer-Verlag.

Hughes, T. (1985) 'The great Cenozoic ice sheets', *Palaeogeography, Palaeoclimatology, Palaeoecology* 50: 9–43.

Hyde, W.T., Crowley, T.J., Kim, K.-Y. and North, G.R. (1989) 'Comparison of GCM and energy balance model simulations of seasonal temperature changes over the past 18,000 years', *Journal of Climate* 2: 864–887.

Ives, J.D. (1987) 'The mountain lands', in M.J. Clark, K.J. Gregory and A.M. Gurnell (eds) *Horizons in Physical Geography*, London: Macmillan Education Ltd.

Janecek, T.R. and Rea, D.K. (1985) 'Quaternary fluctuations in the northern hemisphere trade winds and westerlies', *Quaternary Research* 24: 150–163.

Jarrett, R.D. (1990) 'Paleohydrologic techniques used to define the spatial occurrence of floods', *Geomorphology* 3: 181–195.

Joussaumé, S. (1989) 'Desert dust and climate: an investigation using an atmospheric general circulation model', in M. Leinen and M. Sarnthein (eds) *Paleoclimatology and Paleometeorology: Modern and Past Patterns of Global Atmospheric Transport*, Dordrecht: Kluwer Academic Publishers.

Kiewietdejonge, C.J. (1984) 'Büdel's geomorphology', *Progress in Physical Geography* 8: 218–248, 365–397.

King, C.A.M. (1980) 'Thresholds in glacial geomorphology', in D.R. Coates and J.D. Vitek (eds) *Thresholds in Geomorphology*, London: Allen and Unwin.

Kirkby, M.J. (1989) 'A model to estimate the impact of climatic change on hillslope and regolith form', *Catena* 16: 321–341.

Köppen, W. (1931) *Grundriss der Klimakunde*, Berlin: Walter de Gruyter.

Kraus, E.B. (1973) 'Comparison between Ice Age and present general circulation', *Nature* 245: 129–133.

Kutzbach, J.E. (1985) 'Modeling of paleoclimates', *Advances in Geophysics* 28A: 159–196.

Kutzbach, J.E. and Gallimore, R.G. (1989) 'Pangean climates: megamonsoons of the megacontinent', *Journal of Geophysical Research* 94: 3341–3358.

Kutzbach, J.E. and Guetter, J.E. (1986) 'The influence of changing orbital parameters and surface boundary conditions on climate simulations for the past 18,000 years', *Journal of Atmospheric Science* 43: 1726–1759.

Kutzbach, J.E. and Street-Perrott, F.A. (1985) 'Milankovitch forcings of fluctuations in the level of tropical lakes from 18 to 0k yr BP', *Nature* 317: 130–134.

Kutzbach, J.E. and Wright, H.E. Jr (1985) 'Simulation of the climate 18,000 years B.P.: results for the North American/North Atlantic/European sector and comparison with the geologic record of North America', *Quaternary Science Reviews* 4: 147–187.

Lautenschlager, M. and Herterich, K. (1990) 'Atmospheric response to Ice Age conditions. Climatology near the earth's surface', *Journal of Geophysical Research* 95(D13): 22,547–22,557.

Lloyd, C.R. (1984) 'Pre-Pleistocene paleoclimates: the geological and paleontological evidence; modeling strategies, boundary conditions, and some preliminary results', *Advances in Geophysics* 26: 35–140.

Lorenz, E.N. (1968) 'Climate determinism', *Meteorological Monographs* 8: 1–3.

——(1990) 'Can chaos and intransivity lead to interannual variability?', *Tellus* 42A: 378–389.

McDowell, P.F., Webb III, T. and Bartlein, P.J. (1990) 'Long-term environmental change', in B.J. Turner II *et al.* (eds) *The Earth as Transformed by Human Action*, Cambridge: Cambridge University Press.

Manabe, S. and Broccoli, A.J. (1985) 'The influence of continental ice sheets on the climate of an ice age', *Journal of Geophysical Research* 90: 2167–2190.

———(1990) 'Mountains and arid climates of middle latitudes', *Science* 247: 192–195.

Manabe, S. and Hahn, D.G. (1977) 'Simulation of the tropical climate of an Ice Age', *Journal of Geophysical Research* 82: 3889–3911.

———(1981) 'Simulations of atmospheric variability', *Monthly Weather Review* 109: 2260–2286.

Martonne, E. de (1913) 'Le climat–facteur du relief', *Scientia*: 339–355 (English translation in E. Derbyshire (ed.) *Climatic Geomorphology*, London: Macmillan).

Meehl, G.A. (1984) 'Modeling the earth's climate', *Climatic Change* 6: 259–286.

———(1990) 'Development of global coupled ocean–atmosphere general circulation models', *Climate Dynamics* 5: 19–33.

———(1992) 'Effect of tropical topography on global climate', *Annual Review of Earth and Planetary Science* 20: 85–112.

Meehl, G.A. and Branstator, G.W. (1992) 'Coupled climate model simulation of El Niño/Southern Oscillation: implication for paleoclimate', in H.F. Diaz and V. Markgraf (eds) *El Niño*, Cambridge: Cambridge University Press.

Melton, M.A. (1957) 'An analysis of the relations among elements of climate, surface, properties, and geomorphology', *Department of Geology Technical Report* 11, Columbia University, New York.

Molnar, P. and England, P. (1990) 'Late Cenozoic uplift of mountain ranges and global climate change: chicken or egg?', *Nature* 346: 29–34.

Nicholson, S.E. and Flohn, H. (1980) 'African environmental and climatic changes and the general atmospheric circulation in late Pleistocene and Holocene', *Climatic Change* 2: 313–348.

Ollier, C. (1979) 'Evolutionary geomorphology of Australia and New Guinea', *Transactions of the Institute of British Geographers* NS 4: 516–539.

———(1988) 'The regolith in Australia', *Earth Science Reviews* 25: 355–361.

Passarge, S. (1926) 'Morphologie der Klimazonen oder Morphologie der Land-schaftsgürtel?', *Petermanns Geographische Mitteilungen* 72: 173–175.

Peltier, L.C. (1950) 'The geomorphic cycle in periglacial regions as it is related to climatic geomorphology', *Annals of the Association of American Geographers* 40: 214–236.

———(1975) 'The concept of climatic geomorphology', in W.N. Melhorn and R.C. Flemal (eds) *Theories of Landform Development*, London: Allen & Unwin.

Penck, A. (1910) 'Versuch einer Klimaklassifikation auf physiogeographischer Grundlage', *Sitzungsbericht Preussische Akademie der Wissenschaften, Physikalische-Mathematische Klasse* 12: 236–246.

Petit-Maire, N. (1989) 'Interglacial environments in presently hyperarid Sahara: paleoclimatic implications', in M. Leinen and M. Sarnthein (eds) *Paleoclimatology and Paleometeorology: Modern and Past Patterns of Global Atmospheric Transport*, Dordrecht: Kluwer Academic Publishers.

Phillips, J.D. (1990) 'Relative importance of factors influencing fluvial loss at the global scale', *American Journal of Science* 290: 547–568.

Pielke, R. and Avissar, R. (1990) 'Influence of landscape structure on local and regional climate', *Landscape Ecology* 4: 133–155.

Rea, D.K., Leinen, M. and Janecek, T.R. (1985) 'Geologic approach to the long-term history of atmospheric circulation', *Science* 227: 721–725.

Rind, D. (1986) 'The dynamics of warm and cold climates', *Journal of Atmospheric Science* 43: 3–24.

———(1987) 'Components of the Ice Age circulation', *Journal of Geophysical Research* 92: 4241–4281.

———(1988) 'Dependence of warm and cold climate depiction on climate model resolution', *Journal of Climate* 1: 965–997.

Rind, D. and Chandler, M. (1991) 'Increased ocean heat transports and warmer climate', *Journal of Geophysical Research* 96: 7437–7461.

Rind, D. and Peteet, D. (1985) 'Terrestrial conditions at the Last Glacial Maximum and CLIMAP sea-surface temperature estimates: are they consistent?', *Quaternary Research* 24: 1–22.

Ruddiman, W.F. and Kutzbach, J.E. (1989) 'Forcing of late Cenozoic northern hemisphere climate by plateau uplift in southern Asia and the American West', *Journal of Geophysical Research* 94: 18,409–18,427.

Ruddiman, W.F. and Raymo, M.E. (1988) 'Northern Hemisphere climate regimes during the past 3 Ma: possible tectonic connections', *Philosophical Transactions of the Royal Society of London* B, 318: 411–430.

Salati, E. (1985) 'The forest and the hydrological cycle', in R. E. Dickinson (ed.) *The Geophysiology of Amazonia*, New York: John Wiley.

Sarnthein, M. (1978) 'Sand deserts during glacial maximum and climatic optimum', *Nature* 271: 43–46.

Sarnthein, M., Tetzlaff, G., Koopman, B., Wolter, K. and Pflaumann, U. (1981) 'Glacial and interglacial wind regimes over the eastern subtropical Atlantic and north-west Africa', *Nature* 293: 193–196.

Saunders, I. and Young, A. (1983) 'Rates of surface processes on slopes, slope retreat and denudation', *Earth Surface Processes and Landforms* 8: 473–501.

Schlesinger, M.E. and Gates, W.L. (1980) 'The January and July performance of the OSU two-level atmospheric general circulation model', *Journal of Atmospheric Science* 37: 1914–1943.

Schneider, S.H., Thompson, S.L. and Barron, E.J. (1985) 'Mid-Cretaceous continental surface temperatures: are high CO_2 concentrations needed to simulate above freezing winter conditions?', in E.T. Sundquist and W.S. Broecker (eds) *The Carbon Cycle and Atmospheric CO_2 Natural Variations Archean to Present*, Geophysical Monographs 32, Washington, D.C.: American Geophysical Union.

Schumm, S.A. (1973) 'Geomorphic thresholds and complex response of drainage systems', in M. Morisawa (ed.) *Fluvial Geomorphology, Proceedings of the 4th Annual Geomorphology Symposium*, Binghamton, N.Y.: 299–310.

——(1979) 'Geomorphic thresholds: the concept and its applications', *Transactions of the Institute of British Geographers* NS 4: 485–515.

Schuurmans, C.J.E. and Coops, A.J. (1984) 'Seasonal mean temperatures in Europe and their interannual variability', *Monthly Weather Review* 112: 1218–1225.

Shinn, R.A. and Barron, E.J. (1989) 'Climate sensitivity to continental ice-sheet size and configuration', *Journal of Climate* 2: 1517–1537.

Simmons, A.J., Burridge, D.M., Jarraud, M., Girard, C. and Wergen, W. (1989) 'The ECMWF medium-range prediction models, development of the numerical formulations and the impact of increased resolution', *Meteorology and Atmospheric Physics* 40: 28–60.

Slingo, A. and Pearson, D.W. (1987) 'A comparison of the impact of an envelope orography and of a parameterization of orographic gravity-wave drag on model simulations', *Quarterly Journal of Royal Meteorological Society* 113: 847–870.

Sloan, L.C. and Barron, E.J. (1990) '"Equable" climates in earth history?', *Geology* 18: 488–492.

——(1992) 'A comparison of Eocene climate model results to quantified paleoclimatic interpretations', *Palaeogeography, Palaeoclimatology, Palaeoecology* 93: 183–202.

Starkel, L. (1987) 'Long-term and short-term rhythmicity in terrestrial landforms and deposits', in M.R. Rampino, J.E. Sanders, W.S. Newman and L.K. Königsson (eds) *Climate: History, Periodicity, and Predictability*, New York: Van Nostrand Reinhold.

Stoddart, D.R. (1969a) 'Climatic geomorphology', in R.J. Chorley (ed.) *Water, Earth and Man*, London: Methuen.

———(1969b) 'Climatic geomorphology: review and assessment', *Progress in Geography* 1: 159–222.

Street-Perrott, F.A. and Harrison, S.P. (1984) 'Temporal variations in lake levels since 30,000 yr. B.P. – an index of the global hydrological cycle', in J.E. Hansen and T. Takahashi (eds) *Climate Processes and Climate Sensitivity, Geophysical Monographs* 29, Washington, D.C.: American Geophysical Union.

Stute, M., Forster, M., Frischkorn, H., Serejo, A., Clark, J.E., Schlosser, P., Broecker, W. and Bonani, G. (1995) 'Cooling of tropical Brazil (5°C) during the last glacial maximum', *Science* 269: 379–383.

Tricart, J. and Cailleux, A. (1972) *Introduction to Climatic Geomorphology*, translated by C.J. Kiewietdejonge, London: Longman.

Tsonis, A.A. and Elsner, J.B. (1989) 'Chaos, strange attractors and weather', *Bulletin of the American Meteorological Society* 70: 14–23.

Wallace, J.M. (1983) 'The climatological mean stationary waves: observational evidence', in B.J. Hoskins and R.P. Pearce (eds) *Large-Scale Dynamical Processes in the Atmosphere*, New York: Academic Press.

Warren, A. (1985) 'Arid geomorphology', *Progress in Physical Geography* 9: 434–441.

Washington, W.M. and Meehl, G.A. (1989) 'Cimate sensitivity due to increased CO_2: experiments with a coupled atmosphere and ocean general circulation model', *Climate Dynamics* 4: 1–38.

Webster, P.J. and Streten, N. (1978) 'Late Quaternary ice age climates of tropical Australia: interpretations and reconstructions', *Quaternary Research* 10: 279–309.

Weinert, H.H. (1965) 'Climatic factors affecting the weathering of igneous rocks', *Agricultural Meteorology*. 2: 27–42.

Wilhelmy, H. (1958) *Klimamorphologie der Massengesteine*, Braunschweig: Westermann.

Williams, J., Barry, R.G. and Washington, W.M. (1974) 'Simulation of the atmospheric circulation using the NCAR global circulation model with Ice Age boundary conditions', *Journal of Applied Meteorology* 13: 305–317.

Williamson, D.L., Kiehl, J.T., Ramanathan, V., Dickinson, R.E. and Hack, J.J. (1987) 'Description of the NCAR Community Climate Model (CCM1)', NCAR Technical Note TN-285-STR, Boulder, Colo.).

Wold, C.N. and Hay, W.W. (1990) 'Estimating ancient sediment fluxes', *American Journal of Science* 290: 1069–1089.

Wolman, M.G. and Miller, J.P. (1960) 'Magnitude and frequency of forces in geomorphic processes', *Journal of Geology* 68: 54–75.

10

ON THE LANDFORM HISTORY OF CHORLEY'S WEST SOMERSET

Peter Haggett

Where the landscape in its glory,
Teaches truth to wandering men.
(John Keble)

The first meeting with Richard Chorley that I can recall was in Steyning's Coffee House in the shadow of the parish church of St Mary in the mid-Somerset town of Bridgwater. It was the early summer of 1958. He was back from six years in the United States and was to join the Cambridge department that October as a University Demonstrator. I already held one of these teaching posts and we discussed together a new course of weekly practicals which was to include the then new statistical analysis he was bringing back from Columbia and Brown Universities.[1]

We found we had much in common. This included deep family roots in the same part of England, an illogical support for Somerset's bizarre county cricket team, a love of Gilbertian lyrics, a fascination with the Great War, and a highly developed sense of the comic side of life. (A belief that the euphonium was a musical instrument was Chorley's alone.) We had both been raised in West Somerset and both attended local grammar schools (he at Minehead, I at Bridgwater). We had both gone on from there to Oxbridge[2] and, as undergraduates, had both written our bachelor's dissertations on adjacent parts of the hill country of West Somerset – Chorley on the Brendons and Haggett on the Quantocks.

That meeting was a generation ago and our research and our academic careers, once closely twinned at Cambridge, have gone in different directions since 1966. So, in seeking to write a contribution to a volume on his interests, I've tried to go back to our original common ground in the West Country. Hence this essay on the landforms of Richard Chorley's home area: the hills, valleys and coastline of West Somerset. I call it a reflection since it is over thirty years since I last seriously practised any geomorphology and today's highly skilled professionals will need to read this essay in a forgiving spirit.[3]

PETER HAGGETT

THE WEST SOMERSET LANDSCAPE

Richard Chorley was born, grew up and went to school in the Somerset coastal town of Minehead. The lines of verse which open this chapter (from the Oxford poet and hymnist, John Keble) would have been known to the young Richard.[4] They are carved on the western end of a small 'wind and weather hut' below the crest of Selworthy Beacon, at the western end of North Hill. It was built of local sandstone and erected as a memorial to Sir Thomas Dyke Acland (1722–1785) by his family. The spot was chosen by Acland's youngest surviving son, by then living in Holnicote, New Zealand, to mark the precise destination of family walks. Each Sunday afternoon, Acland, with his children and later his grandchildren, would walk up the combe from the Acland family home at Holnicote in the valley below. In 1944 the family's West Somerset estates including Holnicote and Horner Woods were made over to the National Trust by a descendant, Sir Richard Acland.

Selworthy Beacon (308 metres) is a splendid point from which to view the characteristics of the West Somerset terrain. From the crest above Acland's hut we can see much of the landscape which dominated Chorley's childhood. Eastwards the view along the whaleback ridge of North Hill extends towards the town of Minehead, hidden in its lee with, beyond, the fertile Dunster marshes and the seaward slopes of the Brendon and Quantock ranges. Southwards lies the steep descent into the Selworthy valley and the rise towards Dunkery and the Exmoor uplands. Westwards, the crescent shingle of Porlock Bay and the hump-backed cliffs stretch away towards Lynmouth. Northwards, the steep slopes of Bossington Hill plunge towards the steel-grey waters of the Bristol Channel with the distant Welsh coastline beyond.

The area described here as West Somerset is shown in Figure 10.1. It extends east and west about Minehead over an oblong area, 60 km east–west and 20 km north–south at its widest extent.[5] It is marked on its northern boundary by the coastline of the Bristol Channel and on the other sides by the arbitrary graticule of the National Grid. Given the east–west grain of the topography it represents an area that was just about within the range of a day's return cycle expedition from Minehead. Although it laps over into the county of Devon to the west, it will here for convenience be simply referred to as 'West Somerset'.

The present terrain ranges from 519 metres (Dunkery Beacon on Exmoor) down to sea level. The underlying bedrock spans in age from the Palaeozoic to Recent. The historical sequence of deposition, tectonic movements and erosion gives the structure to this chapter. Thus I look first at the oldest, thickest and spatially most dominant Palaeozoic rocks of the Devonian; then at their disruption by faulting and the Mesozoic sediments that filled the tectonic basins so created. We then jump forward several million years to consider the erosion and peneplanation of the Tertiary period, and the minor

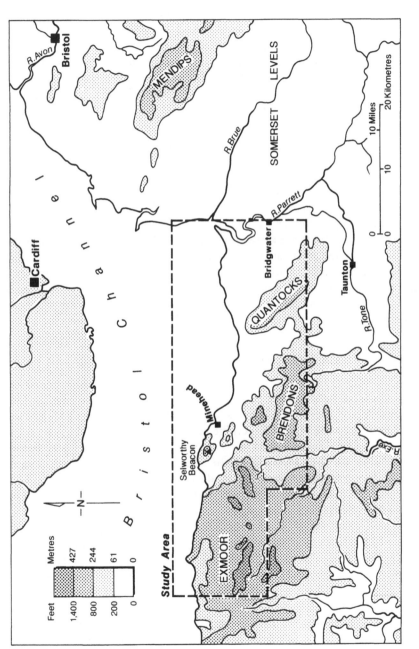

Figure 10.1 West Somerset study area. Location within a wider area of south-west England

modifications of the Quaternary cold periods. Finally, we give examples of those geomorphological processes which are still going on within the West Somerset area today. The emphasis throughout the chapter is on the arguments about the origins, processes and timing of the landforms that dominate this part-fossil, part-active landscape. As such, West Somerset serves as a microcosm of the greater arguments which, at a global scale, have played such an enduring part in Richard Chorley's *History of the Study of Landforms* (Chorley, Dunn and Beckinsale 1964; Chorley, Beckinsale and Dunn 1973; Beckinsale and Chorley 1991).

THE PRIMARY FRAMEWORK

The basic framework of West Somerset is provided by Palaeozoic rocks of Devonian age. These underlie the whole area and outcrop over two-thirds of it to form the attractive uplands of Exmoor, the Brendon Hills and the Quantock Hills (Dineley 1977). The extent of the Devonian outcrop together with its main divisions are shown in Figure 10.2. From north to south the sequence gets younger in age with the Lower, Middle and Upper Devonian beds in sequence. The rocks strike north-east–south-west with a southerly regional dip of 30–40°. We look here at variations in the lithology of the main series, their subsequent folding and dislocation, and the combined effects on present landforms.

Lithological variations within the Devonian

Although Horner published an account of the geology of West Somerset as early as 1816, the preliminary mapping was not completed until the work of Sedgwick, Murchison and de la Beche in the 1830s (Sedgwick and Murchison 1837; De la Beche 1839) and Etheridge and Ussher a generation later in the 1860s (Etheridge 1867; Ussher 1879). It was nearly a century later before Webby subjected the Devonian of the Brendon Hills to detailed modern analysis (Webby 1962). Thus the modern geological maps are a curious palimpsest of well-established local detail and wide areas of sketchy interpolation between exposures.[6]

The main divisions of the Devonian sequence in West Somerset are given in Table 10.1. The Lynton Slates outcrop in the west of the region where they form the core of the Exmoor WNW–ESE-trending anticline and the base of the Quantock anticline. They were formed in shallow seas and comprised mainly slates and siltstones; in landform terms they form weakly resistant beds later exploited by strike streams. The Hangman Grits are more resistant and outcrop more widely. They were formed as coastal plain deposit and consist of buff to reddish brown massive jointed quartzitic sandstones. Individual beds vary considerably in thickness along the strike, and inter-bedded cleaved siltstone and slate are more common towards the contact

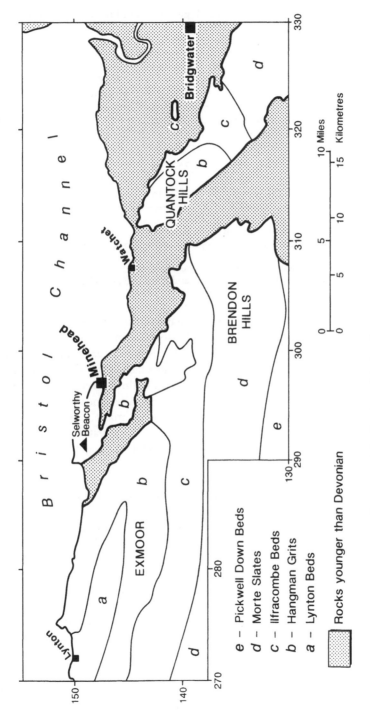

Figure 10.2 Distribution of rocks of Devonian age in West Somerset

Source: Edmonds, McKeown and Williams 1969: Figure 10, p. 27

e – Pickwell Down Beds
d – Morte Slates
c – Ilfracombe Beds
b – Hangman Grits
a – Lynton Beds

Rocks younger than Devonian

Table 10.1 Classification and geomorphological impact of the Devonian rocks of West Somerset

Timing	Beds	Dominant lithology[a]	Estimated thickness[b]	Outcrop	
				Highest elevation[c]	Spatial extent[d] (%)
Upper	Pickwell	Sandstone	90 m		
	Morte	Slates	480 m	430 m	37
	Ilfracombe	Slates	490 m		
Middle	Hangman	Grits	750 m		
				520 m	31
Lower	Lynton	Slates	100+ m		

Notes:
[a] For detailed lithological variation within major beds see text
[b] Based on the generalised vertical section accompanying Sheet 295 'Taunton', 1:50,000 (British Geological Survey 1984).
[c] Highest closed-contour summit in metres within West Somerset
[d] As a proportion of the total area of West Somerset as defined in Figure 10.1

with the Ilfracombe beds. The next formation, the Ilfracombe Slates, are marine sediments with evidence of coral banks in their upper parts. They consist principally of shales, slates and siltstones with occasional limestones. The Morte Slates are also marine deposits and succeed the Ilfracombe Beds conformably. They are dark-coloured slates with almost no limestone beds. The slates are mainly silty in character but occasionally sandy with a flag-like character. Finally, the Pickwell Down Sandstones mark a return to near-terrestrial conditions, being formed in the turbulent waters of rivers and shallow seas. The formation includes red, purple and green continental sandstones with grey-blue shales.

The Devonian rocks can be grouped into two classes on the basis of particle size. First, the mainly sandy rocks with subsidiary siltstones and slates of the Hangman Grits and Pickwell Down series. Second, the fine-grained silty slates and siltstones which dominate the Ilfracombe Beds and Morte Slates. The division is reflected in current soil divisions with coarser podzols on the Hangman Grits and silty gley soils on the slates.

In summary, the Devonian rocks which cover much of the study area were deposited at the southern margin of an Old Red Sandstone continent. Their lithology represents a series of transgressive and regressive cycles in waters of varying depth on the southern shelf of that continent. The Devonian sediments were laid down as the northern part of the compound major synclinorium of Devonian and Carboniferous rocks which constitute much of the south-western peninsula. Hence the oldest rocks exposed in the region, the Lower Devonian, are found outcropping to the far north-east around Foreland Point. South-south-westwards come the Middle and Upper

Devonian and finally the Carboniferous rocks of the Culm Measures south of Exmoor are successively encountered.

Subsequent folding and dislocation

From the late Devonian until the early Permian occurred the deformations of the Variscan orogenic cycle, in which the Devonian sediments just described were subjected to northwards pressure across the Sub-Varisean foredeep (now occupied by the Bristol Channel) against the Archean oldland of the present Wales. This produced an almost continuous east–west line of overthrusting and/or overfolding. To the north of this profound line of disturbance lies the synclinal South Wales coalfield, and to the south, the Culm syncline of central Devon. Our area therefore is represented as the northern ascending limit of the latter syncline, the rocks ageing with distance north up to an anticlinal crest running from east-south-eastwards to Oare, exposing the Lynton Beds in the core.

This severe folding and dislocation in the Variscan orogeny is shown in Figure 10.3.[7] The regional memoir describes the whole series as intensely folded, compressed and cleaved, the southerly dips being generally those of isoclinal folds (Edmonds, McKeown and Williams 1975). Because of the late tectonic activity (mainly Tertiary/Quaternary reactivation), O.T. Jones thought that the regular geological lines drawn across the Exmoor area were partly fictitious: all the exposures of rock on Exmoor he himself had examined 'were heavily affected by pitch-faulting and in an area of pitch-faulting the boundaries did not necessarily run in the direction of the beds' (Jones 1952). His comments have implications for topographic analysis based on these boundaries.

Impact on terrain

Given the complex history of the West Somerset area, it would be difficult to match the contemporary terrain directly to the bedrock geology in any simplistic way: tectonic disruption and later erosional history also play a strong role. However, in terms of its macrogeography, the Palaeozoic rocks of the Devonian form the upland framework of the area. While the Devonian rocks sustain hills at elevations in excess of 250 m in Exmoor, the Brendons and the Quantocks, the prevailing elevation for more recent sedimentary rocks of Mesozoic origin is below 100 m.

Within the main zones of the Devonian itself, the effects on scenery of facies variation can readily be traced. We can illustrate this with reference to the Hangman Grits which outcrop to form the northern half of the Quantock Hills. These are divided into three bands. The thick Trentishoe Grits form the base and coincide with the highest part of the hills (Wills Neck, 384 m). Above these come two narrower bands, Rawn's Shales and the

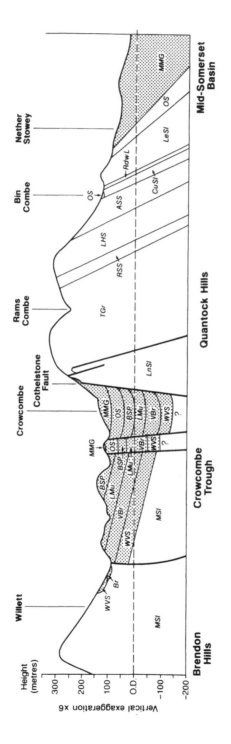

Figure 10.3 Cross-section of the Quantock Hills. Rocks of Mesozoic age are stippled. Initials refer to geological subdivisions. Those mentioned in the text are the Trentishoe Grits (TGr), Rawn's Shales (RSS), Little Hangman Slates (LHS), Vexford Breccias (VBr), Budleigh Salterton Pebble Beds (BSP) and Mercia Mudstones (MMG). Note that the vertical exaggeration is ×6

Source: Modified from British Geological Survey, 1:50,000 series, Sheet 295 'Taunton'

Little Hangman Beds. The shales contain more argillaceous material and have commonly been sought out and followed by streams, while the Little Hangman Beds form uplands and divides (Edmonds and Williams 1985). The north–south alignment of Holford Combe illustrates the relationship. Balchin draws attention to minor dip and scarp topography within the Ilfracombe Beds (Balchin 1952: 455).

The effect of Devonian structures on coastal landforms has been the subject of studies by a father and daughter team, Newell and Muriel Arber (E.A.N. Arber 1911; M.A. Arber 1949). It was in relation to the coast of Exmoor that Newell Arber first introduced the term 'hog's-back cliffs' (and separated them from bevelled and flat-topped cliffs). Muriel Arber thought this form of cliff to be directly related to geological structure. She argued that where the hog's back is developed on the grandest scale (along the Exmoor coast), the regional dip of the Devonian rocks is southerly and thus inland. The seaward slope from the crest down to present sea level is thus seen simply as an escarpment cutting across the edges of the beds; inland from the crest is a gentler dip slope into a valley running more or less parallel with the coast, the opposite (southern) side of the valley being another escarpment.

THE SECONDARY BASINS

The end of the Palaeozoic and the Mesozoic period in West Somerset is marked by three main geomorphological legacies: the formation of tectonic basins, the filling of those basins with sediments and the disruption of the sediments by tectonic activity. We look at each in turn. Together they form the secondary theme in the orchestration of the landscape.

Basin formation by tectonic activity

Most faults in the Palaeozoic rocks were produced during the Variscan orogeny by compressive forces acting from the south.[8] Tensional fractures accompanied early uplift and some faults were probably active at later times under both compression and tension. The pre-Triassic faulting led to the formation of basin-like depressions around the edge of the Devonian highlands. This produced a block-fault system of raised blocks (horsts) and depressed basins (grabens). The highlands of North Hill, Grabbist and the Quantocks fall into the first category; the basins of the Vale of Porlock, the Cleeve, Washford and Williton Lowlands and the Vale of Stogumber into the second. Bordering our area of interest, to the north and east of the Quantocks, the whole of the central Somerset basin extending into the Bristol Channel forms yet another fault-bounded syncline on a still larger scale.

Figure 10.4 shows the faulting pattern around Minehead. A distinction is made between the boundary faults that generally mark a sharp division

223

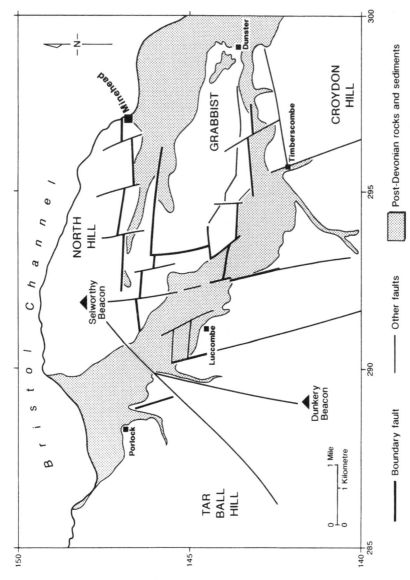

Figure 10.4 Tectonic activity with downfaulted basins in the Minehead area, West Somerset. Main boundary and secondary faults are shown

Source: Thomas 1940: Figure 4, p. 11

between the Devonian and the Trias, and minor faults within the main formations. Inevitably, the map shows only part of the pattern. Exposures only allow some faults to be traced and much of the evidence is buried under later sediments. Around Dunster several steep-sided hills formed of Devonian rock stand as 'islands' above a plain of Red Marl; these later sediments conceal what is probably a highly irregular sub-Triassic surface. Even where evidence is available, the pattern of minor faulting is so dense that only the main lines can be shown on the map: e.g. one single street in Minehead is crossed by nine faults (Thomas 1940: 36).

Most of the fault blocks were subsequently tilted to the north or northeast at angles which vary between 10 and 50°. The step faulting shows regular spacing. In some cases the amounts of vertical movement are considerable. Thomas estimated that the Selworthy strike fault, which forms the southern boundary of North Hill (see Figure 10.4) has a down-throw of at least 700 m (Thomas 1940: 36).

The lowlands which resulted from the pre-Triassic movements provided basins into which later sediments were deposited and in which, in some cases, they remain preserved in the angle between the dip-slope of one fault block and the scarp-faulted edge of the next block.

Basin infilling by Mesozoic sediments

Permo-Triassic rocks were deposited on a very rugged post-Variscan landscape in three major tectonic basins: the large Central Somerset basin to the east of the Quantocks, the Watchet-Stogumber basin between the Brendons and the Quantocks, and the more isolated and smaller Porlock basin to the west. A cross-section of the second basin is given in Figure 10.3. The development of sedimentation was governed partly by the continued subsidence of the basins between the old boundary faults. The sedimentary sequence in the Permian and Triassic period is summarised in Table 10.2.

As the Permo-Triassic sediments accumulated in the basins, so progressively younger strata overlapped the earlier deposits and began to coalesce from one previously isolated basin to another. Thomas has summarised the sedimentary history of this New Red Sandstone period in West Somerset into three epochs: (1) a 'Permian' epoch with deposition restricted to a narrow corridor between the Quantock and the Brendon Hills; (2) a 'Bunter' epoch with a westward transgression of the fluviatile 'Bunter' Beds and formation of the separate Porlock Basin of deposition, with local breccia formation; and (3) a 'Keuper' epoch with a continuation of westerly overlap by lacustrine marls and sandy limestones and uniting of the Porlock basin with the main basin (Thomas 1940: 38).

The accumulation of sediments probably led to synclinal subsidence in the troughs but this remains speculative. For example, despite the cross-section in Figure 10.3 there is no direct evidence of the depth of the Permo-Triassic

Table 10.2 Classification of the Permian and Triassic rocks of West Somerset

Timing	Beds	Dominant lithology[a]	Estimated thickness[b]
Keuper phase	Penarth	Shales and limestones	20 m
	Mercia	Mudstone and sandstones	400 m
Bunter phase	Otter	Sandstone	50 m
	Budleigh Salterton	Pebble beds	20 m
	Littleham	Mudstone	45 m
Permian phase	Vexford	Breccias	30 m
	Wiveliscombe	Sandstones	65+ m

Notes:
[a] For detailed lithological variation within major beds see text
[b] Based on the generalised vertical section accompanying Sheet 295 'Taunton', 1:50,000 (British Geological Survey 1984)

deposits in the Stogumber graben. Estimates of a maximum depth of around 250 m are judgemental. Whitaker has argued that downwarping was more or less continuous from the late Palaeozoic onwards (Whitaker 1973).

Much of the sedimentary evidence suggests that the 'New Red' rocks which filled the basins in Permo-Triassic times were formed under generally continental and arid conditions, probably produced by desert weathering of the Devonian highlands with coarse debris being swept from the barren uplands to form screes and poorly sorted breccias. The evidence of former arid conditions are well shown in the deposits of the Porlock basin. Screes of huge proportions gathered on hillsides to form breccias and breccio-conglomerates, while marls and sandstones were laid down in areas of fresh water. The breccias of the Vale of Porlock appear to have been formed in the way that Norton calls 'Bajada formation' (Norton 1917: 167). He describes this pediment-like landform as a wide slope of rock waste formed of fragments detached by mechanical weathering but accumulated by the action of intermittent streams:

> Its fragments have been more or less water-worn. It forms an imperfect breccia, and yet is far from being a typical conglomerate of well-rounded pebbles. The streams which form the bajada have certain peculiarities which greatly lessen the wear on the stones they carry. The long-accumulated waste on the mountain slope, swept down by spasmodic rains, loads heavily the streams. In the moving mass of the mud-flow, stones are intermingled with the finer waste held in suspension, and are thus protected from mutual abrasion. Hence pebbles remain imperfectly rounded even to the outer edge of the bajada slope.

The bajada is stratified but imperfectly, and the stones may be left in any angle of repose. Calcium carbonate is mentioned as forming part of the matrix, deposited by calcareous evaporating water (Norton 1917).

Later erosion and tectonic disruption of sediments

The contrasts within the Permo-Triassic sediments stand out today as minor topographic variations within the downfaulted lowlands. Within the Stogumber graben, the Vexford Breccias (angular and subangular fragments of sandstone, slate and quartz in a sandy matrix) form a distinct westward-facing scarp on the Brendons side. In the centre of the trough, the Budleigh Salterton Pebble Beds (mainly conglomerates of pebbly sandstone with a strong calcareous and ferruginous cement) mark the base of the Triassic and form the strongest terrain features, with a flat-topped plateau rising to 160 m east of Crowcombe. The later and thicker deposits of Mercia Mudstones (the old Keuper Marls) form a broad plain both east and west of the Quantocks, bounded to the north by an escarpment of Blue Lias limestone. Since the dip of the Liassic strata is consistently to the north, the scarps are all southward-facing, and the strike ridges run east and west.

Tectonic activity continued throughout the period but at a reduced level.[9] Mesozoic strata were affected by tilting and minor faulting. Thus after the deposition of the Lias (how long after is not certain) some Variscan faults moved again, accompanied by new faults following the old trend lines, and caused the emergence of the old fault-blocks. Tilting accompanied these movements, and the tilting of the blocks has been in every case to the north. Figure 10.4 shows a big dip fault at the western end of Grabbist Hill, the Tivington fault, which is shifted by a later strike fault near Wootton Knowle. Again, the southern limit of Grabbist Hill west of Dunster is formed by a big strike fault that is repeatedly shifted by small dip faults.

In the neighbourhood of Watchet the original south-facing Lias escarpment is dislocated and repeated several times by strike faulting. Near Luccombe in the Vale of Porlock a line of knolls represents the breccia escarpment cut into pieces by streams draining from the slopes of Dunkery Hill. At Washford and Vellow strike ridges formed by the conglomerates and breccias of the Bunter period are also disrupted by minor faulting.

THE TERTIARY PENEPLANES

If the Mesozoic in West Somerset was dominated by tectonics and sedimentation, then the Tertiary was a period dominated by erosion. No deposits of Tertiary age are found in the immediate area.[10] In these circumstances, any reconstruction of erosional history is likely to be a hazardous and speculative business. The Tertiary theme is then one of exhumation of the sediments that cloaked much of the primary and secondary structures.

Sub-aerial and marine peneplanes

In a cartographic study of Exmoor and adjacent areas, Balchin recognised a series of eight erosion surfaces (see Table 10.3) (Balchin 1952). The highest Summit surface was presumed to be the remnants of a sub-aerial peneplane at a height of more than 475 m above present sea level. A second sub-aerial peneplane, the Exmoor surface, was ascribed to sub-aerial planation of early Tertiary age.

Below the upland surfaces, Balchin recognised a flight of marine terraces of decreasing height and decreasing age. These ranged in height from a late Miocene Lynton surface tied to a sea level 375 m above the present sea level, to the youngest and lowest surface, the Instow. This was ascribed to the late Pliocene and to a sea level at 85 m.

In the absence of sedimentary deposits, the local evidence for the surfaces is largely cartographic. Flat areas, some of which truncate areas of highly folded and lithologically varied rock types, are interpreted as evidence of planation. Similarly, breaks of slope are seen as evidence of degraded cliff lines at a higher sea level. Fragments of the surfaces at similar heights are aligned and estimates of former coastlines drawn in.

A second source of evidence comes from comparison with other areas. Thus the Lynton Surface is related to the Dartmoor Middle Surface at 412–320 m and planation surfaces of the Dart Basin (Brunsden *et al.* 1964). The most acceptable date for this surface appears therefore to be Middle Tertiary (Oligocene-Miocene). Kidson (1962) has surveyed a detailed thal-

Table 10.3 Peneplane remnants in West Somerset[a]

Surface name	Probable time of formation	Peneplanation process	Height
Instow	Late Pliocene	Submarine	85 m[b]
Georgeham	Pliocene	Submarine	130 m[b]
Buckland	Pliocene	Submarine	210 m[b]
Anstey	Late Miocene	Submarine	255 m[b]
Molland	Late Miocene	Submarine	285 m[b]
Lynton	Late Miocene	Submarine	375 m[b]
Exmoor	Early Tertiary	Sub-aerial	385–392 m[c]
Summit[d]	Sub-Cretaceous (?)	Sub-aerial	475–500 m[c]

Notes:
[a] Mapped by Balchin (1952)
[b] Estimated level of the relevant historical coastline in relation to the present Ordnance Datum
[c] Height range of surface remnants
[d] Height range of hilltop remnants

weg of the Exe and has attempted to link Balchin's work with Green's (1941) studies in south and east Devon.

The discussion which followed Balchin's paper revealed the range of views on this approach to denudation chronology. It was agreed that, if peneplanes existed, then these were likely to be either sub-aerial or marine, as neither glacial nor arid conditions prevailed at this time. Jones was unconvinced by the evidence and ascribed the then current passion among British geomorphologists for recognising multiple erosion surfaces as due to the baleful influence of a visit by Henri Baulig (Balchin 1952).

An alternative view

Balchin's study is typical of the plethora of postwar studies which sought to identify flights of erosion surfaces in different parts of the British Isles. In the absence of ancient shoreline deposits, most of these studies relied on cartographic analysis.

Frey has reworked much of Balchin's data and is unconvinced by his hypotheses.[11] Accordance of summit levels at various heights is highly susceptible to structural control, so that Balchin was on much safer ground to the west of Exmoor rather than to the south, where east–west geological banding produces a step-like form. Even in west Exmoor, however, the effects of 'stream-pinching' across a ridge can lead to many 'flats' in the landscape which can be, and have been, grouped and assigned to a still stand of a falling sea level as shown in Figure 10.5

Stream orientation and pattern provide important evidence on the erosional history of the area. On Exmoor, as throughout the south-west peninsula, streams rise very close to the north coast and flow southwards, often across the low ground of structural depressions, and then cut gorges through some of the highest ground in the peninsula to reach the English Channel. The Exe/Barle system is an excellent example and it is difficult to imagine how the lower ground between Exmoor and Dartmoor could have been excavated to let in Balchin's bench-cutting seas without at the same time totally disrupting these south-flowing rivers. It was indeed these very south-flowing rivers, and their tributaries, which excavated the weaker ground and no doubt created many of the flats in landscape identified as remnants of marine planation. Marine incursions during the Tertiary there must have been, but depositional evidence from our area is frustratingly lacking.

In strong contrast to the striking parallelism of the dominant south-flowing rivers of the south-west peninsula is the threefold pattern shown by the much shorter north-flowing streams. Where the structure is dominated by block faulting the basins obviously carry their own drainage pattern, as in the case of the Porlock Stream and the Vale of Taunton Deane. Second, the predations of the East Lyn along the line of the weaker Lynton beds has gathered in by capture the northerly flow from most of high Exmoor. The

229

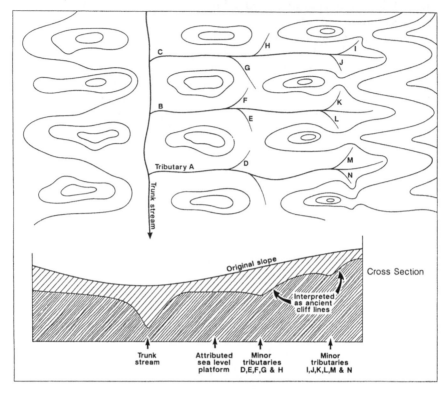

Figure 10.5 A.E. Frey's views of an alternative mechanism for forming the erosional
'flats' on the southern flanks of Exmoor
Source: A.E. Frey, personal communication, 1993

third pattern is of those streams which flow east-north-east of the Quantocks, hypothesised to have been Thames headstreams in the Pliocene, a
direction which can be seen in the vigorous course of the Avill whose
headwaters are in successful combat with the upper course of the south-
flowing Quarme.

Frey sees in the Exmoor-Quantock region evidence of a possible sequence
of stream evolution which would have four stages: (a) south-flowing rivers
as seen over much of southern England; (b) the development of major east–
west components along the many strike lines which follow this direction, so
disrupting an orderly southerly pattern; (c) the opening of the Bristol
Channel by erosion from the west and by probable continued downwarping
to allow the headward growth of streams flowing northwards into the
Channel down often steep slopes with a vigour which enabled their
headwaters and tributaries to encroach into and divert river flow from less
active neighbours (as in the cases of the East Lyn and Avill) and (d) the local

assertion of strong structural lines, as in the downfaulted basins, to produce smaller streams following the dictates of local geology.

THE QUATERNARY FACETS

Although the Quaternary period is very brief in relation to the whole timespan of West Somerset evolution (2 million out of 400+ million years since the early Devonian) it has left significant but essentially small-scale adjustments in the landscape. The two most powerful forces over the period have been the alternation of warm and cold climatic spells and the sea-level changes that accompanied it. A brief summary of the morphological effects are given in Table 10.4, which shows the three main cold stages. We look first at the Anglian and Wolstonian and then at more recent phases.

Anglian and Wolstonian cold phases

Considerable controversy surrounds the exact dating of Pleistocene events. Kellaway thought the Anglian ice might have swept across south-west England but there is no local evidence in West Somerset for this (Kellaway 1971). Other commentators place the earliest deposits as Wolstonian, although earlier local ice caps may be presumed to have existed on high ground (Edmonds and Williams 1985: 49). Most commentators place West

Table 10.4 Probable timing of Quaternary features in the West Somerset area

Period	Epoch	Approximate timing (× 1000 years)	Major landform effects
Recent	Flandrian	10 BP–Present	Alluvial deposits; erosion of head deposits in combes
Late Quaternary	Devensian	115 BP–10 BP	Widespread head accumulation under periglacial conditions
Mid-Quaternary	Wolstonian	290 BP–130 BP	Ice blocks Bristol Channel. Local snow and ice accumulation on Exmoor, Brendons and Quantocks. Combe deepening, overflow channels and stream diversions
	Anglian	540 BP–470 BP	

Source: Based largely on Edmonds and Williams (1985: 49–55), modified by K.C. Crabtree (personal communication)

231

Somerset just to the south of the main ice sheets of the Pleistocene era at their most southerly extent. For example, Mitchell places the ice margin along the north coast of Somerset and Devon during the Gipping period of the Middle Pleistocene (Mitchell 1960). The damming of the Bristol Channel by ice would have led to the establishment of a proglacial lake against the high ground of West Somerset and the present view suggests a very extensive lake, Lake Maw, filling the present Somerset levels and spilling south near Chard towards the English Channel.

Hawkins (1977) has described a number of steep-sided channels in north Somerset related to the presence of ice sheets abutting the present coasts and causing ponding and the overflow of proglacial lakes. Some are interpreted as glacial spillways, others as subglacial channels. Many of the valleys that run down from the Brendons and Quantocks seem very large for their present streams.

Two channels in the West Somerset area have attracted special attention. First, Penning-Rowsell has studied the course of the Hodder/Holford stream as it leaves its Devonian combes in the northern Quantocks and debouches onto the skirt of lower Triassic and Liassic rocks (Penning-Rowsell 1974). As Figure 10.6 shows, gravel spreads mark the earlier 'natural' course of the stream north-east to the coast near Stolford Farm. That earlier course has been abandoned in favour of the present northerly flow along a much more entrenched northerly route to the coast near Kilve. Although the original paper saw the sequence of moves from routes A through B to C as a series of river captures, Penning-Rowsell now interprets the change of course as due to ice north-east of the Quantocks pushing westwards and successively diverting the line of flow.[12] Swelling of the stream by meltwater from a proglacial lake with flow reversal may well have carved the steep-sided gorge in the Liassic escarpment.

A second gorge study is that by Scott-Simpson (1953) of the Valley of the Rocks, west of Lynmouth. He suggests that this steep-sided valley is a continuation of the valley of the East Lyn, which once extended further west but was dismembered by successive retreats of the coastline due to sea erosion. In the light of what is now known about the Wolstonian, an ice-marginal or overflow channel explanation is now more likely.

Later Quaternary phases

Whatever the precise limits of the northerly ice sheets during the Wolstonian, it is probable that West Somerset's climate was very severe, with bare ground and tundra during large periods within the Devensian (see Table 10.4). Two principal effects produced by the severe periglacial environment of the Exmoor region have been recognised by Curtis (1971): first, the removal of weathered material from the summits and the formation of buttresses and clitters; second, the development of some of this material in aprons of

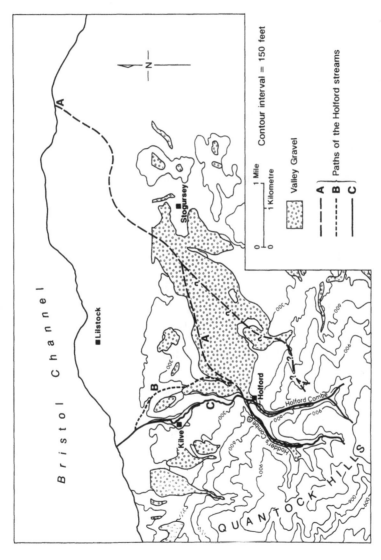

Figure 10.6 Diversion of the course of the Holford-Hodder stream, West Somerset. A marks the original stream with B an intermediate stage and C the present course

Source: Penning-Rowsell 1974: Figure 1, p. 40

solifluction and congelifluction on the flanks of hills and coasts and also the partial filling of valleys with debris.

Head forms an ill-defined mantle of unsorted stony sand and clay covering much of the lower hillside slopes to a depth of 1 to 2 m. It includes the products of weathering during both the Pleistocene and Recent periods and, because of its thinness and ubiquity, is often omitted from drift geology maps.

Most upland valleys dissected into the Devonian uplands have a youthful appearance, with steep valley-side slopes of 20–30° and colluvial accumulations on the lower backslopes. The junction with the surface at the top of the valley sides is usually in the form of a convex slope but can in places be sharp and angular. The state of dissection is rarely sufficiently mature for river terraces to have developed extensively.

Balchin has put forward the view that hog's back cliffs along the Exmoor coast (see earlier discussion, p. 225) were modified during the Quaternary when the climate was far more severe, giving abundant solifluction and the formation of head and when sea level stood lower than it does today relative to the land (Steers 1964: 605).

CURRENT POLISHING

We have argued so far that much of the landscape of West Somerset is a fossil landscape whose main lines were laid down over 300 million years ago and which has been in part exhumed by later erosion. But that does not mean the geomorphological evolution has now stopped. In this last section, we look at examples of change recorded over just the last few decades. We look at those fluvial, slope-forming, and coastal processes that are polishing up the landforms basically shaped in the preceding periods.

Fluvial processes

A recurrent theme in Chorley's writing is the inverse relationship between the magnitude and frequency of geomorphological processes.[13] The great floods at Lynmouth in August 1952 are of particular interest in that they showed the enormous power of what is normally a minor Exmoor stream.

The meteorological background to the event has been described in a number of sources (Bleasdale and Douglas 1952). On 15 August 1952, after a wet period during which the ground became saturated, there was rain nearly all day over Exmoor. The rainfall intensity increased between 8 p.m. and 1 a.m. on the 16th. The precise totals varied spatially but 23 cm was recorded at Longstone Barrow and 19 cm at Challacombe. Kidson calculated that about 300 million tons fell on the 38 square miles drained by the two branches of the Lyn (Kidson 1953; Dobbie and Wolf 1953; Green 1955). For a short period the discharge at Lynmouth was estimated at 670 m^3/sec

(23,000 cusec), or almost as much as the record figure for the Thames. Nearly 40,000 tons of boulders invaded Lynmouth, the small settlement at the mouth of the Lyn, many weighing 10 tons or more. A deltaic deposit on the right bank of the East Lyn was estimated to contain 150,000 m^3 of boulders. Higher up the river there were landslides, and fallen trees temporarily blocked the river. When these dams gave way, the greatest damage was done lower down. Slope failures during the Lynmouth storm have been mapped by Gifford (1953).

Anderson and Calver (1977, 1980, 1982) used the Lynmouth floods to assess the persistence of landscape features formed by a flood with a large recurrent interval. The landscape of the Lyn valley was studied at two dates: following the passage of the 1952 flood and twenty-two years later. The evidence suggests that channel deepening resulting from the flood will survive the mean recurrence interval of the flood event, while channel-erosional scars will not do so: downstream boulder fields deposited at the time of the flood will tend to become a general floodplain feature rather than a specific flood-deposition feature. More generally, it is suggested that the sequence and inter-arrival times between such events, rather than their mean frequency alone, are more significant in their control on landscape form and development than has hitherto been acknowledged.

Slope-forming processes

Innovative studies of soil-water movements in slopes within Bicknoller Combe on the north-west flank of the Quantock Hills have been carried out by Anderson and Burt (1977, 1978a, 1978b, 1978c). An automatically recording tensiometer system was designed and installed to record the spatial variation of soil moisture in a hillslope spur and hollow. Their results demonstrate the significant control of topography on soil moisture conditions and the resulting stream-flow response.

Maximum saturated hillslope flow was shown to coincide with the stream discharge 'throughflow peak'. From the storm event examined and the month-long recession which followed, the hillslope hollow was shown to be significant in both generating a throughflow peak in stream discharge and in maintaining the subsequent base flow.

Coastal processes

Coastal changes represent the most visible forum for observing current processes at work. West Somerset has a wide range of coastal landforms from the dramatic hog's back cliffs (ranging up to summit heights of 350 m within a kilometre of the cliff base) typical of the coast west of Porlock and Lynton, through the wide wave-cut platforms of the Blue Lias rocks near Kilve, to the low shingle ridges and backing marshes of Porlock, Minehead and Stert. The coastal morphology has been described in general terms by Steers (1964). Despite the early work by the Arbers (discussed earlier), there has been little work on cliff recession. Steers noted that, on hog's backs, true cliffs only exist near the foot of the seaward slope near sea level, where they are usually fronted by a beach of coarse boulder. Since only their bases are being cut by the sea, the upper slopes of the hog's back cliffs were probably produced by long periods of sub-aerial weathering including late Quaternary head formation.

By contrast, the evolution of mobile beaches has attracted measurement and research. Kidson (1960) and Carr (1961) have analysed the movement of shingle in some detail for the eastern area between Hinkley Point and Stert Island. Marked material was laid down on the beach but six years later the furthest marked material was found only 2.3 km east of its point of origin. At Hinkley Point, where movement was more rapid than at any other site, the marker that had travelled furthest averaged only 25 m per month. The larger shingle was found to move both at a different speed and locally even in a different direction from smaller shingle. Larger material travels relatively quickly towards the end of the beach ridge or shingle complex of which it is part. There it is held up and the more slowly travelling smaller material begins to overtake it.

EPILOGUE

The present West Somerset landscape appears to have been shaped in its major lineaments many millions of years ago. Indeed, the basic contrast between upland and lowland was imposed around 250 million years ago and, despite subsequent sedimentary burials in the Mesozoic, has been exhumed to form the main framework today. Later processes, despite their intrinsic interest, have only bevelled, infilled and sharpened features at meso- and micro-scales leaving the basic macro-geography little altered.

That huge range of time allows both great periods of inactivity and stillstand and yet enormous change in the environment in which deposition and erosional forces were working. Today's West Somerset environment is only one phase from a history which has switched from mountain range to offshore lagoons, from desert heat to near-glacial cold, from stable block to tectonic splitting. It has, at various times during its long history, looked like an Arizona desert, a Great Barrier Reef shelf, or a Spitzbergen tundra. The fact that we now have no difficulty in postulating such changes emphasises the change in geomorphological science that has gone on since the eighteenth century: our geomorphological forebears would have found such views preposterous or even blasphemous.

Only some of the evidence for this environmental change comes from within the region itself (through fossiliferous rock exposures or diagnostic morphology). Much more comes from setting West Somerset in a wider spatial context. The local Devonian sediments make little sense outside the great syncline that included a much wider area of south-west England. Evidence for periglacial activity would be less compelling did not external evidence suggest the nearness of the Wolstonian ice sheet. Our local interpretations in West Somerset attempt to fit, like a Russian doll, within ever wider spatial interpretations up to the level of global change itself.

All landform interpretations are in some sense provisional. In a modern sense, our interpretations of West Somerset are hardly 180 years old (going back to Horner in 1816). Do the Exmoor erosion surfaces really exist? Did the West Lyn once run through the Valley of the Rocks? Unless we feel that geomorphic science stops in the 1990s, then we must allow that some future generation of geomorphologists will look back in amused disdain at the sense we try to make of the landscapes around us. Today we can only speculate at future interpretive changes for this region but I would be surprised if we didn't find evidence of much greater crustal mobility, and still more landforms credited to the exhumation of desert landforms.

But even if interpretations change, then the challenge of interpreting Keble's 'landscape in its glory' will remain. Whether a boyhood spent in this geomorphological Arcady, one over which so many interpretive debates have arisen, played any part in shaping Richard Chorley's later interests, we shall never know – and perhaps he will never know.[14] I think it is a question best left to some future historian of science when, early next century, surely yet other volumes come to be added to Chorley's greatest work, his *History of the Study of Landforms*.

ACKNOWLEDGEMENTS

Before his, sadly, early death in 1991, Len Curtis (National Park Officer for Exmoor) discussed this chapter with me in its preliminary form. I am grateful to my colleagues Keith Crabtree, Allan Frey, Paul Hancock and Peter Smart at Bristol University (all of whom know the geomorphology of West Somerset much better than I do) for their criticism and advice. The incautious interpretations remain my own. I am grateful to Diana Greenhill who drew the original drafts of the diagrams and Simon Godden who skilfully converted these into text figures.

NOTES

1 The other Demonstrators at that time were Christopher Board (now at the London School of Economics), Michael Morgan (Bristol) and Tony Wrigley (All Souls, Oxford). Six of the undergraduates who suffered at our hands then have survived as contributors to this volume.

2 Exeter College, Oxford (1948–1951); St Catharine's College, Cambridge (1951–1954).

3 My last performance as a geomorphologist was as a replacement for John Kesseli

in a course on 'Introductory geomorphology' given at the University of California at Berkeley in the Summer Semester, 1962.

4　John Keble (1792–1866) was an Anglican priest, theologian and poet who shared with J.H. Newman the leading role in the origins of the Oxford Movement. Keble College, Oxford, was founded in his memory in 1869. During the war years the memorial hut, together with much of North Hill, was an out-of-bounds military area.

5　The topography of the area is covered by three Ordnance Survey Landranger 1:50,000 sheets (180, 181 and 182) and eight Pathfinder 1:25,000 sheets (SS 63/73, SS 64/74, SS 83/93, SS 84/94, ST 03/13, ST 04/14, ST 23/33 and ST 24/34).

6　The solid and drift geology of the area is covered on four 1:50,000 sheets: 278 'Minehead', 279 'Weston-super-Mare', 294 'Dulverton' and 295 'Taunton'.

7　For an outstanding review of the tectonic history of the areas including a specific chapter on south-west England, see P.L. Hancock (ed.) (1983) *The Variscan Fold Belt in the British Isles*, Bristol: Adam Hilger.

8　Much of the discussion is based on Edmonds and Williams (1985: 55–62).

9　This statement is confined to our particular region of interest. In the central Somerset basin to the east and the Bristol Channel to the north, major downwarping in the Cretaceous preserved extensive Jurassic sediments. See Owen (1976).

10　There are, of course, others outside the region, e.g. the floor of the western approaches to the Bristol Channel, the St Erth and St Agnes deposits in Cornwall, the Haldon Hill gravels and Bovey Basin deposits in south Devon.

11　I am grateful to my Bristol University colleague, A.E. Frey, for providing a full exposition of his unpublished research on this area. The section which follows is based, with permission, on his notes.

12　In a letter to the author dated 9 December 1992.

13　See, for example, R.J. Chorley, S.A. Schumm and D.E. Sugden (1984).

14　I have not pursued one obvious source of contemporary information, the shadowy 'Sniffy' Locke, who appears as a Just William character in many of RJC's more outlandish Minehead Grammar School stories. Not all are without scholarly foundation, viz. the tale of the crash of the Junkers Ju-88 on Porlock beach. This certainly took place on 27 September 1940 and is described in Hawkins (1988: 33–34).

REFERENCES

Anderson, M.G. and Burt, T.P. (1977) 'Automatic monitoring of soil moisture conditions in a hillslope spur and hollow', *Journal of Hydrology* 33: 27–36.
——(1978a) 'The role of topography in controlling throughflow generation', *Earth Surface Processes* 3: 331–344.
——(1978b) 'Time synchronized stage recorders for the monitoring of incremental discharge inputs in small streams', *Journal of Hydrology* 34: 141–159.
——(1978c) 'Experimental investigations concerning the topographic control of soil water movement on hill slopes', *Zeitschrift für Geomorphologie* NF 29: 52–63.
Anderson, M.G. and Calver, A. (1977) 'On the persistence of landscape features formed by a large flood', *Transactions of the Institute of British Geographers* NS 2: 243–254.
——(1980) 'Channel plan changes following large floods', in R. Cullingford, D. Davidson and J. Lewin (eds) *Timescales in Geomorphology*, Chichester: John Wiley.
——(1982) 'Exmoor channel patterns in relation to the flood of 1962', *Proceedings of the Ussher Society* 5: 362–367.

Arber, E.A.N. (1911) *The Coast Scenery of North Devon: Being an Account of the Geological Features of the Coast-line from Porlock in Somerset to Boscastle in North Cornwall*, London: Dent.

Arber, M.A. (1949) 'Cliff profiles of Devon and Cornwall', *Geographical Journal* 114: 191–197.

Balchin, W.G.V. (1952) 'The erosion surfaces of Exmoor and adjacent areas', *Geographical Journal* 68: 453–472, discussion 472–476.

Beckinsale, R.P. and Chorley, R.J. (1991) *The History of the Study of Landforms or the Development of Geomorphology*, vol. 3: *Historical and Regional Geomorphology 1890–1950*, London: Routledge.

Bleasdale, A. and Douglas, C.K.M. (1952) 'Storm over Exmoor on August 15th 1952', *Meteorological Magazine* 81: 353–367.

Brunsden, D., Kidson, C., Orme, A.R. and Waters, R.S. (1964) 'Denudation chronology of parts of south-western England', *Field Studies* 2: 115–132.

Chorley, R.J., Beckinsale, R.P. and Dunn, A.J. (1973) *The History of the Study of Landforms or the Development of Geomorphology*, vol. 2: *The Life and Work of William Morris Davis*, London: Methuen.

Chorley, R.J., Dunn, A.J. and Beckinsale, R.P. (1964) *The History of the Study of Landforms or the Development of Geomorphology*, vol. 1: *Geomorphology before Davis*, London: Methuen.

Chorley, R.J., Schumm, S.A. and Sugden, D.E. (1984) *Geomorphology*, London: Methuen.

Curtis, L.F. (1971) *Soils of Exmoor Forest*, Soil Survey Special Publication 5, Harpenden: Soil Survey of England and Wales.

De la Beche, H.T. (1839) *Report on the Geology of Cornwall, Devon and West Somerset*, Memoirs of the Geological Survey of Great Britain, London: Longman.

Dineley, D.L. (1977) 'The Quantock and Brendon Hills, West Somerset and North Devon', in R.J.G. Savage (ed.) *Geological Excursions in the Bristol District*, Bristol: University of Bristol.

Dobbie, C.H. and Wolf, P.O. (1953) 'The Lynmouth floods of August 1952', *Proceedings of the Institution of Civil Engineers* 3: 522–588.

Edmonds, E.A., McKeown, M.C. and Williams, M. (1969, 1975), *British Regional Geology: South-west England*, London: HMSO 3rd and 4th edns.

Edmonds, E.A. and Williams, B.J. (1985) *Geology of the Country around Taunton and the Quantock Hills (Memoir for the 1:50,000 Geological Sheet 295, N.S.)*, London: HMSO.

Etheridge, R. (1867) 'On the physical structure of West Somerset and north Devon and on the palaeontological value of the Devonian fossils', *Quarterly Journal of the Geological Society of London* 23: 568–698.

Gifford, J. (1953) 'Landslides on Exmoor caused by the storm of 15th August 1952', *Geography* 38: 9–17.

Green, G.W. (1955) 'North Exmoor floods, August 1952', *Bulletin of the Geological Survey of Great Britain* 7: 68–84.

Green, J.F.N. (1941) 'The high platforms of East Devon', *Proceedings of the Geologists' Association* 52: 36–52.

Hancock, P.L. (ed.) (1983) *The Variscan Fold Belt in the British Isles*, Bristol: Adam Hilger.

Hawkins, A.B. (1977) 'The Quaternary of the North Somerset area', in R.J.G. Savage (ed.) *Geological Excursions in the Bristol District*, Bristol: University of Bristol.

Hawkins, M. (1988) *Somerset at War, 1939–1945*, Wimborne: Dovecote Press.

Horner, L. (1816) 'Sketch of the geology of the southwestern part of Somersetshire', *Transactions of the Geological Society of London* 3: 338–384.

Jones, O.T. (1952) 'Discussion' in W.G.V. Balchin, 'The erosion surfaces of Exmoor and adjacent areas', *Geographical Journal* 68: 473–474.

Kellaway, G.A. (1971) 'Glaciation and the stones of Stonehenge', *Nature* 232: 30–35.

Kidson, C. (1953) 'The Exmoor storm and the Lynmouth floods', *Geography* 38: 1–9.

——(1960) 'The shingle complexes of Bridgwater Bay', *Transactions of the Institute of British Geographers* 28: 75–87.

——(1962) 'The denudation chronology of the River Exe', *Transactions of the Institute of British Geographers* 31: 43–66.

Kidson, C. and Carr, A.P. (1961) 'Beach drift experiments at Bridgwater Bay, Somerset', *Proceedings of the Bristol Naturalists' Society* 30: 163–180.

Mitchell, G.F. (1960) 'The Pleistocene history of the Irish Sea', *Advancement of Science* 17: 313–325.

Norton, W.H. (1917) 'A classification of breccias', *Journal of Geology* 15: 160–194.

Owen, T.R. (1976) 'Post-Hercynian movements in southwest Britain and their significance in the evolution of Cornubian and Welsh "oldlands"', *Proceedings of the Ussher Society* 3: 361–366.

Penning-Rowsell, E.C. (1974) 'Historical changes in river patterns near Holford, Somerset', *Proceedings of the Somerset Archaeological and Natural History Society* 118: 39–43.

Scott-Simpson, S. (1953) 'The development of the Lyn drainage system and its relation to the origin of the coast between Combe Martin and Porlock', *Proceedings of the Geologists' Association* 64: 14–23.

Sedgwick, A. and Murchison, R.I. (1837) 'On the physical structure of Devonshire and on the subdivision and geological relations of its older stratified deposits', *Proceedings of the Geological Society of London* 2: 556–563.

Steers, J.A. (1964) *The Coastline of England and Wales*, 2nd edition, Cambridge: Cambridge University Press.

Thomas, A.N. (1940) 'The Triassic rocks of south-west Somerset', *Proceedings of the Geologists' Association* 51: 1–43.

Ussher, W.A.E. (1879) 'On the geology of parts of Devon and West Somerset, north of South Moulton and Dulverton', *Proceedings of the Somerset Archaeological and Natural History Society* 25: 1–20.

Webby, B.D. (1962) 'The geology of the Brendon Hills', unpublished Ph.D. thesis, University of Bristol.

Whitaker, A. (1973) 'The central Somerset basin', *Proceedings of the Ussher Society* 2: 585–592.

Part II

ON THEORY AND HISTORY

11

JAMES KEILL (1708) AND THE MORPHOMETRY OF THE MICROCOSM

Geometric progression laws in arterial trees

Michael J. Woldenberg

INTRODUCTION

The purpose of this chapter is to review some of the contributions of James Keill, MD (1673–1719), who introduced geometric progression scaling laws in his morphometric works on the anatomy and physiology of arterial trees (Keill 1708, 1717, 1718, 1738). These scaling laws, based on a *centrifugal, generational* ordering system, are still used in biology (Weibel 1963). These laws assume, unrealistically, that there is a perfect symmetry of branch lengths and diameters at each generation.

Generational ordering anticipates the *centripetally ordered* scaling laws of Horton (1932, 1945) which helped introduce quantification to fluvial geomorphology. Horton's system was modified by Strahler (1952) and applied by his students (Strahler 1964).

Strahler-ordered network analysis has supplanted generational ordering in biology (Jarvis and Woldenberg 1984; Woldenberg 1986) because it is more successful in coping with the observed asymmetrical branching in trees. Perhaps this shortcoming of generational ordering accounts for the fact that it was never used in geomorphology. A more likely reason is the dominance of the Davisian approach in academic geography (Davis 1899; Chorley, Dunn and Beckinsale 1964).

Another centripetal ordering system, invented by Horsfield, produces geometric series scaling laws and is sometimes used by physiologists (Horsfield and Cumming 1968; Horsfield and Woldenberg 1989). The numerical stream ordering systems of Jackson (1834; Goudie 1978) and Gravelius (1914), which can be shown to be identical, cannot lead to scaling laws.[1]

The contributions of James Keill to the morphometry and physiology of tree-like systems have largely been overlooked. A few articles about Keill have been written, chiefly by biographers and historians of science. The historians of science have discussed Keill's work in the context of the application of the chemical and mechanical 'philosophies' to human physiology and the practice of medicine. These applications are known as iatrochemistry (Debus 1977) and iatromechanics (Brown 1968), respectively. One can also speak of iatromathematics; Keill's study of tree morphometry is an excellent example. While many iatrochemical, iatromechanical and iatromathematical ideas have been discussed by historians of science, these scholars do not comment on Keill's introduction of geometric series scaling laws. Aside from a few paragraphs which I wrote (Jarvis and Woldenberg 1984: xi, 9–10, 101) nothing has appeared in the geomorphological literature. The topic of ordering and geometric series scaling laws takes on new significance for geomorphologists because the fractal analysis (Mandelbrot 1977) of fluvial trees has recycled and made use of Strahler network analysis (Tarboton *et al.* 1992).

James Keill's work represents the first major quantitative morphology of a tree. Thus, in the first section of this chapter I will review his contributions that are of interest to geomorphologists and biologists. I will briefly mention his influence on Stephen Hales (1677–1761) and Thomas Young (1773–1829), and I will show how Young extended Keill's analysis. In the second section I argue that Keill was influenced by the ideas of Paracelsus (1493–1541) and his followers, by William Harvey (1578–1657) and by the iatromechanists Lorenzo Bellini (1643–1704) and Archibald Pitcairne (1652–1713). The mathematical, physical and chemical ideas of Isaac Newton (1642–1729) and of John Keill (1671–1721), James's brother, were important. In the last section I attempt to show that geomorphologists such as James Hutton (1726–1797) and William Morris Davis (1850–1934) have been influenced by some of the ideas of Paracelsian science.

KEILL'S SCALING LAWS

In this section I will present a summary of the morphometry found in chapter 1 of Keill's *An Account of Animal Secretion, the Quantity of Blood in the Humane Body, and Muscular Motion* (1708). I will refer to only a few other works in passing because my purpose is to give the reader a continuous exposition of his mathematical analysis. The geomorphologist will see that Keill's analysis presages Horton's geometric progression laws for form (Horton 1932, 1945) and flow (Leopold and Miller 1956), although Keill did not provide 'Horton' diagrams of these relations. The power functions of hydraulic geometry (Leopold and Maddock 1953) can be derived from these laws (Leopold and Miller 1956; Woldenberg 1972). The same derivation is used to calculate the fractal dimension(s) of a tree (Mandelbrot 1977: 42).

As the title suggests, Keill (1708) was making an attempt to understand glandular secretion. As part of the solution to the problem it was necessary to predict the velocity of the blood in various parts of the arterial tree. (I will discuss other aspects of glandular secretion in the section following this one.) Thus the approach taken in this section is purely morphometric. I begin with Keill's use of a generational, centrifugal ordering scheme.

Keill assumed a symmetrically bifurcating tree. An order is assigned to a branch by counting the number of 'divisions' to reach the unbranched trunk of the tree. The trunk has order '0'. Thus the number of branches, N_d, doubles beyond each centrifugally increasing divisional order, d (Keill 1708: 149).

$$N_d = (R_b)^d \text{ or}$$
$$N_g = (2)^{g-1} \tag{1}$$

where R_b is the bifurcation ratio, N_g is the number of branches at generation g and $N_g = N_d$. I have taken the liberty of changing Keill's mathematical notation to conform to that used in the Horton-Strahler type of network morphometry. Unlike the case with divisional orders, the first generational order is assigned to the unbranched base of the tree. The divisional order, d, is equal to $g - 1$. Generations rather than divisions are used in this chapter, again so that the equations conform to traditional stream morphometry.

Cross-sectional area was also assumed to increase by a geometric progression at each division. The value for the base of this geometric series was obtained by direct measurement of wax casts of arteries. (The measurements were reported not as the true area but as a squared diameter. The units were called 'inches' but actually denote square inches.) Squared diameters were summed separately for parent and daughter branches ('trunks' and 'branches' respectively). In every case two daughter branches were recorded for every parent branch and often a daughter branch was included in the parent category as Keill measured successively smaller branches. The grand sum of daughter squared diameters was divided by the sum of parent squared diameters. This yielded an area ratio which for most of the arterial system was approximately 1.252547, but which varied depending on the part of the arterial system that was measured. (In this chapter I have carried out calculations without rounding off to two significant digits. Rounding of values would make it more difficult to follow Keill's argument.) Measured values of the aorta yielded a diameter of 0.73 inches; when squared this becomes 0.5329 square inches (ignoring pi, and the fact that the radius is half the diameter). The law describing the *increase* of squared diameters with generation (the squared diameter law) is as follows:

$$\sum D_g^2 = D_1^2 (R_A)^{g-1}$$
$$\sum D_g^2 = 0.5329 (1.252547)^{g-1} \tag{2}$$

where $\sum D_g^2$ is the sum of all the squared diameters at generation g, R_A is the average *ratio* for the sum of the squared diameters of the two daughters to that of the parent, and D_1 is diameter of the aorta. (R_A is also the ratio of areas, since the constants in numerator and denominator cancel. I will continue to use R with a subscript to denote a ratio.)

Although Keill recognised that two daughter branches did not often have the same diameter, for mathematical modelling purposes symmetrical values were assumed (Keill 1708: 145, 150). Thus the law describing the *decrease* in the squared diameter from parent to a single daughter is given by substituting the ratio $R_A/2 = 0.626273$ in equation (2).

$$D_g^2 = D_1^2(R_A/2)^{g-1}$$

$$2.5 \times 10^{-9} = 0.5329 (0.626273)^{g-1} \tag{3}$$

Keill estimates that capillaries (assumed to have a diameter equivalent to a terminal branch of the arterial tree) have a diameter of 5×10^{-5} inches (1.3 microns). (This figure is too small by a factor of 5. The terminal arteriole has a diameter of about 13 microns; below this value there is some anastomosis (Horsfield and Woldenberg 1989).) Thus, in equation (3) at the last generation, $D_g^2 = 2.5 \times 10^{-9}$ in^2. (Again, Keill ignores *pi* and uses D instead of r.) When equation (3) is solved for the number of generations, the answer is 42. Somehow Keill errs by a factor of 10 and reports: 'the number of branchings (is) above 400' (Keill 1708: 150). In subsequent editions he corrects his error (Keill 1717, 1718, 1738).

Based on the assumption that flows will divide equally at each bifurcation, it is possible to estimate the flows in the branches at each generation, if the flow in the aorta is known. As a first estimate, Keill states that the heart beats 80 times per minute, and each contraction throws into the aorta *at least* one ounce of blood. An ounce of water has a volume of 1.728 in^3. Keill estimates that an ounce of blood has a volume of 1.659 in^3. (He probably generated his estimate of 1.04159 for the specific gravity of blood himself. This is close to the accepted value of 1.05–1.06.) Hence, Keill estimates that the aorta carries a flow of 132.72 in^3/min or 2.17 litres per minute (l/min). (The output of the heart is actually about 5 l/min.) Keill does not concern himself with calculating the flows in branches of the arterial tree other than the aorta; rather his goal is to estimate the *velocity in the terminal branches*.

To estimate this terminal velocity, Keill began with his estimate of the flow in the aorta. He cites (1708: 138) the relation given by his brother John in John's *Lectiones Jo. Keil* (1702: 114; 1720). In modern form

$$Q = AV \text{ or}$$
$$\text{Flow} = \text{Area} \times \text{Velocity} \tag{4}$$

White (1968) cites an earlier work (Keill 1698). Others may have published this relation before John Keill.

Thus the velocity in the aorta is given by dividing the flow by the cross-sectional area. Keill uses a diameter of 0.73 in. to estimate the cross-sectional area at 0.4187 in^2. This yields an average velocity of 26.42 ft/min. However,

Keill says that if the heart throws out two ounces at every systole, then the average velocity is doubled (Keill 1708: 139–140).

Keill next develops a formula for estimating the velocity at different generations. His argument is verbose; an equivalent treatment is given below.

$$A_1 V_1 = A_2 V_2 + A_3 V_3 \tag{5}$$

where A is area, V is velocity and the subscripts 1, 2 and 3 identify the parent and daughter branches. In a symmetrically branching tree, $Q_2 = Q_3, A_2 = A_3$, $V_2 = V_3$. Therefore

$$A_1 V_1 = 2A_2 V_2 \tag{6}$$

$$V_2/V_1 = R_V = A_1/2A_2 \tag{7}$$

$$V_g = V_1 (R_V)^{g-1} \tag{8}$$

Note that R_V is the reciprocal of R_A.

Based on his erroneous estimate of over 400 (instead of 42) generations, Keill makes another error in his estimate of the terminal velocity: the velocity of the blood in the aorta is 10^{38} times greater than in the terminal branches. An improved estimate can be made using the 'correct' number of orders:

$$V_{42} = 26.42 (0.7983732)^{41} = 0.0025845 \text{ ft/min}$$

$$= 0.0013 \text{ cm/sec} \tag{9}$$

Had he not erred in his arithmetic, Keill would have estimated that the velocity of the blood in the aorta is about $26.42/0.0025464 = 10,000$ times faster than in the terminal branches. This estimate is about 600 times greater than the velocity ratio I generate using Keill's estimate of the diameters of the aorta (0.73 in) and a terminal branch (0.00005 in) together with formulas in Horsfield and Woldenberg (1989: 247–249). This analysis used Horsfield's centripetal method of ordering.

In 1717, Keill published an augmented version of his 1708 monograph. One new idea of interest is his attempt to estimate the total time required for blood to travel from the aorta to the terminal branches. Unfortunately, there was a printer's error in the 1717 monograph, and there were several pages missing between pages 142 and 143. The error was corrected in the 1718 edition published in Latin. The corrected version was also republished in English in 1738, along with some other essays. In the 1717 and subsequent versions, Keill takes advantage of some new measurements.

For the mesenteric artery he finds that $R_A = 1.2687$; this yields 36 divisions or 37 generations. He calculates that the velocity of the terminal branches is 1/5261 of the main branch of the artery.

Keill then attempts to calculate the total time it takes for blood to traverse the whole mesenteric artery. To do this he applies the formula for the sum of a geometric series.

$$\sum_{g=1}^{n} T = a \, (R_T^g - 1)/(R_T - 1) \tag{10}$$

where a is the transit time in the aorta, R_T is the ratio of time and n is the number of generations in the series. Since time = length/velocity,

$$R_T = R_L/R_V \tag{11}$$

where R_L is the ratio of lengths from one generation to the next and R_V is the ratio of velocities. But $R_V = 1/R_A$, so

$$R_T = R_L \, (R_A) \tag{12}$$

At this point Keill makes some important errors. He explicitly assumes that $R_L = 1$. Keill did not calculate R_L; he certainly could have made the measurements and calculations. Thus, for Keill, $R_T = R_A$ (1). Apparently, he meant to note that $R_L < 1$, and $R_A > 1$ and that their product, R_T, is very slightly larger than one.

Substituting into equation (10) requires an estimate for the time it takes the blood to traverse the aorta to the mesenteric artery. This is given by the length/velocity. In this discussion Keill (1717, 1718, 1738) estimates systole to take up only one-third the time of a heart beat, and hence the average velocity of 26 ft/min (5.2 in/sec) must be multiplied by 3 to give the average peak velocity. This is an error since the differences between the peak and average velocities diminish in the smaller vessels, as shown in 1733 by Hales (Hales 1733, 1964; Young 1809). The total time required to traverse the tree should have been based on an average velocity in the aorta. Instead, Keill uses a peak velocity of 15.6 in/sec to estimate the transit time in the aorta (with a length of 10 in) as (10 in)/15.6 in $\sec^{-1}$ = 0.64 sec. Using Keill's results, the total time should be given by substituting into eq (10):

$$\sum_{g=1}^{37} (T) = 0.64 \, (1.2687^{37} - 1)/(1.2687 - 1) = 0.64 \, (24828)$$
$$= 15,890 \text{ sec} = 4 \text{ hr } 25 \text{ min} \tag{13}$$

This is an unrealistically long time, even using the peak velocity.

Perhaps this bizarre result may have forced Keill into several errors. He states that the time required to traverse each generation will be constant. This can only be true if the *product* $(R_A)(R_L)$ in eq (11) is very close to 1, not, as Keill would have it, if $R_L = 1$. Keill is very confused and calculates the average transit time by dividing 24825/37 = 670 (actually, 671). Then he multiplies 0.64 seconds by 670 which equals about 7 minutes, a result more to his liking.

According to Young (1809), the length ratio is 1/1.196104. This is almost the reciprocal of $R_A = 1.2687$. In the limit, as $(R_A)\,(R_L)$ approaches 1, ΣT approaches $(a)\,(g)$, which in Keill's case is $0.64(37) = 23.68$ sec. assuming the peak velocity. Had the average aortic velocity been used, the time would be 71.04 seconds.

While Keill made simple mathematical errors, he should still be credited with the first attempted use of the sum of a geometric series to estimate a property of a bifurcating network (in this case total time for blood to flow from the heart to the terminal branches of the mesenteric artery). Although his work contained errors, he certainly pointed the way for Young (1809) and those who directly or indirectly followed both of these men.

It should be noted that Keill made other estimates of R_A based on measurements from different parts of the arterial system. His estimates of the diameter of the aorta (0.73 in) and of a capillary (0.00005 in), which he took to be a terminal branch, remained the same, as did his assumptions of perfectly symmetrical (in terms of length and diameter) dichotomous branching. All of the geometric series equations depend on these values, of which only R_A or its derivatives are variable. The number of generations also varies only with R_A. Table 11.1 gives a list of the values of R_A and R_L found in Keill's works and in Thomas Young's 1809 paper.

Before moving on to a discussion of James Keill's antecedents, I should mention the influence of his work on two very well-known scientists, Stephen Hales and Thomas Young (both had been students at Cambridge University). (Young was a physician, contributor to the theory of light, the physiology of the eye, the physics of elasticity, and the translation of the Rosetta stone.) Stephen Hales, one of the greatest animal and plant physiologists, was a contemporary of Keill's and specifically quotes Keill's *Tentamina . . .* (Keill 1718; Hales 1733, 1964). In contrast to Keill's approach, Hales's work is based on many very skilful experiments. Mathematical deductions, though important, take a secondary role. Hales did not use Keill's scaling laws to model the morphometry or physiology of arteries or veins. Using experiments and more accurate measurements, Hales follows up

Table 11.1 Values for R_A and R_L from Keill and Young

Location	R_A	R_L	Reference
Aorta and first branches	1.0454		Keill (1708: 144, 147–148)
Beyond the first branches of the aorta	1.25247		Keill (1708: 150)
Same	1.23887		Keill (1717: 72, 75)
Mesenteric artery	1.2687		Keill (1718: 92)
Mesenteric artery		1.0	Keill (1718: 94)
Arteries	1.2586	0.836048	Young (1809: 4–5)

on many issues addressed by Keill, including the velocity of the blood in the aorta, the resistance of the blood in various parts of the tree, and the role of capillary attraction in glandular secretion and arterial and venous blood pressure. (Instead of trying to measure or predict blood pressure, Keill (1738) attempted to calculate the force of the heart.)

Thomas Young recapitulated and extended the geometric series laws of the arterial tree. I will discuss some of his results.

Keill's motivation for developing his scaling laws was to determine the velocity of the blood in order to explain glandular secretion. It was Young's purpose to calculate the resistance to flow in the arterial system, first considering the arteries as inelastic tubes, and then investigating the effects of the pulse and the elasticity of the vessel (Young 1809). These notions are found explicitly, but non-quantitatively, in Hales's *Haemastaticks* (Hales 1733, 1964).

Young delivered two lectures related to hydraulics and physiology to the Royal Society (Young 1808, 1809). The Croonian Lecture (Young 1809) is much better known to physiologists than is the work of Keill, although it is clearly based on Keill's work. While Young gives Keill credit for making some measurements, the reader is not told that the topological and mathematical methods and several relations that Young uses are based on the work of Keill (Young 1809: 4).

Aside from some slight differences in the measurements taken and numbers derived, Young uses the same approach as Keill does but goes further. In his first extension of Keill's analysis he calculates a length ratio (R_L) and the total length of the arterial tree. The calculation of length and area ratios allows Young to determine the volume ratio (R_{Vol}). The volume of the tree can be calculated from the volume of the aorta and R_{Vol} using the formula for the sum of a geometric series.

Young develops an equation for resistance in order to estimate the pressure loss associated with the flow of blood in the arterial tree. He calculates generational scaling laws to deduce ratios for the terms in his resistance equation. From these ratios he creates a 'resistance ratio' and uses this to calculate the relative resistance for proximal and distal parts of the arterial tree.

In his paper on hydraulic investigations, Young (1808) attempts to devise formulas for velocity and resistance in turbulent and laminar flow in rivers and pipes. He starts by reviewing the work of DuBuat and Gerstner, who developed unsatisfactory formulas for velocity and friction. Young proposes a two-component formula for determining the pressure head loss (P) due to friction (1808: 166, 172).

$$P = b(V^2)L/D + 2c(V)(L/D) \qquad (14)$$

The first part is for large pipes (turbulent flow), while the second part is for small pipes (laminar flow). The coefficients b and c are highly complex and depend on D. The expression becomes much simplified for 'pipes' the size of the arteries (Young 1809: 6). Young's expression for this loss of pressure due to friction is given by

$$P = 2.1126 \times 10^{-7} (L)(V)(D)^{-3.5} \tag{15}$$

The constant is dependent on the units used (inches). In contrast, an analogous expression for P using the Hagen-Poiseuille equation is given by

$$P = (C)(L)(V)(D)^{-2} \tag{16}$$

where C is a constant which includes the dynamic viscosity.

Using equation (15) as a model, the dimensionless pressure loss ratio is given by Young as follows:

$$R_P = (R_L)(R_V)(R_D)^{-3.5} \tag{17}$$

Since Young finds that $R_L = 1/1.1961014 = 0.8360495$, $R_V = 0.7945185$ and $R_D = 0.7932919$, then the *loss* in pressure from generation (g) to $(g + 1)$ *increases* by a ratio R_P of 1.49425 (Young 1809: 6).

Given equation (15) and assuming the heart is pumping water, not blood, the pressure drop in the aorta (with $V = 8.5$ in/sec, $D = 0.75$ in, $L = 9$ in), measured in inches of water pressure, can be evaluated as 4.42348×10^{-5}. Thus, the pressure drop at a given generation (g) is given by:

$$P_g = 4.42348 \times 10^{-5} (1.49425)^{g-1} \tag{18}$$

The total pressure drop due to resistance, including the aorta to generation g is:

$$\sum_{g=1}^{n} P = (\text{Aortic drop}) (R_P^{g-1} - 1)/(R_P - 1) \tag{19}$$

or

$$\sum_{g=1}^{n} P = 4.42348 \times 10^{-5} (1.49425^{g-1} - 1)/(0.49425) \tag{20}$$

Young estimates that the resistance of the arteries consumes a pressure equivalent to a column of 15 inches of water. The capillary veins (*sic*) consume another 5 inches. Using equation (20), Young points out that most of the resistance is concentrated in the last five generations of the 30-generational system. Young has given a deductive, quantitative justification for a conclusion reached by Hales on the basis of experiment. A similar result can be achieved using the Poiseuille relation (equation 16) and ratios based on a Strahler-ordered tree.

In this paper, Young is the first to calculate correctly the sum of a geometric series to estimate a property of a tree. While Keill tried to use the same technique in his estimate of transit time for the blood, he failed because of a misunderstanding of the meaning of the terms in the formula. However, he is to be credited with realising the potential application for the sum of the series in the analysis of tree properties, and for making the first attempt to use this mathematical tool in this context. It was not until Horton's paper in 1945 that the sum of a geometric series was used to describe a property of a river network (the sum of all stream lengths in the network).

PHYSIOLOGY: PARACELSIAN AND NEWTONIAN THEMES

From the Greeks to Paracelsian science

Readers who are geomorphologists deserve some explanation as to why an historical discussion of work primarily related to human physiology has any relevance to our field. The answer lies in the common history all the sciences share in the search for models, analogies or metaphors to give order to our perceptions of the material world (Chorley and Haggett 1967; Haggett and Chorley 1969). There is a long history of the attempt to conceive a cosmos out of chaos. At least since the time of Plato (428–348/7 BC) (Adams 1954: 60) natural philosophers and scientists from western, Arabian and oriental traditions have investigated the relations of the macrocosm to the microcosm. In their search for explanations relating to phenomena on the Earth, and within the solar system and beyond, on the one hand, and also within the human (or animal) body on the other, they proposed that human anatomy and physiology (and, by extension, that of animals) could serve as a miniature model for the cosmos, and vice versa.

I will mention some concepts which have been used at the macro- and microcosmic scales (Ellenberger 1972, 1973). The first is the theme of the cycle, including energy cycles, exemplified by the diurnal and annual seasonal cycles; material cycles, exemplified by the hydrologic or life cycles (Tuan 1968); and cycles of geological time (Gould 1987). Implied in biological and other material cycles is the idea of growth, decline and renewal. The concepts of steady state, autoregulation and dynamic equilibrium seem to have been introduced only after classical thermodynamics and Le Chatelier's negative feedback principle were invented in the nineteenth century (Gilbert 1877; Bertalanffy 1950; Strahler 1950; Hack 1960; Chorley 1962).

The second theme is the application to the human body of the physical and chemical laws derived from an analysis of the inanimate world. The purpose of this was to improve the understanding of physiology and the practice of medicine. At the beginning of the eighteenth century this included applications from mechanics, hydraulics, mathematics (including a mathematical and physical understanding of harmony) and alchemy.

The reader will recognise that these themes and some of the applications are very old, going back to the ancient Greeks, including Plato, Aristotle and

Pythagoras. All of these ideas were given a renewed emphasis by Paracelsus (1493–1541), an extraordinary alchemist and physician, who advocated an experimental and observational approach to studying natural processes and who attempted to apply this knowledge to find 'a method of cure for man's bodily ills' (Debus 1965: 23; 1977). Though he was personally abrasive, his ideas and his cures gained him an important following among alchemists and physicians.

Among the followers of Paracelsian alchemy and medicine in England, the most prominent was Robert Fludd (1574–1637). Fludd was a physician and colleague of William Harvey (1578–1657), the discoverer of the circulation of the blood, published in 1628 (Harvey 1949).

Harvey's wonderful discovery was based first on qualitative and then quantitative estimates of the flows in the major vessels from and to the heart (Kilgour 1954). In the middle of this discussion, Harvey includes the Aristotelian and Paracelsian notions of his colleague Fludd (Debus 1972: 75) to support his anatomical observations (Harvey 1949: 70).

> This motion [of the blood] may be called circular in the way that Aristotle says air and rain follow the circular motion of the stars. The moist earth warmed by the sun gives off vapors, which, rising, are condensed to fall again moistening the earth. By this means things grow. So also tempests and meteors originate by a circular approach and recession of the sun. . . .
>
> So the heart is the center of life, the sun of the Microcosm, as the sun itself might be called the heart of the world. The blood is moved, invigorated, and kept from decaying by the power and pulse of the heart. It is that intimate shrine whose function is the nourishing and warming of the whole body, the basis and source of all life.

The discovery of the circulation of the blood revolutionised physiology and made a great contribution to the methodology of science. The *quantitative elaboration* of the circulation by James Keill predates fluvial network morphometry (Horton 1932, 1945) and even hydraulic geometry (Leopold and Maddock 1953; Leopold and Miller 1956; Woldenberg 1972) by over 200 years.

Iatromechanism and Newtonian ideas

In the next few paragraphs, I will outline how and why the mathematical description of the geometry and flow of the arterial tree by James Keill came to be. The story involves several Scottish scientists who were part of Isaac Newton's circle.[2]

Archibald Pitcairne (1652–1713) was born in Edinburgh, and earned a medical degree from Reims in 1680. Pitcairne was a close friend of David Gregory (1661–1708), professor of mathematics in Edinburgh. They had studied at the university together and both had interests in mathematics and in medicine, especially iatromechanics. Gregory corresponded often with Isaac Newton, eventually meeting him in 1691. So great was Newton's respect for him that shortly afterwards he had Gregory appointed Savilian Professor of Astronomy at Oxford (Gjertsen 1986).

During the late 1680s Pitcairne and Gregory had been reading the *Principia* of Newton, as well as works by Borelli, the father of iatromechanism, and Bellini, who described the microstructure of the kidney. Bellini had argued that fevers were caused by a fault in the motion, quantity or quality of the blood (Brown 1968: 207). Bellini stated that an increased flow and velocity produces fevers. The quality of the blood is affected by the secretions of the glands. This implied that problems of disease could be understood, given a knowledge of how glandular secretion was controlled by hydraulic and chemical principles.

In 1692 Pitcairne called on Newton, who was the Lucasian Professor of Mathematics in Cambridge. Newton gave Pitcairne his unfinished manuscript, 'De natura acidorum' (Newton 1961), to which Pitcairne appended some notes of their conversation.

In this important paper Newton extended the idea of an attractive force operating in the macrocosm to one that operates at the atomic and molecular (microcosmic) scale. Newton proposed that atoms and/or molecules are caused to join by a short-range but strong attractive force, analogous to, but different from, gravity. The interatomic force was inferred from the dissolving properties of acids; gravity had been inferred from the behaviour of planets and falling bodies. In Pitcairne's notes of his conversation with Newton, Newton envisages that the atoms and molecules would form a kind of hierarchy of matter, moving from the simple to the complex.

Just after visiting Newton, Pitcairne left for Leiden to accept a medical professorship. He remained only a year and a half, from 1692 to 1693. In one of his lectures, 'On the circulation of the blood through the minutest vessels of the body', he states that although the circulation of the blood was the key to life, the production and distribution of glandular secretions are of equal importance. Furthermore, most diseases are due to disorders causing too much or too little production of these secretions (Guerrini 1987: 75). The impact on health of hydraulics and the mechanisms of glandular secretion, suggested by Bellini, had become a central problem for Pitcairne (Brown 1968).

Having seen spectacular models of the tree networks in tissues and organs injected with wax and mercury in Leiden and Amsterdam (Cole 1926), it became apparent to Pitcairne that the animal body was an infinitely complicated web of vessels and fluids flowing in them (Brown 1968: 222). It was clear to him that a mathematical analysis of the hydraulics of the circulation in these vessels would lead to important advances in understanding the causes and cures of diseases.

In the meantime, John Keill (1671–1721) and his brother James (1673–1719) had begun their studies in Edinburgh in 1688. James studied philosophy and John (and perhaps James) were students of mathematics under David Gregory. Gregory introduced them to Newton's *Principia*. It is possible that the Keills knew Pitcairne, who taught medicine in Edinburgh, and who lived with Gregory during part of this period. After Gregory had gone to Oxford as Savilian Professor of Astronomy, John Keill followed him in 1694 to lecture on Newtonian natural philosophy. From these lectures he produced a classic text (Keill 1702). Eventually, he too was appointed the Savilian Professor of Astronomy in 1710. John Keill is remembered for his rather infamous role in attacking Leibniz in the dispute with Newton over the invention of the calculus (Westfall 1980; Gjertsen 1986; Brown 1987).

James Keill left Edinburgh for Paris, where he studied chemistry and anatomy; in this he seems to have followed in the footsteps of Pitcairne, who discovered an interest in medicine in Paris after starting law in Edinburgh. Keill matriculated in Leiden in 1696 but did not receive a degree. He returned to England, where he served as an unofficial anatomy instructor at Cambridge and Oxford. He wrote a popular anatomy textbook in 1698. In 1699 he purchased an MD degree from Aberdeen, and in 1705 received an honorary MD from Cambridge. He practised medicine in Northampton, where he died (Valadez and O'Malley 1971).

John made two important contributions which were used by James in his analysis of the circulation. First of all, he wrote a paper elaborating Newton's idea of short-range attraction between molecules, proposed in 'De natura acidorum' (John Keill 1708). Second, James quotes his brother on the relationship of flow or discharge to area and velocity (Keill 1702; White 1968).

James Keill had kept abreast of the ideas of Archibald Pitcairne, as well as other iatromechanists, and he was determined to create a mathematical theory which would make concrete Pitcairne's hydraulic model, and show how it related to the model of glandular secretion. As Bellini and Pitcairne had suggested, if these processes were better understood, then diseases caused by excesses or deficiencies in these secretions could be addressed.

In his 'De natura acidorum', Newton also had proposed that a hierarchy of complex molecules could be produced by varying combinations of a few simple molecules and atoms (Newton 1961; Guerrini 1985: 251). To understand the glandular secretions it was necessary to propose a mechanism which allowed the complex molecules in the secretions to form. Once formed, the secretions had to be taken out of the blood by the glands.

How could these secretions be formed in the blood? Following the suggestions of Bellini and Pitcairne, James Keill believed that the component atoms and molecules in the blood are in constant motion, bouncing off each other and off the arterial walls. The attractive force between the molecules falls off with distance raised to the third, fourth or higher power, according to his brother (Jo. Keill 1708). Thus the force becomes very weak as the particles separate. High velocities increase the energy of the collisions so that the rebounding force usually overcomes the attracting force. Only the particles with the greatest mutual attraction can overcome rebounding forces and combine to form complex molecules. These molecules can than enter the arteries of a gland and be sieved out of the bloodstream.

It was understood that the blood velocity decreases in the smaller arteries, and this implies a greater distance (and a greater number of bifurcations) from the heart and the aorta. Since different secretions can only be formed at different velocities, a gland would be located where the velocity was appropriate for the formation of its secretion. (If the complex molecule formed in some other part of the arterial system at the correct velocity, it would be eventually broken up in the aorta, or by the mechanical force of the lungs.)

Keill's geometrical progression laws

A crucial problem, then, was to find a way to calculate the blood velocity in various parts of the arterial tree. James Keill attempted to solve the problem by using a generational system of ordering which would allow him to develop a co-ordinated group of geometrical progression scaling laws.

It is possible that James Keill may once have read an article reporting an unusual 'starfish', published in the *Philosophical Transactions of the Royal Society* in 1670. (This was the only scientific journal published in England.) The author was John Winthrop (1606–1676), governor of Connecticut, and the first North American member of the Society. Winthrop noted the unusual multiple branching arms of this 'starfish', and sent a specimen to the editor, Mr Henry Oldenburg. Oldenburg was so fascinated by this animal that he appended his own italicised remarks to Winthrop's note, and had an artist create a detailed illustration. Oldenburg labelled the divisions of the branching structure with consecutive numbers, beginning with 1 at the first division point above the base of the tree. He points out that the number of branches *doubles* at each divisional order. He also notes that the branches beyond each bifurcation are not equal in length, one being 1.2–1.25 times longer than the other. The artist rendered the drawing to indicate a constant percentage decline in diameter and length with divisional order, a point which Keill might have noticed. In current parlance, this is a generational ordering system and is the first ordering of a tree known to me. (The 'starfish' was *Astrophyton agassizii* (Lyman 1878). Lyman wrote of the asymmetry of the branches.)

It is also possible that Keill independently reinvented this generational system; if not, he is still the first person to use this ordering system explicitly to generate a group of geometric series laws expressed in mathematical form.

I will speculate on why Keill assumed that the arterial tree was organised around geometric progressions. If one accepts an ideal model of perfect symmetrical bifurcation, then the number of branches doubles with generational (centrifugal) order. For purposes of creating a simple model, the diameters of, and flows and velocities within, the two daughter branches at each bifurcation are assumed to be symmetrical (even though paired daughter diameters are usually not equal). Since the flow of each daughter is always one-half that of the parent, the daughter/parent area and velocity ratios are also assumed constant. Support for this assumption comes from Keill's discovery that the average ratio of the summed daughter areas to parent area was about 1.26. This number ($2^{1/3}$) is of significance to modern physiologists (Roy and Woldenberg 1982).

The mathematical analysis of these laws requires the use of logarithms. John Napier published the first table of logarithms in 1614. After conversations with Napier, Henry Briggs, the first Savilian Professor of Geometry at Oxford, created a table of logarithms to the base 10 in 1617 and extended the table in 1624 (Boyer 1968).

In an article on the harmonic roots of Newtonian science, Gouk (1988) reviews the mathematics of musical scales and harmonies. The mathematics of scales requires the use of logarithms. She points out that in the seventeenth century, Oxford and Cambridge students were taught astronomy, geometry, arithmetic and music as part of their mathematics curriculum. Gouk shows that Newton had pondered the mathematics of scales as a student, and later used the diatonic scale as a model to define and name the seven colours in the spectrum. In this he was motivated to use music theory to describe the phenomena of light. This is an application of Pythagorean and Paracelsian ideas.

Keill's motivation to use geometric series and logarithms stems from the nature of the problem. However, it is no accident that James Keill had a knowledge of logarithms. It is known that both Keills studied mathematics with Gregory, and John published a well-received essay on the use of logarithms. It is also possible that familiarity with the method of calculating intervals in an equal-tempered scale provided James Keill with a mathematical and philosophical model.

GEOMORPHOLOGY: PARACELSIAN AND NEWTONIAN THEMES

We return to the themes of the Paracelsians: the macrocosm and the microcosm; cycles; alchemy; and their applications to physiology and medicine. How do these relate to geomorphology? I can give only a brief and partial discussion.

I have mentioned the direct line of descent from Harvey through Bellini, Pitcairne and Newton, to James Keill (1708), which led to Keill's development of imagined, deductive, centrifugally ordered geometric series scaling

laws. Over two hundred years later Horton (1932, 1945) developed empirically verified centripetally ordered geometrical series laws. The reversal of
the direction of ordering was the key to the modern success of the Hortonian
model because this allowed a good correspondence of branch size and order.
Keill also understood that since joint tributary cross-sectional area increased,
the velocity decreased geometrically with generational order. He therefore
had insights which anticipate some of the findings of hydraulic geometry
(Leopold and Maddock 1953; Leopold and Miller 1956).

Finally, like Pitcairne before him, Keill had a concept of the infinite
regression in size in a fractal tree:

> ... it is evident to the naked Eye, and agreed on by all Anatomists, that
> [the coats of the blood vessels] are composed of Myriads of Veins and
> Arteries.... [The more powerful the microscopes] we use, still the more
> vessels we discover.... Whoever is acquainted with the preparations of
> the curious Dr. Ruysck [Ruysch] would be apt to believe that the whole
> Body, and all its Fibres were nothing but Blood-Vessels.
>
> (Keill 1708: 124–125)

For a description of Ruysch's exquisite work, see Cole (1921).

There is another line of descent based on a theme of cycles, from Harvey
through Keill to Hutton and even to Davis. James Hutton's medical
dissertation, written in Leiden, was entitled *The Blood and the Circulation
of the Microcosm* (Hutton 1749). The scales of the macrocosm and the
microcosm, of cycles and of chemical renewal were obviously much on
Hutton's mind when he wrote this work. Like Harvey and many others,
Hutton believed that the blood, driven in a circuit by the heart, is 'renewed'
along the way in the lungs and in the digestive system by chemical and
mechanical processes. He sees a hierarchy of cycles in the body, including the
nutritional circulation in the bones and muscles, a circulation of glandular
secretions, and a circulation of highly refined juices from the brain. He
speaks also of the cycle of life as a renewal process. He comments on how
the nutrition of plants is related to the nutrition of man, calling attention to
the subcycles in the macrocosm (Donovan and Prentiss 1980: 8–10).

Ellenberger (1972, 1973) argues that the same themes are found in
Hutton's *Theory of the Earth*. Hutton compares a river network to a venous
tree which returns the blood to the heart (Hutton 1795, II: 533). Just as the
circulation of the blood repairs and renews the body of an animal, so also do
material cycles and subcycles renew the earth. 'This earth, like the body of
an animal, is wasted at the same time that is repaired. It has a state of growth
and augmentation; it has another state, which is that of diminution and
decay' (Hutton 1795, II: 562; Gould 1987: 83).

Hutton's cyclic ideas are not found in Newton's chemistry. While he was
aware of Newton's chemical ideas he preferred to emphasise the Paracelsian

ideas of a hierarchy of cycles within the macrocosm and the microcosm (Donovan and Prentiss 1980: 17).

We are indebted to James Hutton for his discovery of geological cycles and 'deep time' and the principle of uniformitarianism. These discoveries are an outgrowth of the application of the ancient idea of cycles to rocks and landscapes in the macrocosm (Hutton 1795). It should not be forgotten that Hutton began with an examination of cycles in the microcosm (Hutton 1749).

Geomorphologists have returned to the idea of the microcosm as a model for the macrocosm at various times. Recently, General System Theory, originally based on the study of animals (Bertalanffy 1950), was applied to larger systems, such as rivers (Strahler 1950; Chorley 1962).

I interpret Davis's geographical cycle (Davis 1899) as an earlier use of the microcosm (the life cycle of an animal) to explain the macrocosm. Stoddart (1986) argues that Davis took his idea of the life cycle of the landscape from Darwin, especially Darwin's interpretation of the origin of atolls as the last stage of a developmental sequence beginning with a coral reef fringing a volcanic island which gradually subsides while it erodes. Darwin's interpretation served Davis as a model for Davis's geographical cycle. Davis expressed the developmental sequence of a landscape in terms of the life cycle of an animal.

The notion of the substitution of space for time may have had its inspiration from palaeontology. Just as a palaeontologist may reconstruct stages in the life cycle for an individual from different fossils of the same species at different stages in life, so also did Darwin reconstruct the life cycle of an atoll and Davis the life cycle of a landscape from landforms in different stages of development. (It should be remembered that Davis was a student at Harvard's Museum of Comparative Zoology founded by Louis Agassiz. Here Davis studied palaeontology as a student of Nathaniel S. Shaler.)

Davis (1899) also compared the veins of a leaf to a river network, again using a microcosmic model of the macrocosm. Of course the explicit use of a tree comparison was not new. In 1636 Harvey defended his discovery of the circulation of the blood in a letter to a sceptical Caspar Hofmann. In this letter he compared the venous tree to the tributaries of the Rhine River (Ferrario et al. 1960: 19). In another letter sent in 1652 to Robert Morison, Harvey compares the venous system and portal vein to the roots and trunk of a tree (Harvey 1958: 84). Undoubtedly such comparisons were made even earlier and this kind of thinking continues to the present. Indeed, the study of trees lends itself to making connections between the microcosm and the macrocosm (Woldenberg 1968, 1986; Woldenberg et al. 1970).

In the same spirit, Thompson's On Growth and Form (1942) investigates form and process in biology and makes many comparisons between life forms, mechanical systems and ideal geometrical shapes. This book has served as an inspiration for quantitative geographers since at least the 1960s (Bunge 1966).

259

We have seen a Noachian flood of literature on fractals, beginning with Mandelbrot (1977). This flood has overtopped many natural and artificial disciplinary levees. It illustrates a continuation of the search for description and explanation across scales and disciplines, themes which quantitative geographers have emphasised (Chorley and Haggett 1967; Haggett and Chorley 1969).

Just as the principle of uniformitarianism is invoked to unify the processes at work in the past and the present, the notion of the macrocosm and the microcosm implies in some sense that similar processes work at different scales.

ACKNOWLEDGEMENTS

The author wishes to thank Professor Gordon Cumming, head of the Midhurst Medical Research Institute (West Sussex) for his excellent hospitality and encouragement during my sabbatical in 1981–1982. I also wish to thank Mrs Karen Bicker (Secretary) and Mr Norman Harris (Librarian) of the same Institute for their kind help. I thank Professor Theodore M. Brown, who introduced me to several important sources.

I wish to thank Professor Richard Chorley for his encouragement over the years, for his kind hospitality at Cambridge in the spring of 1988, and for the excitement of his ideas.

NOTES

1   The Jackson and Gravelius systems recapitulate the convention of naming streams upstream from the mouth. Instead of using names, consecutive numbers are used so that the name or identification number n persists beyond a confluence along the major daughter branch, while the the number $n + 1$ is assigned to the minor daughter branch (Woldenberg 1968; Haggett and Chorley 1969). The major daughter is the stream entering a junction with the greatest magnitude or flow. Where two streams of equal magnitude join (e.g. two first-order streams) the major daughter is the stream most in line with the axis of the parent and so receives the same number as the parent branch.

2   Selected biographical details for this section have been taken from Brown (1968, 1987), Guerrini (1983, 1985, 1987), Gjertsen (1986) or from the *Dictionary of Scientific Biography*, e.g. Debus (1972).

REFERENCES

Adams, F.D. (1954) *The Birth and Development of the Geological Sciences*, New York: Dover.

Bertalanffy, L. von (1950) 'The theory of open systems in physics and biology', *Science* 111: 23–28.

Boyer, C.B. (1968) *A History of Mathematics*, New York: John Wiley.

Brown, T.M. (1968) 'The mechanical philosophy and the "animal oeconomy" – a study in the development of English physiology in the seventeenth and early eighteenth century', unpublished Ph.D. thesis, Deparment of History, Princeton University.

——(1987) 'Medicine in the shadow of the *Principia*', *Journal of the History of Ideas* 48: 629–648.

Bunge, W. (1966) *Theoretical Geography*, 2nd edition, Lund: Gleerup.

Chorley, R.J. (1962) 'Geomorphology and general systems theory', *U.S. Geological Survey Professional Paper* 500-B: 1–10.

Chorley, R.J. and Haggett, P. (eds) (1967) *Models in Geography*, London: Methuen.

Chorley, R.J., Dunn, A.J. and Beckinsale, R.P. (1964) *The History of the Study of Landforms*, vol. 1: *Geomorphology before Davis*, London: Methuen.

Cole, F.J. (1921) 'The history of anatomical injections', in C. Singer (ed.) *Studies in the History and Method of Science* 2: 285–345.

Davis, W.M. (1899) 'The geographical cycle', *Geographical Journal* 14: 481–504.

Debus, A.G. (1965) *The English Paracelsians*, London: Oldbourne.

——(1972) 'Robert Fludd', *Dictionary of Scientific Biography*, New York: Charles Scribner's Sons, 4: 75.

——(1977) *The Chemical Philosophy*, New York: Science History Publications.

Donovan, A. and Prentiss, J. (1980) 'James Hutton's medical dissertation', *Transactions of the American Philosophical Society* 70, part 6: 1–57.

Ellenberger, F. (1972) 'Les origines de la pensée huttonienne: Hutton étudiant et docteur en médecine', *Comptes Rendus Hebdomadaires des Séances de l'Académie des Sciences, Paris* 275: 69–72.

——(1973) 'La thèse de doctorat de James Hutton et la rénovation perpétuelle du monde', *Annales Guébhard-Séverine* 49: 497–533.

Ferrario, E.V., Poynter, F.N.L. and Franklin, K.J. (1960) 'William Harvey's debate with Caspar Hofmann on the circulation of the blood', *Journal of the History of Medicine and Allied Sciences* 15: 7–21.

Gilbert, G.K. (1877) *Report on the Geology of the Henry Mountains*, chapter 5, 'Land sculpture', Washington, D.C.: Government Printing Office.

Gjertsen, D. (1986) *The Newton Handbook*, London: Routledge & Kegan Paul.

Goudie, A.S. (1978) 'Colonel Julian Jackson and his contribution to geography', *Geographical Journal* 144: 264–270.

Gouk, P. (1988) 'The harmonic roots of Newtonian science', in J. Fauvel, R. Flood, M. Shortland and R. Wilson (eds), *Let Newton Be*, Oxford: Oxford University Press.

Gould, S.J. (1987) *Time's Arrow Time's Cycle*, Cambridge, Mass.: Harvard University Press.

Gravelius, H. (1914) *Flusskunde*, vol. I, Berlin and Leipzig: Goschenesche Verlag.

Guerrini, A. (1983) 'Newtonian matter theory, chemistry, and medicine, 1690–1713', unpublished Ph.D. thesis, Department of the History and Philosophy of Science, Indiana University.

——(1985) 'James Keill, George Cheyne, and Newtonian physiology, 1690–1740', *Journal of the History of Biology* 18: 247–266.

——(1987) 'Archibald Pitcairne and Newtonian medicine', *Medical History* 31: 70–83.

Hack, J.T. (1960) 'Interpretation of erosional topography in humid temperate regions', *American Journal of Science* 258A: 80–97.

Haggett, P. and Chorley, R.J. (1969) *Network Analysis in Geography*, London: Edward Arnold.

Hales, S. (1727) *Vegetable Staticks*, London: W. & J. Innys at the West End of St. *Paul's*; and T. Woodward, over against St. *Dunstan's* Church in *Fleetstreet*.

——(1733) *Statical Essays: Containing Haemastaticks*, London: W. Innys & R. Manby, at the West End of St. *Paul's* and T. Woodward, at the *Half Moon* between the Temple-Gates, *Fleetstreet*.

——(1961) *Vegetable Staticks*, London: Oldburne.

——(1964) *Statical Essays: Containing Haemastaticks*, introduction by A. Cournand, New York: Hafner.

Harvei, Guilielmi Angli (1628) *Exercitatio Anatomica de Motu Cordis et Sanguinis in Animalibus*, Francofurti: Sumptibus Guilielmi Fitzeri.

Harvey, W. (1949) *An Anatomical Study on the Motion of the Heart and Blood in Animals*, 3rd edition, translated by C.D. Leake, Springfield, Ill.: Charles C. Thomas.

——(1958) 'The third letter in reply to Robert Morison, M.D., of Paris', in *The Circulation of the Blood: Two Anatomical Essays by William Harvey together with Nine Letters Written by Him*, translated by K.J. Franklin, Springfield, Ill.: Charles C. Thomas.

Horsfield, K. and Cumming, G. (1968) 'Morphology of the bronchial tree in man', *Journal of Applied Physiology* 24: 373–383.

Horsfield, K. and Woldenberg, M.J. (1989) 'Diameters and cross-sectional areas of branches in the pulmonary arterial tree', *Anatomical Record* 226: 245–251.

Horton, R.E. (1932) 'Drainage-basin characteristics', *Transactions of the American Geophysical Union* 13: 350–361.

——(1945) 'Erosional development of streams and their drainage basins: hydrophysical approach to quantitative morphology', *Bulletin of the Geological Society of America* 56: 275–370.

Hutton, J. (1749) *Dissertatio Physico-Medica Inauguralis de Sanguine et Circulatione Microcosmi* [Inaugural Physico-Medical Dissertation on the Blood and the Circulation of the Microcosm], Leyden: Wilhelm Boot.

——(1795) *Theory of the Earth with Proofs and Illustrations*, Edinburgh: William Creech.

Jackson, J. (1834) 'Hints on the subject of geographical arrangement and nomenclature', *Journal of the Royal Geographical Society* 4: 72–88.

Jarvis, R.S. and Woldenberg, M.J. (eds) (1984) *River Networks*, Stroudsburg, Pa.: Hutchinson Ross.

Keill, Jacobus (1718) *Tentamina Medico Physica ad quasdam Quaestiones, quae Oeconomiam Animalen spectant Accomodata. Quibus Accessit Medicina Statica Britannica*, London: George Strahan.

Keill, James (1708) *An Account of Animal Secretion, the Quantity of Blood in the Humane Body, and Muscular Motion*, London: George Strahan.

——(1717) *Essays on Several Parts of the Animal Oeconomy*, 2nd edition, London: George Strahan.

——(1738) *Essays on Several Parts of the Animal Oeconomy. The Fourth Edition to Which is Added, a Dissertation concerning the Force of the Heart, by James Jurin, M.D., F.R.S. with Dr. Keill's Answer and Dr. Jurin's Reply. Also Medicina Statica Britannica, or Statical Observations, Made in England, by James Keill, M.D., Explained and Compared with the Aphorisms of Sanctorius, by John Quincy, M.D.*, London: George Strahan.

Keill, John (1698) *An Examination of Dr. Burnet's Theory of the Earth, together with Some Remarks on Mr. Whiston's New Theory of the Earth*, Oxford: The Theater.

——(1702) *Introductio ad veram Physicam, seu Lectiones Physica: Habita in Schola Naturalis Philosophia Academia Oxoniensis: quibus accedunt Christianii Hugenii Theoremata de Vi Centrifuga & Motu Circulari Demonstrata*, Oxonia: E. Theatro Sheldoniano, impensis T. Bennet.

——(1708) 'Epistola ad cl. virum Gulielum Cockburn, M.D. In qua leges attractionis aliaque physices principia traduntur', *Philosophical Transactions of the Royal Society of London* 26: 97–110.

——(1720) *An Introduction to Natural Philosophy; or, Philosophical Lectures Read in the University of Oxford, Anno Dom. 1700: to which are added, the Demonstrations of Monsieur Huygens's Theorems, concerning the Centrifugal Force and Circular Motion*, translated from the last edition of the Latin. London: Henry Woodfall, for William and John Innys and John Osborn.

Kilgour, F.G. (1954) 'William Harvey's use of the quantitative method', *Yale Journal of Biology and Medicine* 26: 410–421.

Leopold, L.B. and Maddock, T. Jr (1953) 'The hydraulic geometry of stream channels and some physiographic implications', *U.S. Geological Survey Professional Paper* 252: 1–57.

Leopold, L.B. and Miller, J.P. (1956) 'Ephemeral streams – hydraulic factors and their relation to the drainage net', *U.S. Geological Survey Professional Paper* 282A: 1–36.

Lyman, T. (1878) 'Mode of forking among Astrophytons', *Proceedings of the Boston Society for Natural History* 19: 102–108.

Mandelbrot, B.B. (1977) *Fractals: Form, Chance, and Dimension*, San Francisco: W.H. Freeman.

Newton, I. (1961) 'De natura acidorum' (1692) ('On the nature of acids'), in H.W. Turnbull (ed.), *The Correspondence of Isaac Newton Volume III (1688–1694)*, Cambridge: Cambridge University Press.

Roy, A.G. and Woldenberg, M.J. (1982) 'A generalization of the optimal models of arterial branching', *Bulletin of Mathematical Biology* 44: 349–360.

Stoddart, D.R. (1986) *On Geography and Its History*, Oxford and New York: Basil Blackwell.

Strahler, A.N. (1950) 'Equilibrium theory of erosional slopes approached by frequency distribution analysis', *American Journal of Science* 248: 673–696, 800–814.

——(1952) 'Hypsometric (area–altitude) analysis of erosional topography', *Bulletin of the Geological Society of America* 63: 1117–1142.

——(1964) 'Quantitative geomorphology of drainage basins and channel networks', in V.T. Chow (ed.) *Handbook of Applied Hydrology*, New York: McGraw-Hill.

Tarboton, D.G., Bras, R.L. and Rodriguez-Iturbe, I. (1988) 'The fractal nature of river networks', *Water Resources Research* 24: 1317–1322.

Thompson, D.W. (1942) *On Growth and Form*, 2nd edition, Cambridge: Cambridge University Press.

Tuan, Y.F. (1968) *The Hydrologic Cycle and the Wisdom of God*, Toronto: Toronto University Press.

Valadez, F.M. and O'Malley, C.D. (1971) 'James Keill of Northampton, physician, anatomist and physiologist', *Medical History* 15: 317–335.

Weibel, E. (1963) *The Morphometry of the Human Lung*, Berlin: Springer Verlag.

Westfall, R.S. (1980) *Never at Rest: a Biography of Isaac Newton*, Cambridge: Cambridge University Press.

White, G.W. (1968) 'John Keill's view of the hydrologic cycle, 1698', *Water Resources Research* 4: 1371–1374.

Winthrop, J. (1670) 'An extract of a letter written by John Winthrop, concerning some natural curiosities … especially a very strange and curiously contrived fish', *Philosophical Transactions of the Royal Society of London* 4: 1151–1153, Figure 1, facing p. 1142.

Woldenberg, M.J. (1968) 'Hierarchical systems: cities, rivers, alpine glaciers, bovine livers and trees', unpublished Ph.D. thesis, Department of Geography, Columbia University.

——(1972) 'Relations between Horton's Laws and hydraulic geometry as applied to tidal networks', *Office of Naval Research Report* (AD 744043), National

Technical Information Service, US Department of Commerce, Springfield, VA 22151 (*Harvard Papers in Theoretical Geography* 45).

——(1986) 'Quantitative analysis of biological and fluvial networks', in A.S. Popel and P.C. Johnson (eds) *Microvascular Networks: Experimental and Theoretical Studies*, Basle: Karger AG.

Woldenberg, M.J., Cumming, G., Harding, K., Horsfield, K., Prowse, K. and Singhal, S. (1970) 'Law and order in the human lung', *Office of Naval Research Report* (AD 709602), National Technical Information Service, US Department of Commerce, Springfield, VA 22151 (*Harvard Papers in Theoretical Geography* 41).

Young, T. (1808) 'Hydraulic investigations', *Philosophical Transactions of the Royal Society of London* 98: 164–186.

——(1809) 'The Croonian lecture on the functions of the heart and arteries', *Philosophical Transactions of the Royal Society of London* 99: 1–31.

12

THEORY, MEASUREMENT AND TESTING IN 'REAL' GEOMORPHOLOGY AND PHYSICAL GEOGRAPHY

Keith Richards, Susan Brooks, Nicholas Clifford, Tim Harris and Stuart Lane

> Whenever anyone mentions theory to a geomorphologist, he instinctively reaches for his soil auger.
>
> (Chorley 1978: 1)

INTRODUCTION

The well-known quotation above encapsulates a perennial problem in a discipline rooted in the reality of 'the field'; how to strike an acceptable balance between concrete empiricism and abstract theorising. The incommensurability of those with data and no model, and those with a model but no data, was throughout the late 1970s and early 1980s typical of meetings of the British Geomorphological Research Group (BGRG). One strategy which has attempted to resolve this dichotomy, espoused by Haines-Young and Petch (1986), has been to emphasise a critical rationalist methodology in which a hypothesis-testing framework for research is encouraged (for example, in a workshop for beginning research students organised by the BGRG). This has long been a tradition in geographical field study. Eversden (1969: 67), in recommending teaching through fieldwork, argued that we

> begin with a problem it is desired to solve ... decide the hypothesis ... which is to be tested in the field as a solution to the problem ... decide what information is needed to test the hypothesis ... results can then be considered and the original hypothesis can be rejected or accepted.

Taken to extreme, this approach can lead to narrow scientism, and in general, too normative an emphasis ignores the fact that critical rationalism is itself only a model of the process of scientific enquiry. To convert it into a proscriptive recipe for research method is potentially limiting, perhaps

265

especially in field-based research into systems that are intrinsically 'open'. It was for this reason that an Editorial and subsequent discussion in *Earth Surface Processes and Landforms* (Richards 1990, 1994) have sought to suggest a 'realist' antidote to the hypothesis-testing model. This chapter attempts to consider in more detail some of the implications of such a methodological shift for the *practice* of research in geomorphology.

There are, of course, well-known objections to the 'hypothesis–experiment–test–falsify' procedure of critical rationalism. First, it has its origins in empirical positivism, and continues to rely on observation and measurement as the essence of its (experimental) method. In environmental and social research, the reliability of measurement requires continual scrutiny, and a proscriptive methodology which relies on observable entities is therefore liable to result in debatable results and potentially inadequate understanding. Second, the falsification central to critical rationalism is as unlikely to succeed as a methodological procedure as the verification principle it sought to displace. If a theory fails to generate successful predictions, there is no fundamental reason to reject it completely, since it might provide a better account (better in terms of predictive success) in another time and place. Furthermore, tests are not conclusive sources of re-evaluation when they fail, because what really resulted in failure is rarely obvious (Beck 1987), either because the theory is invariably a set of complex statements any one of which may be invalid, or because there may be a failure of the 'faithful measurement postulate' when the correspondence rules that connect concepts to their measurable entities are inadequate.

The persistence of this essentially positivist methodology in environmental and social research probably reflects the illusion of rigour provided by the appeal to falsification, with its assumption of an experimental practice dominated by critical testing prior to theory development. In physical geography it has deep roots in a functional, systems-oriented geomorphology (Chorley and Kennedy 1971), and an associated emphasis on statistical methods (Chorley 1966), and was part of the search for scientific respectability which led to the quantification movement in several disciplines in the 1960s (Gibbon 1987). Deep-seated objections in human geography have led to an apparent methodological schism between physical and human geographies which has made it difficult for geography to respond to the current concerns with the relationships between humanity and environment (Chorley 1965). One attraction of realism is therefore that it has a potentially broad appeal in the geographical sciences, reflecting the fact that beyond the discipline it has adherents in both the natural (Harré 1970) and the social (Bhaskar 1989) sciences. Furthermore, in the context of problems that are often global in scale and involve both natural and social processes (Stoddart 1987), and are the current concern of the environmental sciences, an experimental science tradition may have little to offer. Adherence to a research tradition which divides problems into elements, rather than adopting some kind of holistic or integrative approach (Phillips 1992), perhaps accounts for some of the failures to address larger issues, as bemoaned by Kennedy (1993).

REALISM

In one form of realist approach to scientific enquiry, a key argument is that there are three levels into which any phenomenon can be structured: the fundamental generating mechanisms (underlying structures with causal powers), the events which they may or may not produce depending on the circumstances, and the empirical observations of those events made by humans (Table 12.1). Observations are only made of events if capable observers are present – that is, if technology has created the means whereby observation and measurement of the particular kinds of events generated by a given set of mechanisms are possible. Also, events do not always arise when appropriate sets of causal mechanisms are in place, because the local space–time context must also be appropriate to trigger the mechanism. Furthermore, when multiple causal mechanisms exist together in an 'open' system, their effects may be self-cancelling in certain contexts, with the result

Table 12.1 A summary of the basic tenets of realism

1 Reality is structured into three levels:		
The real	*The actual*	*The empirical*
Mechanisms	*Events*	*Observations*
(Intransitive)	*(Cognitive)*	*(Experimental)*

2 The real mechanisms, generating systems, or structures may be time and space invariant, but the events that are observed are not, because:

3 The events are realisations of the causal powers of mechanisms acting in particular contingent conditions.

4 Thus, in the complex 'open' systems of nature and society, a particular mechanism may not always produce the same kinds of events because at different times and places those conditions vary – indeed, interacting mechanisms may prevent the occurrence of events.

5 This means that identification of the laws of nature through conventional positivist experimental activity cannot be expected to occur, because such experimentation creates specific 'closed system' conditions, and any 'constant conjunctions of events' that are observed reflect the existence of those conditions.

6 Furthermore, positivist methods assume that a choice between competing theories can be made on the basis of their relative predictive success; indeed, that a form of explanation is achieved when predictions are proved correct. In realism, an asymmetry exists between prediction and explanation, and the former does not imply the latter.

that no events occur. Thus there can be no simple equivalence between an empirical observation and the identification of the causal mechanism underlying it. Any positivist approach which fails to acknowledge such a structure, and collapses the three levels into one, is likely to oversimplify the research process whereby an understanding of 'causes' is created by an observational procedure.

The practice of positivist science involves a reliance on experimental closure in order to assist in the identification of the 'laws' of nature. This closure permits manipulation of causal relationships, so that observations can be made of the constant conjunctions which lead to identification of the empirical relationships that represent those laws. The realist response to this form of experimentation is to note that the constant conjunctions observed in such closed experiments are themselves a product of the experimental closure, leading to the absurd conclusion that the laws of nature are created by experimental activity (Bhaskar 1989). This is itself an indication that a model of scientific activity that requires experimental closure, observation and measurement to develop understanding and explanation requires careful evaluation. Experimental activity has a wide variety of purposes (Harré 1983), and identifying explanatory causal relationships may not even be the most important one.

A further well-known implication of realism (Table 12.1) is therefore that a strong asymmetry exists between prediction and explanation. In critical rationalism, hypotheses are used to make predictions. Explanatory theories underlying the hypotheses are accepted for as long as the predictions are successful, and are modified when they fail. However, successful prediction does not of itself provide a basis for developing an explanatory account. Time series modelling can be undertaken which allows successful short-term predictions to be made, but the models themselves provide no explanation or understanding of the mechanisms underlying the variation in the series; this is derived from other forms of investigation. Realism is therefore a search for a model of the method by which explanation rather than prediction is achieved, and the criteria whereby one explanatory account may be preferred over another (criteria which may appear less rigorous than those which measure predictive success, but at least in part because the rigour tradition-ally imputed to the latter is exaggerated).

The quotation from Chorley (1978) at the head of this chapter, from a paper in which he explicitly addressed these varied bases for development of explanation, suggests that geomorphologists in particular, and physical geographers and environmental scientists more generally, have tended to eschew these theoretical considerations. It is therefore useful to consider some practical questions that arise during research, and evaluate them in relation to the realist positions outlined above and in Table 12.1. It may seem inductive to attempt to understand the method by which knowledge is acquired in the environmental sciences through the experience of doing research, but an informed view on scientific method necessarily follows empirical activity.

MEASUREMENT

Ironically, empirical positivism and critical rationalism appear to have neglected the issue of measurement, perhaps because they are so rooted in a belief in observable entities. However, a perspective which is governed by belief in the three-level reality outlined above immediately raises questions about the relationships among those levels which affect our ability to observe and measure. Relationships between the real and the actual, and between the actual and the empirical, involve processes which have been termed, respectively, translation (from concept to entity) and transduction (from entity to fact).

Translation: concepts and entities

Geography is replete with vague and imprecise concepts (equilibrium, thresholds, locality, system). Concepts such as these must be clearly defined before they can be translated into entities that are susceptible to measurement, and to representation as facts. This translation can be quite simple, particularly in cases where no more than classification occurs. Surprisingly, the principles of classification have been neglected in the geographical literature since Grigg's review of 'regions as classes' (1967). Indeed, rejection of quantification in some areas of human geography seems even to include refusal to acknowledge the value of classifying, or 'naming', as a weak form of measurement that lends itself to quantification and statistical analysis via the enumeration procedures of categorical data analysis (Wrigley 1985). In physical geography, with a stronger taxonomic background, a debate has already been joined and resolved in the literature about the consequences, for classification and its subsequent use in explanation, of imprecise translation of concepts to entities. This arose in relation to the idea of equifinality.

Chorley (1962) originally suggested that equifinality could arise when a landform appears to have been formed by several different processes or process histories. Haines-Young and Petch (1983) rejected this conclusion on critical rationalist grounds, arguing that the assumption that more than one causal hypothesis may be equally acceptable violates the requirement that one should search for a critical test. It is difficult to isolate a critical test if the notion of equifinality lurks in the background, explicitly allowing non-discrimination between competing hypotheses. This apparent difference of view in part reflects the distinction between developing a general account of a class of phenomena, and searching for a deductive explanation of an individual case. Nevertheless, it is reasonable to reject the view that equifinality does occur, on the grounds that when it appears to apply, it is usually because the concept under examination has been translated with a lack of precision into an entity. Thus the concept of a landform consisting of exposed, joint-bounded blocks may be classed as a 'tor', but this is such a vague specification as to admit inevitably of several possible explanations. Equifinality, and equal validity of competing hypotheses, may therefore be attributed to improper translation between an initial general concept and a specific entity that might be the subject of explanatory examination. If tors may be explained by both Linton's two-stage hypothesis (weathering and regolith stripping; 1952) and Palmer and Nielsen's one-stage hypothesis (free-face formation and destruction; 1962), this is likely to be because the examples of tors explained by these competing hypotheses were in fact different entities (for example, tors distinguished by their hilltop and valley-side locations, and by their lithological differences). This general argument even undermines Culling's (1987) appeal to chaotic and bifurcating behaviour in non-linear systems as a basis for justifying a belief in equifinality. These fashionable mathematical models have yet to accommodate the representation of complex phenomena whose character is multivariate.

Transduction: entities and facts

A functional, systems-analytical approach to geomorphology (Chorley and Kennedy 1971) requires operational definitions which permit meaningful conversion from concept to entity, and from entity to measurable quantity. This latter process has been referred to as transduction by Pawson (1989), using a term derived from the 'transducer', a device familiar to physical geographers for transforming one form of energy into another (for example, the piezo-ceramic transducer that converts hydrostatic pressure to electrical current, and is used to record variations in water level in stream gauging stations). In Pawson's (1989) analysis, transduction includes the questionnaire used in social sciences. The fact that the transducer analogy can be employed in social science research methods, where measurement is frequently regarded with suspicion, suggests that there is no reason to suppose either that measurement is relevant to physical geography alone because only here is it feasible, or that transduction is any more straightforward in physical than in human geography. In fact, because in many (but not all) of the environmental sciences there is only a weak tradition of instrument development, there are similar problems in the control and interpretation of the transduction process.

A critical question concerns the circumstances in which transduction can be believed, and its solution requires the progressive application of technology to science in order that a variety of different systems of transduction can be brought to bear on a particular theoretical entity. A theoretical 'thing' can become a 'real' phenomenon when there is coincidence between the results of several different measurements (Hacking 1983). Cells are believable when optical and electron microscopy both reveal structures that are interpretable in terms of the same theory of the cell. Each of these observation devices embodies its own theories, and it is reasonable to suppose that it would be improbable that these different theories (about the process of representation) could reveal the same structure in a phenomenon under observation if that structure did not exist in reality, and had precisely the properties associated with it in the theory developed to explain its properties.

An example of observation and measurement problems in geomorphology is provided by the increasing interest in the assessment of sediment transport in relation to fluid turbulence. The study of turbulence is actually necessitated by a realist investigation of sediment transport, because the fundamental underlying mechanisms of fluid entrainment and suspension are related to the process of momentum exchange in turbulent flow. Turbulent flow structures and the short-term Reynolds stresses are physically more relevant than time-averaged velocities and shear stresses derived from the phenomenological model of the law of the wall. This is an illustration of the principle that understanding is provided by observation and measurement undertaken at the scales appropriate to the physical behaviour of the process being examined. However, research by Lapointe (1992) has shown that measurement problems present a barrier to the development of theory in this field. Turbidity records reveal short-term fluctuations in suspended sediment concentration that may be correlated with turbulent velocity variation (Clifford et al. 1995), but analysis of sediment fluxes requires the calibration of turbidity; and this can only be achieved by extracting pumped samples from the flow over timescales in excess of those of the turbidity variations being monitored.

At a deeper level, the question arises as to whether the available instruments are capable of revealing the nature of turbulence. In field-oriented fluvial geomorphology, for example, the robustness of the design of electromagnetic current meters (EMCMs) encourages their use in studies of the structure of velocity variation at frequencies $> 0.1 \mathrm{Hz}$. Two EMCMs in common use are the Colnbrook-Valeport (CV) and the Marsh-McBirney (MM). Experiments in which these two instruments have been used to examine turbulent velocity variation are reported by Lane, Richards and Warburton (1993). These included measurements at low and high ambient flow velocities in a flume and the River Sence in Leicestershire, with velocities being data-logged at $10 \mathrm{Hz}$. The structure of the velocity series obtained by the two EMCMs can be examined by Box-Jenkins autoregressive-moving average (ARMA) modelling. Correlograms are shown in Figure 12.1, and some details of the analyses are summarised in Table 12.2. In the table, the lag-one autocorrelation coefficient is shown in column 4, followed by the number of significant terms in the autocorrelation (ACF) and partial autocorrelation (PACF) functions. Also listed is an indication of the general form of the ACF and PACF, which is an indication

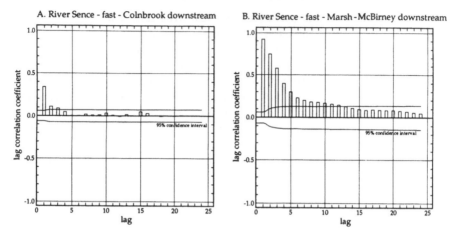

Figure 12.1 Correlograms of time series of the downstream velocity components measured at a constant discharge at the same point in a selected vertical within a rapidly flowing riffle section of the River Sence, Leicestershire. The series were generated by two different electromagnetic current meters logged at 10Hz

of the appropriate ARMA model required to decribe the series. Usually, an autoregressive model is suggested by an ACF which decays exponentially and a PACF which cuts off, the order of the model being defined by the lag of this cutoff. A moving average model is suggested by an ACF that cuts off and an exponential PACF. In this series of comparative experiments with two different EMCMs used at the same location, where the structure of velocity series would be expected to be constant, it is clear that the instruments provide different information on the nature of that structure. The CV meter produces less smooth series (low values of r_1, and more rapid decay of the ACF to values not significantly different from zero). The MM meter produces series whose PACFs decay exponentially, while the CV series have PACFs which cut off. This suggests that the MM sensor induces a moving average component in the signal which it generates. From these results it is difficult to avoid the conclusion that the 'turbulent structures' recovered by 10Hz sampling of EMCM signals can be considerably distorted by instrumental artefacts. Since EMCMs were initially developed to allow orthogonal mean flow components to be measured, their unquestioned and unchecked application to studies of turbulence may not be justifiable. There is a need to adopt the principle of coincidence noted above, using ultrasonic and laser doppler anemometric methods to measure the characteristics of turbulence at the same sites as electromagnetic instruments, in order to separate the natural signal from the instrumental noise more clearly.

Table 12.2 Properties of downstream (x) and vertical (y) velocity series obtained in the flume and in the River Sence in Leicestershire, using a Colnbrook-Valeport (CV) discoidal and a Marsh-McBirney (MM) spherical electromagnetic current meter logged at 10Hz with a Campbell data-logger

Data source	Mean velocity	Series	r_1	ACF	Form	PACF	Form
Flume	Low	CVx	0.60	2(8)	Exp	2(7)	Cutoff
Flume	Low	CVy	0.76	1	Exp	1	Cutoff
Flume	Low	MMx	0.96	3	Exp	3	Exp
Flume	Low	MMy	0.91	4	Exp	4	Exp
Flume	High	CVx	0.66	3	Exp	3	Cutoff
Flume	High	CVy	0.72	2	Exp	2	Cutoff
Flume	High	MMx	0.90	4	Exp	4	Exp
Flume	High	MMy	0.88	4(6)	Exp	4(6)	Exp
Sence	Low	CVx	0.76	8	Exp	1(2)	Cutoff
Sence	Low	CVy	0.87	12	Exp	2(5)	Cutoff
Sence	Low	MMx	0.96	9	Exp	2(14)	Cutoff
Sence	Low	MMy	0.91	8	Exp	2	Cutoff
Sence	High	CVx	0.34	3	Exp	1(3)	Cutoff
Sence	High	CVy	0.69	2	Exp	1	Cutoff
Sence	High	MMx	0.92	12	Exp	5	Exp
Sence	High	MMy	0.86	9	Exp	5	Exp

Notes: The lag-one serial correlation is given in column 4. Column 5 gives the number of statistically significant terms (95%) in the autocorrelation function (ACF), with an indication of any outlying terms which are also significant in parentheses; all ACFs are exponential in form. Column 7 gives information on the number of significant terms in the partial autocorrelation function (PACF); the form of the PACF appears to differ between current meter types.

DEMARCATION CRITERIA

A second practical issue concerns how to 'validate' the theoretical findings of research programmes typical of realist science. Critical rationalism itself developed as a response to the methodological limitations of logical empiricism, and involves emphasis on epistemological considerations (methodology). As a result it is well endowed with criteria for choice between competing hypotheses, particularly those involving predictive success. However, such successes can often only occur if a form of reductionism is followed in which isolation and experimental closure reduce extraneous influences. This differs from a realist reductionism, in which explanation is 'reduced' to another level (not always at a smaller scale, although this is often assumed). Sediment transport may be explained by a reduction to the level of turbulence, but mountain landscapes require an appeal to plate tectonics (N.J. Cox, personal communication).

Realism demands explanatory power of its theories, not exclusively predictive ability. In cases such as solute transport, which requires information on mineral stability and chemical kinetics (Richards 1990), or sediment transport, which requires consideration of turbulent flow processes, this invokes reductionist uncovering of ontological depth. A phenomenon is probed at increasingly smaller scales, until the physically appropriate scale is reached. By invoking relationships with other phenomena, the criterion of explanatory power also frequently demands a strongly interdisciplinary approach to research, as in the examples outlined below. These characteristics imply that the explanation developed for a phenomenon at a particular scale should be mutually consistent with, and reinforcing of, accounts of other related phenomena at both the same and adjacent space–time scales. This combines the notions of 'structural corroboration' and 'referential adequacy' considered by Phillips (1992) as two demarcation criteria which provide alternatives to that of predictive success. Both may be criticised: a swindler's story may involve mutual support from a wide range of evidence, and an invalid theory may nevertheless help us to appreciate other phenomena. However, combined with 'multiplicative replication', 'consensual validation', and 'the search for negative evidence', these criteria provide a 'soft' form of the conventional scientific method, which is liable to be more suited to research in open systems, and which is also probably a closer approximation to the true practice of science.

However, this does not imply that a change in emphasis from predictive to explanatory demarcation criteria will not affect the research practice. In fact, the demarcation criteria themselves impose constraints on the nature of research questions, and the style of research. In particular, explanatory criteria demand a research process in which several related issues are addressed together. An example of the reinforcing explanation required of a realist research project is a study of glacial hydrology undertaken at the Haut Glacier d'Arolla, in Valais, Switzerland (Richards *et al.* 1996). It is clear that an understanding of the subglacial hydrology of a glacier is unlikely to be attained by any simple observational process; instead, a variety of linked experiments must be performed which successively and in concert reveal the nature of the process whereby daily patterns of surface melt are transformed into different hydrograph forms, as the drainage process changes daily and seasonally to delay and attenuate that transformation to varying degrees.

An initial requirement for this research was that the surface and bed topographies of the glacier were established by conventional surveying and radio-echo sounding. Digital terrain models based on gridded elevation data were then to be used to provide a first approximation to the subglacial drainage network structure, by plotting the contributing area draining through each grid square over the subglacial potential surface (Figure 12.2; Sharp *et al.* 1993). Dye-tracing experiments (over 500 in this case) then make it possible to check this drainage system structure. In addition, the time of travel of the dye from the moulins into which the injections were made to the meltwater stream portal varies seasonally, and it is possible to distinguish the

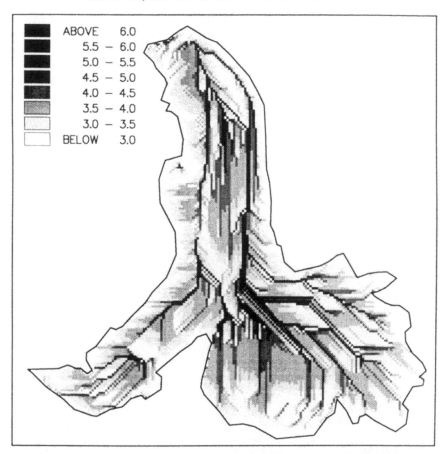

ABOVE	6.0
5.5 –	6.0
5.0 –	5.5
4.5 –	5.0
4.0 –	4.5
3.5 –	4.0
3.0 –	3.5
BELOW	3.0

Figure 12.2 The up-glacier contributing area on the hydraulic potential surface of the Haut Glacier d'Arolla, Valais, Switzerland, illustrating that this provides a means of identifying the probable subglacial drainage network. The digital elevation model is based on a 20 m grid, and the key defines classes of the common logarithm of contributing drainage area in square metres

dye return curves of (a) rapid throughput in a conduit system, and (b) delayed flow in a slower distributed drainage system. This evidence is then evaluated against hydrochemical data from the glacier melt stream, with the cation sum, pH, and $p(CO_2)$ being used to discriminate between open and closed system waters, which can be interpreted in terms of both access to atmospheric CO_2 and the chemical kinetics during the mixing of quickflow and delayed flow. The water balance of the glacier is also monitored so that periods of net water storage and drainage can be identified, to correlate with the evidence of open and closed system conditions. This can be assessed using experiments to measure basal water pressure by lowering pressure transducers into moulins, and survey data capable of resolving vertical and horizontal components of glacier sliding motion. Finally, the data can be integrated by a physically based numerical modelling strategy. In this, spatially distributed melt across the glacier surface is simulated on the basis of a surface energy balance sub-model driven by continuous meteorological data and incorporating components to define the effects of surface albedo (ice or snow) and diurnal variations in topographic shading (Figure 12.3), coupled with a conduit flow sub-model which transfers the hourly moulin input through the drainage network structure defined by the initial topographic and dye-tracing data. This highly integrated programme of data collection and analysis provides a multivariate, multiple-process database, the evidence from which links together to create a realist interpretation of the glacier's hydrology, through a mutually reinforcing explanatory system

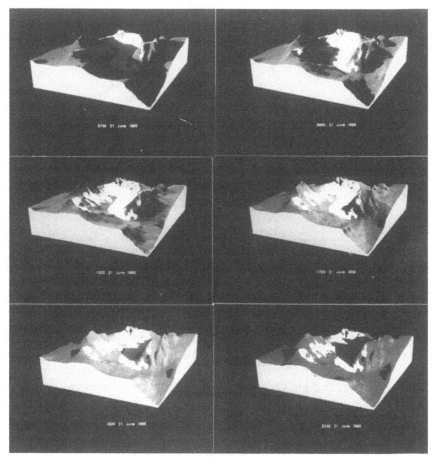

Figure 12.3 Three-dimensional computer maps of the Haut Glacier d'Arolla with topographic shading as calculated for a melt model based on the surface energy balance

rather than a series of prediction exercises.

A second example is a study of the relationships among bedload transport, suspended sediment concentration, and water quality in a lowland gravel-bed stream (Harris and Richards, in press). Here, the objective is to consider the ways in which fine sediment stored in inter-gravel pore spaces influences water chemistry when released from that store following a bed-disrupting flow. This research aim was best achieved through parallel analyses of the three components of river load, and by a combination of field and laboratory study of both the interactions between coarse and fine bed sediments, and between bed sediments and the transporting fluid. Conventional field monitoring of the transport processes was undertaken in the River Kennett, near Newmarket, using an automated, continuously recording bedload trap, a turbidity meter, and a conductivity meter, and these continuously logged determinands were supplemented by pumped samples for suspended sediment and chemical analysis. However, insights into the details of the processes of interaction were then provided by flume experiments on sediment mixtures, which revealed the role of fine sediment in destabilising a gravel bed through the alteration in shear stress partitioning between grain and form drag (cf. Iseya and Ikeda 1987), and laboratory experiments on fluid–sediment chemical interaction which demonstrated the varied rates of solution of different solute species that occur when vigorous mixing takes place of bed sediment and stream water.

Such examples of 'interdisciplinary' research as these suggest that realist research is best undertaken as an in-depth case study at a single site, with multiple interacting processes being the explicit focus of attention, and without reliance on a single mode of experimentation. This issue is explored further in the next section. It also hints that a realist research programme runs counter to the intellectual structures established during a dominantly positivist twentieth century. Academic institutions are organised on departmental lines that can inhibit the development of parallel research into related structures in order that reinforcing explanatory accounts are constructed. This has implications for the ability of existing intellectual structures to address the larger environmental problems alluded to at the end of the Introduction.

EXPERIMENTAL DESIGN: FIELD WORK

Any research project requires a precise definition of the problem to be investigated at each stage, and this is likely to involve a cascade of increasingly specific (potentially reductionist) 'researchable' questions as the ontological depths of the phenomenon are exposed. This progressive research design involves both extensive and intensive phases (Sayer 1984; Sarre 1982), both of which are commonly necessary to provide description and explanation. The extensive research phase may involve hypothesis testing and quantitative (statistical) generalisation, and reveals the background patterns evident in the products of the underlying processes. The intensive research phase may involve a detailed case study whose objective is to uncover the nature of those processes. In geomorphological (and geographical) research, investigation has often stopped at the stage when functional relationships are identifiable, and before a full understanding of mechanisms has been pursued (Yatsu 1992).

In geomorphology and other environmental sciences, the intensive research phase which discloses these mechanisms often cannot be based on the closed system experiments of positivism. The realist experiment in glacial hydrology described above is, typically, based on intensive case-study research. There has been a resurgence of interest in case-study research in recent years (Ragin and Becker 1992), reflecting its importance in intensive realist investigation. A case study is 'a detailed examination of an event (or series of related events) which the analyst believes exhibits (or exhibit) the operation of some identified general theoretical principle' (Mitchell 1983: 192). The objective of a case study is to understand mechanisms, and generalisation from its results to other cases is not through empirical extrapolation using statistical inference: 'the validity of extrapolation depends not on the typicality or representativeness of the case but upon the cogency of the theoretical reasoning' (Mitchell 1983: 207). Since observable events arise when generating mechanisms interact with specific sets of contingent circumstances, intensive, open-system, realist experiments often involve 'longitudinal' (historical) or 'comparative' (place-based) studies (Pawson 1989). An essential objective of realist experiments is therefore recontextualisation of the circumstances within which mechanisms might act, rather than isolation of an individual mechanism. The local contingent conditions in time and place have to be understood fully in order that the nature of the events, created by particular mechanisms acting in specific circumstances, can be postulated. Hence, the geography of a problem – the contingency of place – must be described in terms of the potentially significant, local, time–space influences. Such experiments are necessary in the environmental and social sciences where experimental closure is either impossible or creates such unrealistic conditions as to inhibit explanation of reality. The 'field area' is thus the main laboratory for realist research into environmental and social processes, and its selection and description requires a full account of the regional and local contingencies within which the (general) mechanisms operate, and which are to be disclosed by the research process.

Realist case-study experiments require that field investigations provide the context for mechanisms to be observed in action via the events that they create. This requires identification of the significant attributes of the field area, the choice of which is therefore critical. For example, Richards *et al.* (1993) have considered channel patterns in part of the Gangetic Plain, arguing that in this tectonic foredeep basin, the conversion of linear sedimentation aligned along the major river axes to areal sedimentation is a necessary process to ensure that the whole basin is uniformly filled in the long term. The result is that avulsion should dominate the long-term behaviour of the rivers, and this appears to be the case in the north-central plain. However, in the south-central plain, the Ganga itself appears to be associated with incision and terrace formation, driven by Quaternary climatic change. Thus the field area chosen, which affects the sedimentation rate in this case, determines the relative importance of different underlying mechanisms, and the kinds of events that are observed. Quite different conclusions about the control of long-term river behaviour can be reached in different parts of the Gangetic Plain, because the background rates of sedimentation determine which controlling mechanism dominates. The choice of field area is thus critical to the scientific interpretation, and in the absence of laboratory control, experimental control must be provided by careful field site/area selection (effectively, purposive sampling). If a particular mechanism is being investigated, field sites must be chosen whose character-istics allow that mechanism to trigger events that can be observed. This places considerable demands on the researcher, who must have a deep understanding of both the mechanism and the field site in order that the experiment yields observable and logically interpretable events. The choice of field site is thus a critical element of experimental design, and should not be a matter of arbitrary convenience in a properly conducted realist experiment. The primacy attached to identification of general laws in conventional science must be balanced by a much stronger consideration of the space–time context in open-system experimentation.

This emphasis echoes the recent resurgence of interest in place-based research in geography (a new regional geography), although it suggests that it carries with it certain dangers. In physical geographical research a place-based methodology was championed by Pitty in 1979. This was not specifically concerned with the comprehension of integrated but diverse phenomena, such as for example characterises traditional regional geography, but involved the deliberate choice of an area for study because its distinctive characteristics enabled particular phenomena to be better understood. He summarised this geographical approach as a method which 'utilizes the variability at the earth's surface as a source of natural selectors in its investigations, and depends on the exploration for, and discovery of, localities naturally simplified by the marked presence or absence of selectors of particular interest' (1979: 279). Wider conclusions were then to be drawn from *comparative* studies of the same phenomena in other areas. This place-centred investigation is substantively systematic and problem-oriented, but is firmly rooted in detailed description of the local boundary conditions. The problem for geography is that such a research framework is, however, by no means peculiar to geographical research, but is part of a more generally applicable realist methodology.

Furthermore, contingent factors have also always been required by the deductive-nomological model of critical rationalist explanation, in which the explanandum is deduced from the combination of a general covering law and a set of statements defining the initial conditions (the geographical and historical context for the event to be explained). It is equally necessary for this kind of explanation that critical properties of the field area are defined. Richards (1987) provides an example: that of channel pattern change from braided to meandering in the South Platte River, where a plausible account is initially based on (i) a generalisation about the threshold stream power distinguishing braided from meandering rivers, and (ii) historical evidence of river impoundment, flow regulation, and reduced stream power in the South Platte River. This account is only partial, however, because the description of the geographical context neglects relevant additional evidence of increased groundwater levels in floodplain environments receiving irrigation return flows, and the associated increases in riparian vegetation and stabilisation of floodplain sediments (Nadler and Schumm 1981). This example hints that a positivist search for explanation, based on supposedly law-like and empirically justified quantitative regularities, tends to narrow the contextualisation so that a regularity can be recognised. A realist account of the adjustment of the South Platte River is more likely to invoke, perhaps qualitatively, a set of interacting physical mechanisms of channel pattern change, and in doing so to allow a wider and deeper contextualisation. The detailed character of the field area is therefore more significant to a realist environmental scientist. This is well expressed by Rouse (1989), who argues:

> not that scientific knowledge has no universality, but rather that what universality it has is an achievement rooted in local know-how.... The empirical character of scientific knowledge is the result of an irreducibly local construction of empirical reference, rather than the discovery of abstract universal laws that can be instantiated in any local situation ... much of scientific knowledge involves preparing the situation to make laws applicable to it, and learning how to describe it in terms to which laws can attach. Such preparation and description always constitute a form of local knowledge.

Realist, case-study, place-based research therefore demands detailed representation of those properties of a field area relevant to the mechanisms it is hoped to disclose (through observation of the events they trigger in a particular field site).

The precise roles of the field area and fieldwork in research have rarely been considered in depth, by comparison with the extent to which the role of experimentation in the laboratory has been evaluated (Woolgar 1988). However, fieldwork is typically organised to reinforce a particular model of enquiry – evident in Eversden's (1969) view of the nature of fieldwork quoted above. One critical approach to the significance of the 'field' in environmental science involves adoption of Markus's (1987) hermeneutic strategy to evaluate the manner in which the *Field Area* becomes an element in the writing of environmental science. This in essence requires a deconstruction of the sections, in selected published papers, that present background information on the field area, in order to establish whether the details provided yield the contextual data required for satisfactory explanation. This could lead to a critique comparable to that of Markus (1987) of the meaning of *Materials and Methods* sections in scientific writing (he views them as legitimising devices, rather than as a true provision of the means whereby replication of experiment can occur).

An example chosen to illustrate the importance of textual detail about the field location, in papers reporting environmental research, is provided by a comparison of the studies of meander planform shape and migration process by Hooke (1984) and Carson (1986). It should be noted that this comparison is between a general review and a specific case study, and is therefore somewhat unfair to the former; however, it illustrates the general principle. Figure 12.4 depicts a number of different styles of meander bend migration as classified by Hooke (1984). A fundamental problem with the interpretation of these is that little detailed information is presented on the precise field contexts within which each example of meander migration was observed. The result is that it is impossible to judge the degree to which a particular migration style might reflect hydrodynamic processes (the general underlying generating mechanisms), as opposed to the external influences of the (local) boundary conditions. Since the observed rivers are relatively small, there is a high probability that floodplain sediment heterogeneity, or bank vegetation, or random bank collapse events, will have influenced the pattern of migration of individual bends, and without detailed information on these field contexts it is impossible to judge either the uniqueness of a particular migration style, or the circumstances in which each style might occur. Hooke (1984) attempts to identify the circumstances in which different patterns of migration occur, but via an appraisal of literature which itself often lacks the necessary depth of contextual detail. The result is that the distributions of meander forms and patterns of change are related to generalised catchment or floodplain characteristics; 'anomalous' meander patterns are interpreted in relation to 'factors' that may be responsible for their occurrence and behaviour, such as inconsistent floodplain sediment composition, or 'locally soft banks, resistant lenses, various non-alluvial formations, geological structure, engineering structures, and stream confluences' (1984: 490). Hooke's paper is, of course, an attempt to generalise about meander habit, and this style of argument is therefore inevitable. Apparent anomaly in individual bend behaviour might, however, be less noteworthy with emphasis on the manner in which individual morphologies interact with hydrodynamic processes to give rise to subsequent evolution in morphology. This indicates the dangers of empirical classification based on the *products* of

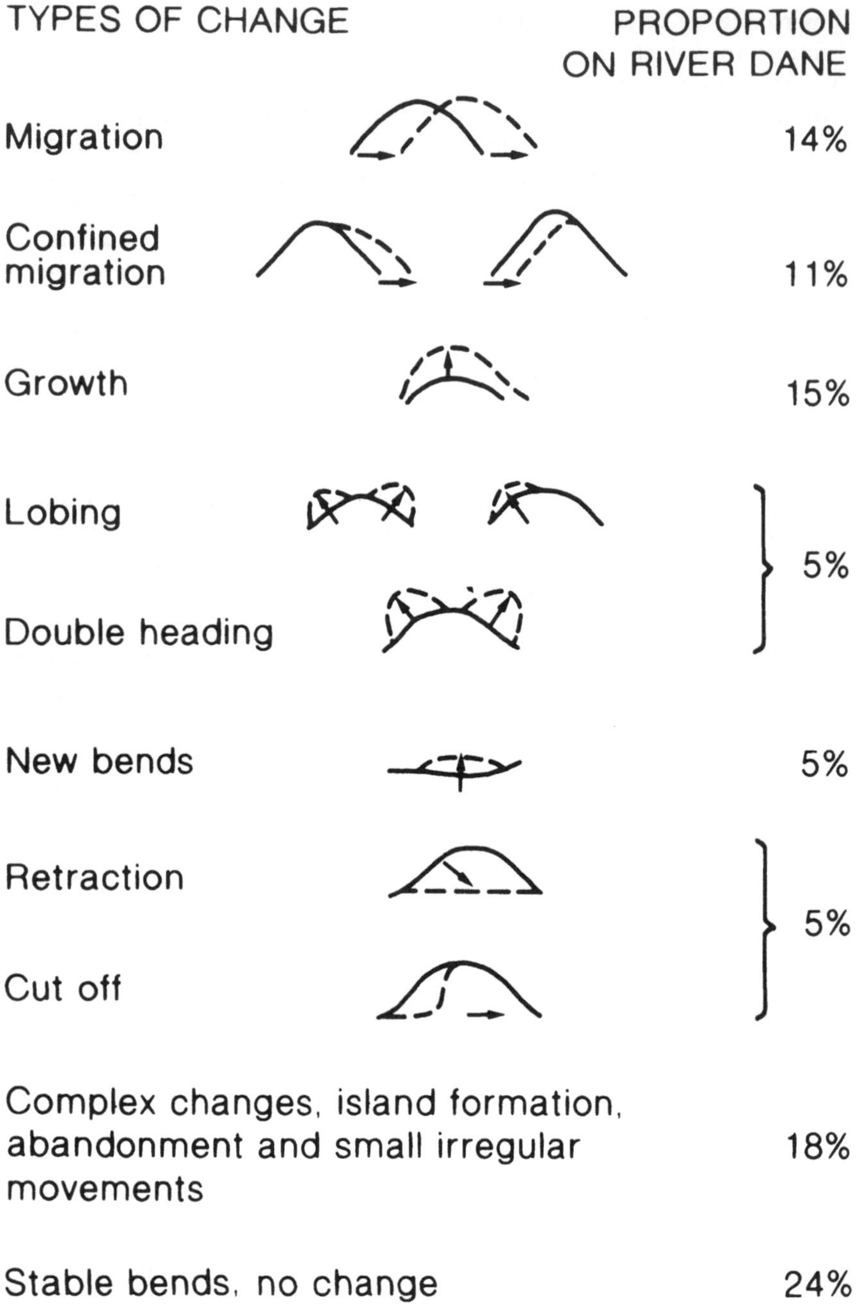

TYPES OF CHANGE		PROPORTION ON RIVER DANE
Migration		14%
Confined migration		11%
Growth		15%
Lobing		
Double heading		} 5%
New bends		5%
Retraction		
Cut off		} 5%
Complex changes, island formation, abandonment and small irregular movements		18%
Stable bends, no change		24%

Figure 12.4 A typology of river meander migration, based on observations of the
River Dane in Cheshire
Source: After Hooke 1984

processes (the observed form of bend migration in this case), rather than the processes themselves. The varying local contingent circumstances will inevitably randomise the 'products' and obscure the generalisation, unless their effects can be stripped out.

By contrast, Carson (1986) studied the 'anomalous' meander bend shapes of steep, high-power gravelly rivers on the Canterbury Plains in New Zealand, an example of whose prematurely inflected forms is shown in Figure 12.5. Through a combination of meticulous field observation of sediments, transport paths, flow dynamics, and channel form, and very detailed qualitative description of their interdependence, Carson was able to propose a convincing model for the shape and migration habit of bends of this type. The premature inflection displayed in these bends is characteristic of the high energy conditions of these gravel bed rivers, where the high momentum of flow and bedload inhibit development of an inward current at the bed across the point bar. The flow instead constructs an over-widened point bar as decelerating longitudinal flow diverges across the bar surface. The flow then converges and plunges into a deep scour pool adjacent to the outer bank; where banks are erodible, this is often associated with an undercut notch. However, particularly where the outer bank is more resistant, the flow then diverges and shoals rapidly on exiting from the scour pool, deposits a gravel bar at the outer bank, and is deflected towards the up-valley bank where bank erosion causes up-valley bend migration. This pattern of behaviour is particularly associated with high-energy, bedload-transporting rivers; asymmetric 'delayed inflection' meander bends are more characteristic of suspended or mixed load rivers, whose interaction between form and process behaves differently (Carson and Lapointe 1983).

Carson (1986: 890) refers to the 'myth of widespread inward deflection of bed flow' and attributes it to the theoretical notion of uniform circular motion common in mathematical approaches to the meander problem, but 'rarely verified in nature'. The emphasis on (numerical) generalisation has obscured the richness of behaviour in river channels that is best understood by a combination of (perhaps qualitative) deep process understanding, together with detailed field observation at levels from general reach properties to local conditions ('lupins recolonize the point bar ... their stems break, snap forward, become ... an additional indicator of local flow direction at the bed' (Carson 1986: 887). This can yield real insight into the richness of river behaviour. Furthermore, this assessment also demonstrates a point made by Hooke (1984) in quoting Hickin (1983): 'many river reaches may be in a transient state permanently'. This transient state reflects the fact that the flow is continually adapting to a continually changing 'container' (the channel). The overemphasis on 'uniform circular motion' in physical-mathematical approaches to meandering reflects simplification of boundary conditions in such approaches, which is now no longer necessary given the capacity to model depth-averaged flow in reaches of complex boundary shape (Whiting

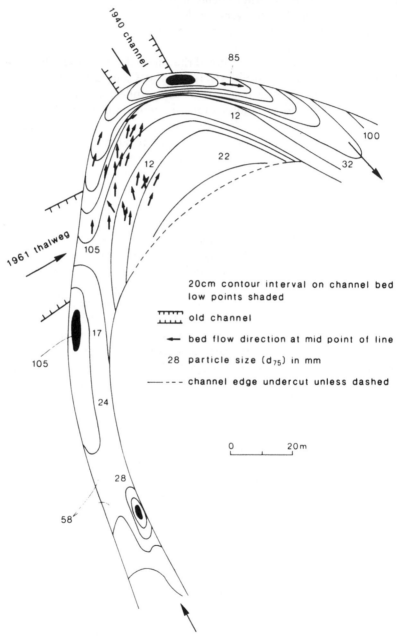

85

1940 channel

12

100

12

22

32

1961 thalweg

105

20cm contour interval on channel bed
low points shaded

old channel

bed flow direction at mid point of line

28 particle size (d₇₅) in mm

channel edge undercut unless dashed

17

105

24

0   20m

28

58

Figure 12.5 Bed topography, bed material sizes, and bed flow directions at
bankfull stage in an over-widened bend of the Waireka Stream on the
Canterbury Plains, New Zealand
Source: After Carson 1986

and Dietrich 1991; Lane, Richards and Chandler 1993, Figure 12.6). The richness of behaviour in rivers thus reflects the way spatially distributed feedback occurs between the flow and sediment transport processes, and the complex and evolving channel form (Richards 1988). This is why recontextualisation of the local conditions is necessary in each study of meander behaviour, and why generalisation is often premature if such recontextualisation has not occurred in other environmental studies.

MODELLING

Finally, the roles of modelling in scientific investigation need to be re-examined in the context of the realist perspective on the nature of scientific enquiry, given the increasing importance of modelling in the description, explanation and prediction of the behaviour of environmental systems. A traditional assumption is that models represent devices for making predictions (George 1967). This relates both to the critical rationalist methodology, and to technological rather than scientific objectives. For example, predictive models in hydrology are used in engineering hydrology for the purposes of estimating flood magnitudes prior to the design of structures, while physical hydrology is concerned with scientific modelling in which physical understanding is represented mathematically. Close links exist between key issues in modelling, and the issues discussed above in relation to realism: the problem of achieving accurate, precise and meaningful measurement of environmental variables; the explanatory objectives of realism; the frequently non-viable and untestable nature of predictions; and the importance of detailed observation of the local boundary conditions.

In geomorphology, scientific modelling plays an important role in accelerating time, it being impossible to monitor landform change given the timescales commonly involved. As Rhoads and Thorn (1993) note, in empirical, positivist geomorphology *abduction* rather than deduction is commonly necessary, in which the 'result' (the observed landform) and the (process-based) 'law' are used to reconstruct the former, unobservable, environmental 'initial conditions'. However, an alternative approach is to *simulate* landform development, by embodying in a numerical model a mathematical representation of the key processes (Kirkby 1984; Ahnert 1987). If the simulated landscapes resemble real landscapes, the scientist can have some confidence in the manners in which both the processes are embodied in the model's conceptual structure, and their mathematical representations are parameterised, and can even use the model time steps to estimate the real time required to produce the landscape being simulated (as in Ahnert's study of the Kall valley). Models may, however, be most usefully used in sensitivity analyses and the assessment of system behaviour under different scenarios, in which cases the consequences for landform development of alternative parameter values are evaluated. For example, a model-based assessment of the Holocene occurrence of shallow translational earthslides on slopes in Scotland is reported by Brooks *et al.* (1993). This couples a finite-difference model of infiltration with a shallow-slide stability analysis, and investigates the interaction between evolving podsol hydraulic properties and different rainstorm scenarios for periods of the Holocene, to identify the rainfall amounts and intensities likely to have triggered failure at different stages of soil development (dependent on soil thickness, horizon structure, density and texture).

It is now widely recognised in hydrology that model predictions cannot readily be validated, because of the problems of space–time discretisation, and spatial variability (Beven 1989). Although processes operate in nature continuously, their numerical representation (in finite difference or finite element formulations) must divide time and space into discrete increments. The result is that the model averages properties across these time and space steps, which may not accord in scale with the scale at which those properties are normally measured. Furthermore, attempts to measure these properties in the real world are confounded by their small-scale spatial variability, which implies that a single point measurement within an area represented by a model grid cell cannot be assumed to approximate the average for that cell. The result is that models are often best applied as tools for probing understanding, through sensitivity analyses and scenario simulation. Questions of which parameters most exercise the greatest influence on system behaviour under different conditions can be assessed, and comparison of simulation results leads to improved understanding, although not necessarily any greater capacity to predict, because of residual uncertainty about parameter values and the conceptualisation of process. An illustration is the study by Jain *et al.* (1992), which shows how a physically based model can be parameterised relatively readily for an 800 km^2 basin on the basis of about two weeks' fieldwork and available maps. However, the uncertainty surrounding rainfall inputs precludes prediction of specific hydrograph events. Instead, the model is used to examine scenarios, and sensitivity analyses are conducted in order to identify the key parameters, and the rate of change of system response to changes in specific parameters. These examples demonstrate modelling in its general rather than its specific role; it is an experimental device for probing understanding, rather than a method of obtaining specific predictions.

Spatially distributed simulation models are increasingly important: models that incorporate spatially distributed parameters, and simulate spatial feedbacks. Two consequences flow from the application of such models. First, they place considerable demands on detailed fieldwork to ensure that the spatially distributed boundary conditions are fully described. For example, Figure 12.6 required high resolution digital terrain models of the channel bed and water surface (obtained by a combination of analytical terrestrial photogrammetry and tacheometry), details of bed material size

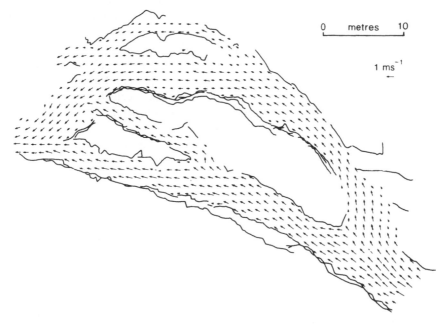

Figure 12.6 Depth-averaged flow vectors in a complex braided reach of the gravel-bedded meltstream of the Haut Glacier d'Arolla, simulated using a finite difference (volume) numerical solution of the Navier Stokes equations, with a rigid lid approximation and a two-equation k-ε turbulence model. The water surface was fixed photogrammetrically, and a detailed digital elevation model of the bed topography was obtained by similar means in conjunction with rapid tacheometric bed survey using a total station. Roughness estimates were based on surface clast counts, and a detailed inflow velocity distribution was measured by Braystoke current metering at multiple sites across the upstream section

distributions in order to provide bed roughness estimates, and inflow velocities across the upstream section. The rigorous application of a numerical model of a geomorphological process thus forces the researcher's attention to the detailed contextualisation required of a realist case study. Furthermore, key applications of modelling involve comparison of system response for differing assumed boundary conditions, and manipulation of spatially distributed boundary conditions to test the effects of their changes (for example, land-use changes affecting hydrological response, or roughness changes affecting the distribution of flow velocity, shear stress, and transport capacity). Second, it is clearly a waste of the sophistication of such a model if it is simply used to provide an output prediction (of the hydrograph at the basin outlet, in the case of a hydrological model); the model's internal state variables should also be interrogated. Indeed, the predicted system output can appear to be successful although the simulated internal system behaviour is incorrect – the right (output) answer for the wrong (internal) reasons. This is illustrated by Anderson and Burt (1992), in the case of hydrological models which give reasonable output hydrographs, but fail to simulate the correct spatial pattern of soil moisture. Thus, validation of the simulated velocity pattern in Figure 12.6 was based on comparison of several point measurements of resultant velocity made with two-component EMCMs, with the simulated resultant velocities in the appropriate grid cells (a correlation of r = 0.78 being obtained, with points distributed uniformly about the line of equality between observed and predicted velocities).

Spatially distributed sensitivity analyses are also desirable, because predictions of system behaviour at a single point may appear to be counter-intuitive when spatial feedbacks occur. This can again be illustrated with reference to the simulation in Figure 12.6. Increasing the bed roughness coefficient in a sensitivity analysis results in reduced velocity over the upstream bar head, but apparently anomalous *increased* velocity elsewhere (for example, along the upstream right-hand bank). This is because of spatial feedback in which extra flow is diverted laterally by the increased bar head resistance. Quite different sensitivities may therefore occur in different locations as the flow is redistributed spatially; however, these patterns may still be consistent with understanding of the physical processes incorporated in the model, and therefore add to confidence in the model's performance. However, although distributed models produce distributed predictions, and it is therefore possible to evaluate process responses in their correct spatial context, this generates large quantities of (simulated) data which are difficult to assimilate and interpret, and new approaches are required to handle this. Beven and Binley (1992) suggest qualitative comparison of simulated patterns of response with field-based understanding, and argue that visualisation techniques on graphics workstations may allow rapid display of distributed data for qualitative examination. An increasing need in the study of dynamic behaviour of environmental processes is for computer animation to permit visualisation of complex processes which unfold in space–time on scales that are otherwise difficult to comprehend. Studies of oceanic and atmospheric motion provide obvious examples, where the visual observation of complex patterns is a guide to the realism of circulation models. Thus the results of physical-mathematical modelling, based on physical understanding, may be judged not by quantitative significance testing, but by qualitative appraisal.

These observations demonstrate the links between current preoccupations in hydrological and hydraulic modelling, and the realist epistemological considerations of the preceding sections. It would appear that the roles of modelling are looser, broader and more complex than is often implied by critical rationalists. These aims are more than to make predictions; they are to improve and deepen understanding, and to contribute to the ability to explain the spatially complex nature of environmental processes. It is therefore part of the pluralistic epistemology of a realist science that is sceptical about measurement and prediction, but explicit about the importance of proper description of local space–time contingencies as a precursor to explanation of causal mechanisms.

CONCLUSIONS

Two broad aspects of the realist perspective particularly commend it to geographers; its emphasis on 'local' circumstances, and its interdisciplinarity. Renewal of a form of regional geography, or at least, place-centred research, might imply a preference for a realist approach to the explanation of events in particular places through the interaction of local circumstances with causal mechanisms operating at a range of scales (both beyond and within the 'place'). However, a discipline that has recently rediscovered the region may need to beware of assuming this to be a uniquely geographical method; other open-system sciences also evidently require thick description of regional context in their attempts to disclose environmental or sociological mechanisms. The region is therefore unlikely to provide a uniquely geographical structure for investigation, and in realist science is revealed as a methodological rather than a substantive device.

Chorley commended interdisciplinarity to geographers in 1969 when he edited *Water, Earth and Man*. A potential, but under-exploited, strength of physical geography has always been its concern with dynamic interactions among environmental phenomena (climatic, biotic, physical and chemical). In geography as a whole, the dynamic interaction between human beings and the physical environment perhaps needs to be rediscovered as the centre ground (as opposed to the static conception of resource management rejected as a basis of integration in geography by Johnston (1983)). Environmental problems present good examples of the kind of integrated place-centred research to which geography could aspire. For example, surface water acidification in upland catchments (and its effects on aquatic ecology) has been variously explained as the consequence of land-use change (particularly by coniferous afforestation), of long-term postglacial soil acidification, and of acid precipitation. In different catchments these contributory processes may vary in their relative importance. In a catchment rich in calcium carbonate sources, the natural buffering capacity of the soil may limit the degree to which acid precipitation is effective in causing acidification of surface waters. When the competing claims of interested groups of scientists and policy-makers about the causes of acidification are being evaluated, it is important that the local catchment conditions are evaluated which allow some mechanisms to operate and inhibit others. If a programme of liming is proposed to restore the pH of acidified waters, other environmental processes and phenomena must be understood in order that this strategy can be optimised, and it also becomes a matter for economic assessment, for consideration of the externalities introduced by the demand for limestone, and for place-sensitive policy formulation (in short, a problem for dynamic appraisal of environmental management options involving continually adjusting physical and human geographical issues). Such problems seem likely to be most successfully addressed by physical geographers through adoption of the realist research practices which have been discussed in this chapter.

ACKNOWLEDGEMENT

The Natural Environment Research Council is acknowledged for providing Research Studentships, all tenable in the Department of Geography at Cambridge, to K. Richards (1970–1973, supervised by R.J. Chorley), and to S. Brooks, N. Clifford, T. Harris and S. Lane (between 1984 and 1994, and supervised by K. Richards). It is hardly necessary to add that the influence of R.J. Chorley continues throughout this line.

REFERENCES

Ahnert, F. (1987) 'Approaches to dynamic equilibrium in theoretical simulations of slope development', *Earth Surface Processes and Landforms* 12: 3–15.

Anderson, M.G. and Burt, T.P. (1992) 'Subsurface runoff', in M.G. Anderson and T.P. Burt (eds) *Process Studies in Hillslope Hydrology*, Chichester: John Wiley & Sons.

Beck, M.B. (1987) 'Water quality modelling: a review of the analysis of uncertainty', *Water Resources Research* 23.

Beven, K.J. (1989) 'Changing ideas in hydrology', *Journal of Hydrology* 105: 157–172.

Beven, K.J. and Binley, A. (1992) 'The future of distributed models: model calibration and uncertainty prediction', *Hydrological Processes* 6: 279–298.

Bhaskar, R. (1989) *Reclaiming Reality: A Critical Introduction to Contemporary Philosophy*, London: Verso Press.

Brooks, S.M., Richards, K.S. and Anderson, M.G. (1993) 'Shallow failure mechanisms during the Holocene: utilization of a coupled slope hydrology-slope stability model', in D.S.G. Thomas and R.J. Allison (eds) *Landscape Sensitivity*, Chichester: John Wiley & Sons.

Carson, M.A. (1986) 'Characteristics of high energy "meandering" rivers: the Canterbury Plains, New Zealand', *Geological Society of America, Bulletin* 97: 886–895.

Carson, M.A. and Lapointe, M.F. (1983) 'The inherent asymmetry of river meander planform', *Journal of Geology* 91: 41–55.

Chorley, R.J. (1962) 'Geomorphology and general systems theory', *U.S. Geological Survey, Professional Paper* 500-B.

——(1965) 'A re-evaluation of the geomorphic system of W.M. Davis', in R.J. Chorley and P. Haggett (eds) *Frontiers in Geographical Teaching*, London: Methuen.

——(1966) 'The application of statistical methods to geomorphology', in G.H. Dury (ed.) *Essays in Geomorphology*, London: Heinemann.

——(ed.) (1969) *Water, Earth and Man: A Synthesis of Hydrology, Geomorphology and Socio-economic Geography*, London: Methuen.

——(1978) 'Bases for theory in geomorphology', in C. Embleton, D. Brunsden and D.K.C. Jones (eds) *Geomorphology: Present Problems and Future Prospects*, Oxford: Clarendon Press.

Chorley, R.J. and Kennedy, B.A. (1971) *Physical Geography: A Systems Approach*, London: Prentice Hall.

Clifford, N.J., Richards, K.S., Brown, R.A. and Lane, S.N. (1995) 'Scales of variation of suspended sediment concentration and turbidity in a glacial meltwater stream', *Geografiska Annaler* 77A: 45–65.

Culling, W.E.H. (1987) 'Equifinality: modern approaches to dynamical systems and their potential for geographical thought', *Transactions, Institute of British Geographers*, NS 12: 57–72.

Eversden, J. (1969) 'Some aspects of teaching geography through fieldwork', *Geography* 54: 64–73.

George, F.H. (1967) 'The use of models in science', in R.J. Chorley and P. Haggett (eds) *Models in Geography*, London: Methuen.

Gibbon, G. (1987) *Explanation in Archaeology*, Oxford: Basil Blackwell.

Grigg, D. (1967) 'Regions, models and classes', in R.J. Chorley and P. Haggett (eds) *Models in Geography*, London: Methuen.

Hacking, I. (1983) *Representing and Intervening*, Cambridge: Cambridge University Press.

Haines-Young, R.H. and Petch, J.R. (1983) 'Multiple working hypotheses: equifinality and the study of landforms', *Transactions, Institute of British Geographers*, NS 8: 458–466.

——(1986) *Physical Geography: Its Nature and Methods*, London: Harper & Row.

Harré, R. (1970) *The Principles of Scientific Thinking*, London: Macmillan.

——(1983) *Great Scientific Experiments: Twenty Experiments That Change Our View of the World*, Oxford: Oxford University Press.

Harris, T.R. and Richards, K.S. (in press) 'Bed-stored fines release and water quality variation in a lowland stream', *Earth Surface Processes and Landforms*.

Hickin, E.J. (1983) 'River channel changes; retrospect and prospect', in J.D. Collinson and J. Lewin (eds) *Modern and Ancient Fluvial Systems*, Oxford: Basil Blackwell.

Hooke, J.M. (1984) 'Changes in river meanders: a review of techniques and results of analyses', *Progress in Physical Geography* 8: 473–508.

Iseya, F. and Ikeda, H. (1987) 'Pulsations in bedload transport rates induced by a longitudinal sediment sorting: a flume study using sand and gravel mixtures', *Geografiska Annaler* 69A: 15–27.

Jain, S.K., Storm, B., Bathurst, J.C., Refsgaard, J.C. and Singh, R.D. (1992) 'Application of the SHE to catchments in India. Part 2. Field experiments and simulation studies with the SHE on the Kolar sub-catchment of the Narmada River', *Journal of Hydrology* 140: 25–47.

Johnston, R.J. (1983) 'Resource appraisal, resource management and the integration of physical and human geography', *Progress in Physical Geography* 7: 127–146.

Kennedy, B.A. (1993) '"... no prospect of an end"', *Geography* 78: 124–136.

Kirkby, M.J. (1984) 'Modelling cliff development in South Wales: Savigear revisited', *Zeitschrift für Geomorphologie* 28: 405–426.

Lane, S.N., Richards, K.S. and Chandler, J. (1993) 'Within-reach spatial patterns of process and channel adjustment', in E.J Hickin (ed.) *River Geomorphology*, Chichester: John Wiley & Sons.

Lane, S.N., Richards, K.S. and Warburton, J. (1993) 'Comparison between high frequency velocity records obtained with spherical and discoidal electromagnetic current meters', in N.J. Clifford, J.R. French and J. Hardisty (eds) *Turbulence: Perspectives on Sediment Transport*, Chichester: John Wiley & Sons.

Lapointe, M. (1992) 'Burst-like sediment suspension events in a sand-bed river', *Earth Surface Processes and Landforms* 17: 253–270.

Linton, D.L. (1952) 'The problem of tors', *Geographical Journal* 121: 470–487.

Markus, G. (1987) 'Why is there no hermeneutics of natural sciences? Some

291

preliminary theses', *Science in Context* 1: 5–51.

Mitchell, J.C. (1983) 'Case and situation analysis', *Sociological Review* 31: 187–211.

Nadler, C.T. and Schumm, S.A. (1981) 'Metamorphosis of South Platte and Arkansas rivers, eastern Colorado', *Physical Geography* 2: 96–115.

Palmer, J. and Nielsen, R.A. (1962) 'The origin of granite tors on Dartmoor, Devonshire', *Proceedings of the Yorkshire Geological Society* 33: 315–340.

Pawson, R. (1989) *A Measure for Measures: A Manifesto for Empirical Sociology*, London: Routledge.

Phillips, D.C. (1992) *The Social Scientist's Bestiary: A Guide to Fabled Threats to, and Defences of, Naturalistic Social Science*, Oxford: Pergamon.

Pitty, A.F. (1979) 'Conclusions', in A.F. Pitty (ed.) *Geographical Approaches to Fluvial Processes*, Norwich: GeoBooks.

Ragin, C.C. and Becker, H. (eds) (1992) *What Is a Case? Exploring the Foundation of Social Inquiry*, Cambridge: Cambridge University Press.

Rhoads, B.L. and Thorn, C.E. (1993) 'Geomorphology as science: the role of theory', *Geomorphology* 6: 287–307.

Richards, K.S. (1987) 'Rivers: environment, process and form', in K.S. Richards (ed.) *River Channels: Environment and Process*, Oxford: Basil Blackwell.

——(1988) 'Fluvial geomorphology', *Progress in Physical Geography* 12: 435–456.

——(1990) '"Real" geomorphology', *Earth Surface Processes and Landforms* 15: 195–197.

——(1994) '"Real" geomorphology revisited', *Earth Surface Processes and Landforms* 19: 277–281.

Richards, K.S., Chandra, S. and Friend, P.F. (1993) 'Avulsive channel systems: characteristics and examples', in J.L. Best and C.S. Bristow (eds) *Braided Rivers*, Geological Society of London Special Publication.

Richards, K., Sharp, M., Arnold, N., Gurnell, A., Clark, M., Tranter, M., Nienow, P., Brown, G., Willis, I. and Lawson, W. (1996) 'An integrated approach to modelling hydrology and water quality in glacierised catchments', *Hydrological Processes* 10: 479–508.

Rouse, J. (1989) *Knowledge and Power: Toward a Political Philosophy of Science*, Ithaca: Cornell University Press.

Sarre, P. (1982) 'Realism in practice', *Area* 19: 3–10.

Sayer, A. (1984) *Method in Social Science: A Realist Approach*, London: Hutchinson.

Sharp, M.J., Richards, K.S., Arnold, N., Lawson, W., Willis, I., Nienow, P. and Tison, J.-L. (1993) 'Geometry, bed topography and drainage system structure of the Haut Glacier d'Arolla, Switzerland', *Earth Surface Processes and Landforms* 18: 557–571.

Stoddart, D.R. (1987) 'To claim the high ground: geography for the end of the century', *Transactions, Institute of British Geographers*, NS 12: 327–336.

Whiting, P.J. and Dietrich, W.E. (1991) 'Convective accelerations and boundary shear stress over a channel bar', *Water Resources Research* 27: 783–796.

Woolgar, S. (1988) *Science: The Very Idea*, Chichester: Ellis Horwood.

Wrigley, N. (1985) *Categorical Data Analysis for Geographers and Environmental Scientists*, London: Longman.

Yatsu, E. (1992) 'To make geomorphology more scientific', *Transactions, Japanese Geomorphological Union* 13: 87–124.

13

OPEN SYSTEMS – CLOSED SYSTEMS

A Portuguese vignette

Robert J. Bennett

One of Richard Chorley's most significant contributions to geomorphology was to draw attention to the interaction between geomorphological and human systems. In his 1973 comments he quotes George Perkins Marsh's nineteenth-century recognition of man's ability to dominate the landscape (Chorley 1973: 157–159). Later, in *Environmental Systems*, he argued against ecological models that place society in too subordinate and ineffectual a role (Bennett and Chorley 1978). He described the interaction between human and environmental systems as one of 'interfacing'.

Chorley saw the analysis of social and geomorphological interfaces as fundamentally bound up with issues of stability, reaction and relaxation times. As in so much of his work he was naturally attracted to systems theory to unravel these complexities. In one of his first contributions on general systems theory (Chorley 1962), he was concerned to distinguish cyclic time, graded time and steady time. Lying behind this was his attempt to relate research on short-term channel adjustments from the great traditions of Gilbert, whom he considered to have been undervalued by many other geomorphologists, to the Davisian and Penckian concepts of long-term landform evolution. Chorley's use of general systems theory was the main anchor of his early argument. But general systems theory, like the Davisian cycle, also led to concepts that emphasised steady state and equilibrium over disequilibrium. As a result, although emphasising dynamic and open systems, most general systems views were ironically fundamentally closed. Like the developments in the subject of economics, an overemphasis on the concept of equilibrium diverted attention from the key factors that prevented equilibrium from occurring. Chorley recognised this problem. In the progress from his work on general systems theory through *Models in Geography* to *Physical Geography* and *Environmental Systems*, Chorley increasingly recognised the importance of the forces leading to disequilibrium as well as equilibrium solutions. This led to his wider conceptualisation of open systems as disequilibrium systems. This naturally led him to the

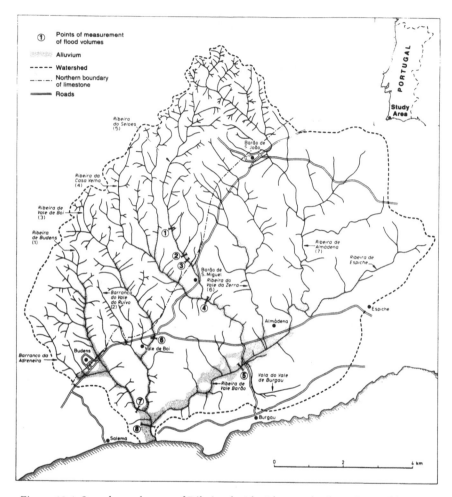

Figure 13.1 Sample catchment of Ribeira de Almádena and tributaries, and location in southern Portugal. The eight points of measurement of flood volumes are marked

increasing recognition of the influence of human actions on geomorphological development (see Chorley 1962, 1967; Chorley and Kennedy 1972;
Bennett and Chorley 1978).

Chorley was not alone in recognising these important conceptual developments, but he was one of the first to grasp the implications of their
significance for the whole subject of geomorphology. Nowadays it is routine
to look at disequilibrium, thresholds and the environment–landform–human
interface. This was not routinely the case in the 1960s or even 1970s. Richard
Chorley was one of the first to pursue these ideas and disseminate them
through his highly successful writings and teaching. His undergraduates and

doctoral students have become an important force in the subsequent development of geomorphological research. This chapter examines how some of Chorley's ideas on interfacing apply in the context of one specific region, the Algarve of southern Portugal. The main catchment that is examined in this chapter within the Algarve is shown in Figure 13.1. The research on which the chapter is based is drawn, in part, from the fieldwork of students who were strongly influenced by Chorley's teaching.

CONCEPTS OF GEOMORPHOLOGICAL DEVELOPMENT OF SOUTHERN PORTUGAL

Southern Portugal has been an area for vigorous debate and rival hypotheses as to the forces that dominate landform evolution. The debate has reached a similar stage of development as that for south-east England in the 1960s when Chorley opened his case for an open-systems perspective. The traditional view has been based on an account of landscape evolution developed by Feio (1951) and Zbyszewski (1958) who have sought to recognise a series of surfaces stretching from the Iberian meseta into the southern Portuguese mountains and littoral. An important benchmark for this chronology is a 'Pleistocene' marine deposit. This deposit, dated to the Lower Calabrian by largely circumstantial evidence, covers a wide area over much of the land above 160 m. After the Calabrian, however, there is little in the way of datable depositional materials until more recent archaeological and preserved vegetation materials give some indications of dates of some deposits. Hence for most of the period between about 2 million years and about 8000 years BP there are only landform features of terraces and surfaces from which to develop a chronology.

Using this fragmentary evidence Feio and Zbyszewski, who published 281 papers on the subject between 1933 and 1979, have sought to identify and correlate marine and aeolian sediments on littoral sites. From these Zbyszewski extrapolated inland, largely using altimetric correlation of river terraces linked to the chronology of interglacial episodes from northern Europe, the relationship of which to the glacial history of northern Portugal has been inferred by Feio (1946, 1949, 1951). Zbyszewski's (1958) Quaternary sequence, shown in Tables 13.1 and 13.2, is thus largely still a morphological chronology which has to assume an altitude sequence as a surrogate for time, i.e. higher deposits are assumed to be older.

There is undoubtedly considerable merit in the Feio–Zbyszewski research and it has been accepted uncritically in the geological survey of the area (see for example Rocha *et al.* 1979, 1983) and in Embleton's study (see Sala 1984). The research certainly provides a baseline for subsequent study. However, the morphological and altimetric evidence alone is clearly unsatisfactory and reflects many of the concerns that Chorley ridiculed in much of the research on south-east England which had been used by Wooldridge and Linton

Table 13.1 Association of surfaces and recent geological events in southern Portugal

Age	Height of hypothesised surface / terraces (m)	Stage	Southern Portugal
Pleistocene	5–8 10–15 20–35 50–60 80–90	Grimaldrian II Grimaldrian I Tyrrhenian Millazian Sicilian	Progressive regression forming terraces at successively lower levels. Oscillations of sea in glacial periods
	120–130 160–170	Upper Calabrian Lower Calabrian	Marine sands of Lower Calabrian. The 160 m surface is the most significant in the serra of west Algarve
Pliocene		Astian Plaisantian Pontian	A period of aridity with deposition of the continental deposits of 'raña' or 'calhausmal rolados'. Subtropical climate. Very important period of tectonic unrest along old Hercynian trends. Mesquita and San Marcos Faults. Becoming more important at end of period with graben of Aljezur
Miocene		Sarmatian Tortonian	Regression and continued planation. Beginning of exhumation of Monchique massif
		Helvetian	Marine deposits. End of the major planation of Lower Alentejo at a height of 240 m at Évora and 200 m at Beja

Source: After Feio 1951

to produce a 'staircase' of evolution since the Cretaceous.

The analysis of interfacing of current processes and their relative role in sediment transport in the past is completely absent in the Feio–Zbyszewski work. More recent research in southern Portugal has made some attempts to move forward by dating some of the deposits using ^{14}C and other methods. Devereux (1983a), for example, was able to relate some of the valley fill alluvium in the western part of the area to the end of the last (Würm) glaciation through a dated deposit at 7450 ± 90 years BP. Other dates in Devereux's work relate to valley fills in a dispersed range of sites. Dates range from 1750 ± 90 years BP to 780 ± 50 years and 520 ± 60 years BP. A related

Table 13.2 Quaternary chronology for southern Portugal

Event	Marine series	Continental series	Archaeological assemblages	Date (years BP)
Flandrian	Silting estuaries	'Muddy sands' Palaeodunes in the centre and south-west	Upper Palaeolithic and Languedocian	6000–1800
Würm (glacial)	Max. marine regression to –100 m			
Tyrrhenian III	Beaches 5–8 m	Terraces 5–8 m	Mousterian, Languedocian and 'Final' Acheulian	120,000–6000
Tyrrhenian II	Beaches 12–25 m	Terraces 12–20 m	Languedocian	200,000–120,000
Riss (glacial)		'Gravels & pebbles'		
Tyrrhenian I	Beaches 30–35 m	Terraces 20–40 m	Levalloisian Tayacian Acheulian upper (Micoquian) middle lower	
Mindel (glacial)		'Gravels & pebbles'		600,000–300,000
Sicilian II	Beaches 50–60, 60–70 m	Terraces 45–50, 60–70 m	Abbevillian	700,000–600,000
Sicilian Ib	Beaches 70–90 m	Terraces 70–80 m		
Gunz		'Gravels & pebbles'		1.2 m–700,000
Sicilian Ia	Beaches 90–100 m	Terraces 90–100 m	Abbevillian?	2.0 m–1.5 m
Calabrian	Platforms at: 115–130 m 150–160 m 180–190 m			

Source: After Zbyszewski 1958, modified from Feio 1951

valley alluvium in the central mountain area of the Algarve has been dated to 2690 ± 70 BP by Chester and James (1991). These dates do indicate active erosion and deposition in channels in the Algarve littoral and mountain valleys up to 500 years ago. Devereux, like Chester and James, also attempts to relate the ^{14}C dates to a more general sequence of valley filling, drawing analogies with the work of Vita-Finzi (1969). It is not entirely clear whether this analogy is valid. Vita-Finzi was working primarily in the eastern Mediterranean whereas Portugal faces the Atlantic and has been subject to a different recent tectonic history. There is also considerably less archaeo-logical material available in Portugal than that available to Vita-Finzi, so that any general dating is not possible. Certainly there is considerable morpho-logical evidence of fill in many valleys, but whether this relates to sequences of fill–erosion–fill seems doubtful. There is also some evidence that the upper valley areas close to, but beneath, the level of the Lower Calabrian surface may have been affected by periglacial processes, which is conjectured as reworking some of the higher-level pre-Calabrian deposits (Daveau 1973). Chester and James (1991) also conjecture on the interrelation of cut and fill sequences to social activity, associated particularly with late Roman and Moorish settlement up to about 700 BP, followed by valley incision during a period of agricultural abandonment, to be further followed by more alluvial fill as a result of rapid soil erosion during the Salazar expansion of cereals in the mountain area from 1930 to 1970. The last twenty years they recognise as one of further land-use change as a result of abandonment of the mountain cereal cultivation and invasion by scrub garrigue.

Chester and James's analysis is closer to the conception of the open-system dynamics that affect landscape erosion. But they too tend to fall prey to the trap of monocausal explanation by over-emphasising the role of human intervention over geological processes. The debate in southern Portugal, therefore, has tended to neglect the full detail of the field area that must be exhaustively examined for all possible influences before a full open-systems perspective can be embraced. The crucial gap in these previous analyses has been assessment of the interaction of human interference with the main bedrock of the Algarve littoral, which is limestone. This interaction is examined below.

An extensive area of limestone forms the main bedrock of the Algarve littoral. This is an important feature which can be expected to act in a fundamental way to control the erosional development of the area. This influence has been neglected in all earlier research on the geomorphological evolution of southern Portugal. The hypothesis which is investigated below is that the limestone belt has acted as a form of control system regulating fluvially dependent evolution. The limestone outcrops in a belt 2–50 km wide across the southern littoral of the area. It frequently covers the whole littoral from just below the Lower Calabrian surface to the sea. It covers all of the area surveyed by Devereux in his 'eastern valleys', most of the reaches of his 'western valleys', and the lower reaches of Chester and James's analysis of the Odelouca, Arade and Enxerim valleys. In a more pluvial period a limestone area may allow water tables to rise, which allows river erosion and sediment transport, but in more arid times, such as the present, the capacity for transportation is severely reduced as a large proportion of the water disappears underground. Thus the limestone controls a form of on/off switch for fluvial erosion. The importance of this control is reflected in survey research, which shows that typically channel widths and depths reduce by 20% and channel slopes by 30% over the first 1.5 km of flow over limestone bedrock (see Table 13.3). This pattern is of course highly variable depending on the extent of alluvial fill from more impermeable material, derived from up-river, deposited in the valley floor, as well as the height of the local water table, which depends on the local relative relief.

The patterns of alluvial fill and the variable transport capacity of the channels controlled by the limestone water table suggest that there is a very complex erosional system operating in this area in which small changes in the overall rainfall/evaporation balance may have disproportionately large influences on the capacity of channels, slope processes, and the development of landform features. The limestone can act as a system control device which regulates a threshold between active or inactive fluvial erosion dependent on quite small changes in the height of the water table. As a result, very inactive periods may be punctuated by highly active phases of erosion. If the controlling parameters are close to the threshold the on/off erosion and

Table 13.3 Comparison of channel geometry above and below the limestone junction for four tributaries in the Ribeiro de Almádena catchment[1]

Tributaries		Upstream	Downstream	% Difference
Rib. de Budens	width	418	272	−34.9
	depth	44	34.4	−21.8
	slope	0.4	0.4	0
Rib. de Seloes	width	630	629	−0.2
	depth	252	180	−28.6
	slope	1.5	1.1	−26.7
Rib. de Vale da Zerra	width	870	672	−22.8
	depth	245	280	+14.3
	slope	1.8	1.2	−33.3
Rib. de Almádena	width	559	324	−42.0
	depth	148	95	−35.8
	slope	2.1	1.5	−28.6

Note:
[1] Mean of five measurement points, at 150 m intervals immediately above, and from 0.75 km to 1.50 km below the limestone junction (width and depth in cm, slope in degrees)

transport can be regulated by only slight shifts in the water balance as a result of changes in rainfall or human interference (as, for example, observed in other areas by Sweeting 1972). On the ground this would be recognised by frequent alternate channel filling and incision as queues of sediment material respectively accumulate or disperse. This conceptualisation in other environments has been developed by Schumm (1979) and Brunsden and Thornes (1979) into a wider model of sediment queuing and threshold controls. Such a conceptualisation also fits closely to Chorley's open-system ideas. It appears more relevant to the Portuguese area than the concept of a single time arrow of development giving a 'staircase' in the form of an altitudinal sequence of landforms, as hypothesised by Feio and Zbyszewski, or indeed by Devereux using comparison with Vita-Finzi's research. It also widens the control parameters considered in the Chester and James analysis of anthropogenic factors. We investigate below how this on/off erosion switch has operated as a result of human interference.

INVESTIGATION OF THE ON/OFF EROSION MODEL

A specific region of the western part of southern Portugal is used in the discussion below to illustrate how Chorley's disequilibrium and threshold concepts based on the control systems idea can be applied to geomorphological evolution. The region discussed is shown in Figure 13.1, together with the features of the main catchment within which most fieldwork was undertaken. The fieldwork was developed over 1984, 1985 and 1990. The area covered focuses on the littoral belt below the Lower Calabrian plateau and mountain area. The littoral varies from a 30 km-wide belt west of Silves to 15–20 km between Lagos and Portimâo, narrowing to 10–15 km north of Burgau where the study catchment is located. It disappears between Burgau and Vila do Bispo where the Lower Calabrian 'surface' reaches the coast, often in a formidable line of cliffs. The northern edge of this littoral zone is usually founded on relatively impermeable Carboniferous slates and shales. This is followed by a narrow band of impermeable Triassic marls with occasional dolerite intrusions. The upper reaches of the channels all derive from this slate or marl area. The river systems then transverse a wide area of Jurassic limestones. Many of these are massive-bedded and dolomitic, giving rise to classic karstic features. Other limestones are thin-bedded and in places grade into shaly limestones and sandstones. However, the sequence is dominated by highly permeable limestone or sandy limestone except where a cover of alluvial fill occurs. The geology is often complex and modified by faulting. The boundary of the limestone belt is shown in Figures 13.1 and 13.2.

The littoral is covered by extensive areas of limestone plateaus (the 'Barrocal'). But between plateaus there are steeply incised valleys, often with wide alluvial floors. Whilst some of the immediate coastal valley floors evidence marine and estuarine infill, of Flandrian age, in general most of the valley-floor deposits display cut and fill graded to terrestrial processes. Thus, since the Lower Calabrian, active incision and fluvial deposition have occurred. This is further supported by the low relative relief of most of the area under investigation. In the area under investigation the local relative relief is generally 30–60 m between plateaus and valley floors. As a result the water table is generally close to, or at, the surface in the valley floors. A few major channels have perennial flow, whilst most have flow in a 3–6-month period of the winter–spring.

Whilst considerable incision has occurred in the past, there is some evidence of more recent acceleration of valley and channel infill. Using air photograph coverage of the area in 1947 and 1972 at a scale of approximately 1:10,000, and fieldwork measurements of the same sites in 1981, Devereux (1983) was able to examine long-term channel characteristics. The method of calculation of channel widths and depths used a calibrated stereometer, whilst field measurements attempted to identify the same sites in the field. The accuracy of the photographic interpretation is open to some doubt since the depth of channels used (60–300 cm), and widths (600–1600 cm), stretch the tolerance of the equipment and the operator to the limit, given the original scale of the photographs. Also the tolerances from the air-photographic work do not necessarily align with those in the field measurements. There are also difficulties of definition arising from the frequent occurrence of channel-in-channel phenomena. However, the overall consistency of Devereux's results indicates some useful general conclusions.

The general pattern of channel development observed by Devereux was for channel infill over the period 1947–1981. He argued that erosion is currently minimal, with the only significant erosion occurring in upstream reaches and western valleys between 1947 and 1972 and equalling about 1.1 cm per year from channel banks. He found sedimentation between channel banks to be far more important, with rates varying between about 12 cm per year in upstream areas to about 6 cm per year in the downstream sections of the western Algarve valleys. In the 1972–1981 period in the western Algarve he argued that channel-bed sedimentation had increased to more than 25 cm per year in upstream sections and more than 14 cm per year in downstream sections, while channel bank accretion was 10.5 cm per year and 4.5 cm per year respectively in the two reaches.

Devereux accounts for the changes in channel morphometry in terms of subtle climatic shifts during the past thirty years. Analysing precipitation data he notes that there was an increase in the mean annual total from 1950 to 1981, together with a shift in the seasonal pattern, particularly in terms of the increased contribution to annual totals by the months of June and August. With an increase in precipitation in these normally arid months there could be a widespread increase in channel vegetation in the Algarve. In

ROBERT J. BENNETT

addition, in the western Algarve the same climatic data show a decrease in evaporation rates for most months of the year and a consequent increase in effective rainfall in May, July and August. Devereux dates the 'sudden' decrease in evaporation rates in the western Algarve to 1972, following which rapid sedimentation and increasing channel vegetation growth occurred. In addition, the channels are predominantly stony which increases stream bed friction and so aids sedimentation.

Whilst there is thus some evidence of changed precipitation regimes on channel morphology, the speed and extent of changes appear to be too great to have occurred solely as a result of the rather minor changes in precipitation that Devereux discusses. This leads to a need to search for other potential influences, the most obvious one of which is human interference, a view echoed by Chester and James. There are three chief sources of such influence. First is the dramatic increase in demand for water as a result of the rapid increase of tourism and second homes in the area. Second is the changing pattern of land use. Third is the influence of the large network of small and large dams built in the area.

The demand for water in this area has increased rapidly since the 1960s. Population in the Lagos and Portimão administrative districts (concelhos) has increased by 150% and 80% respectively between the Censuses of 1950 and 1990. A large part of this increase has occurred in the most recent years, 30% in Lagos and 45% in Portimão concelhos between 1971 and 1987. Water supply relies heavily on aquifers but the total increase in abstraction has not been fully measured. Local public supply systems in Lagos and Portimão have increased water abstraction at a rate of 300–500% between 1950 and 1990, greater than the rate of population growth because of increased demand per head (from swimming pools and increased development and mains water supply) and increased seasonal tourist demand. But in addition to public supply systems many large water-users, such as major hotels, and scattered rural houses have developed their own boreholes to tap the aquifer. This has been largely uncontrolled. Illegal wells are believed to far outnumber licensed ones. Groups of farmers and householders have often formed co-operatives to drill large-capacity wells into deep aquifers. The result has been a rapid lowering of water tables across the whole littoral, particularly in the immediate 1–3 km coastal zone that has attracted most tourist development.

Changing land-use patterns among farmers has been a further feature increasing the demand on water. A substantial shift from the traditional agricultural systems that were based on winter rains has occurred. Instead of winter wheat and 'dry' tree crops such as figs, olives and almonds, extensive areas have been developed for irrigated vegetables, often under glass, as well as 'thirsty' citrus and other fruit trees. Agriculture is estimated to take 90% of all water resources in the area (MHOPT 1982) and its demand has increased more rapidly than that of the population because of the expansion

302

of the cultivated area for exporting foodstuffs.

Dams have also been constructed to provide additional non-aquifer water supplies. Two large schemes constructed in the area in the 1960s at Bravura and Silves have a combined capacity of over 40 million cubic litres. This has reduced the flow of the two major rivers (Odeaxere and Arade) to a trickle. In addition, a wide range of smaller dams and reservoirs has sprung up as a result of both local government and private sector activity. Although none of them affect the Almádena catchment, dams have cut off or reduced the flow in many other catchments since the late 1970s, they affect the adjacent catchment of the Odeaxere and hence must draw down some of the water table in the Almádena, and the area is also widely affected by boreholes, particularly for Burgau and Luz.

As a result of these influences, river flow in most catchments, including the Almádena, has been reduced in winter months to very small volumes and has ceased in the summer. The water tables have been lowered in all areas. In inland areas a comparison of 1968 and 1990 levels of the Vale Fuzeiros valley near Silves is possible because of historical data available for that valley: it shows the water table measured in wells to have been lowered by 10–20 m. In the coastal margins the water table has been lowered sufficiently in many cases to allow entry of the saline wedge. Public authority records show that it is up to 100 m lower than in 1970, for example around Lagos, which is close to the Almádena.

Land-use changes in the area have also been extensive. Figure 13.2 shows a comparison of the 1951 land-use surveys of a sample area and a 1990 physical survey. Comparison of the 1990 position with earlier surveys in 1984 and 1985 demonstrate the process of land-use change to have been one that is steadily developing. The pattern of change can be divided into two main areas. First, the northern half of the sampled catchment, on the carboniferous slates north-west of the littoral boundary, has suffered extensive abandonment of cultivation with resulting expansion of scrub – a net effect of 28% of the surface area. This, combined with expansion of tree cultivation, is likely to have significantly reduced runoff in the upper part of the catchment. Second, in the southern half of the area, where the bedrock is a mixture of limestones, shales, marls and alluvium with a much lower drainage density, the area has experienced a net abandonment of cultivation of 13.7% of the surface area, but significant intensification of irrigation as well. This has increased water demand for irrigation from the water table. Overall the catchment has experienced extensive land-use change which is likely to have significantly reduced total runoff. The pattern of change is summarised in Table 13.4.

The effect of changing land uses was tested by measuring the infiltration rate on comparable soils and topographic conditions in cultivated and uncultivated conditions at fourteen sample sites. The results of this test, shown in Table 13.5, demonstrate that the abandonment of cultivation on the

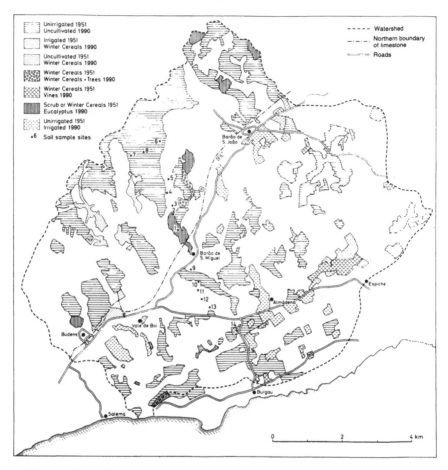

Figure 13.2 Land-use change in the Ribeira de Almádena catchment, 1951–1990.
Locations of sites for measurement of infiltration rates are also shown

slate soils and alluvial fills has greatly increased their infiltration capacity, by
2 to 6 times, in all cases except in the slate valley floors. In contrast, increases
in cultivation in the limestone littoral result in increases in infiltration rates.
Overall the balance of increased over decreased infiltration rates indicates
that about 20% of the catchment surface area has increased infiltration
significantly, whilst approximately 8% has decreased infiltration rates.

Reduction in normal channel flows would not in itself indicate that the
erosion potential of the rivers had been reduced. Although the erosion
system appears to have been more inactive for a larger proportion of time
than in the past, it may be that when a major rainfall event occurs channel
erosion and transport may be more readily reactivated and its effects may be
more significant. However, this view can be discounted on two counts. First,

Table 13.4 Land-use changes in the Ribeira de Almádena sample catchment, 1951–1990

Land use		% of catchment area changed		
1951	*1990*	*Slate serra*	*Limestone littoral*	*Total*
1 Cultivated	Uncultivated	31.1	13.7	22.2
2 Irrigated	Winter cereals	–	0.3	0.1
3 Uncultivated	Winter cereals	3.1	–	1.5
4 Winter cereals	Winter cereals & trees	1.1	1.1	1.0
5 Winter cereals	Vines	–	0.2	0.1
6 Scrub or winter cereals	Eucalyptus	2.1	0.5	1.3
7 Unirrigated with crops	Irrigated	–	4.2	2.2
Total gross area affected by change		37.4	20.0	28.4

Table 13.5 Infiltration rate (mm/hr) averaged over sample sites in the Ribeira de Almádena catchment; 1990 fieldwork[1]

Bedrock	Uncultivated			Cultivated		
	Interfluve	*Hillslope*	*Valley floors*	*Interfluve*	*Hillslope*	*Valley floors*
Slates (Av. unused = 790) (Av. used = 610)	1195	436	738	521	Winter cereals: 225 Eucalyptus forest: 945	1084
Limestone littoral (Av. unused = 181) (Av. used = 327)	135	167	240	386	279	315
Alluvium (Av. unused = 1289) (Av. used = 204)			1289			204

Note:
[1] Rates given are the steady-state rates averaged over 14 dispersed sample sites in each geologic and topographic situation. The location of sample sites is shown in Figure 13.2

the manner of channel fill has been such that flow velocities have been severely reduced. Considerable channel and bank vegetation has occurred in almost all rivers. A channel-in-channel phenomenon has developed widely as a result of reduced flows. This has allowed strongholds of bushes, grass and weeds to establish within the channel whilst leaving a small flow-way clear for normal discharges. Often the clear flow-way is less than 25% of the total former channel width. The Manning coefficient has therefore increased from 0.04 to 0.07 or greater in most cases. This results in a halving of flow velocity under bankfull conditions. In turn this reduces the Reynolds and Froude numbers and the sediment transport capacity. It appears that the sediment transport capacity of the rivers in this area has been at least halved since the aerial photographic surveys of 1947. Also because of the influence of the limestone bedrock, the slower flow has allowed greater time for bed and bank infiltration so that reduced velocity reduces transport capacity to a greater extent downstream than the normal predictions of the Reynolds and Froude numbers. The reduced heights of the water table have generally increased the potential for this infiltration considerably.

A second source of evidence is measurement of the effect of recent flood events. The year 1989 provided a major rainfall event. Over a 48-day period of October–December 1989 924 mm of rain fell across the region under investigation and four daily rainfall maxima exceeded 60 mm. This represents at least a 90-year event according to available historical records of the Portuguese meteorological office. The 30-year annual average precipitation is 450 mm. The main event, however, was a storm on 14 October 1989, of 158 mm measured at Faro 80 km away, and believed to be of larger magnitude in the sample catchment, where unfortunately no records are maintained. In the catchment area depicted in Figures 13.1 and 13.2 it was possible to measure maximum flood volume as a result of this storm in eight locations marked in Figure 13.1. These gave flood heights ranging from zero to 1.9 m above bankfull.

The flood discharges estimated from sampling of the extensive flood debris are shown in Table 13.6. The most notable feature is the small increase in flood discharge between points 3, 4 and 5 on the same channel, the Ribeira do Seloes. Point 3 is on alluvial fill just before the limestone littoral. Point 4 is below the joining of a major tributary that doubles the catchment area, but is 0.5 km into the limestone area. Point 5 is after the joining of three major rivers that approximately triple the catchment area and is on alluvial fill across the limestone. The minimal increase in flood discharges between these measuring points shows the strong influence of the limestone bedrock allowing infiltration into the aquifer. As a result the four rivers to the west of the catchment (Barranco da Adreneira, Ribeira de Budens, Ribeira de Vale de Boi and Barranco do Vale do Ruivo) supply approximately three-quarters of the flood discharge at the mouth of the catchment as it enters the sea. This occurs despite the fact that they drain only one-quarter of the catchment area. This is the result of their having a much narrower line of limestone to cross: only 2 km compared to the 5–10 km that the eastern rivers of the catchment have to traverse. The limestone thus appears to exert a strong depressing influence on the erosive potential of extreme event flood discharges.

Table 13.6 Flood discharge for the storm of 14 October 1989 and channel characteristics at 8 sample sites in Ribeira de Almádena catchment[1]

Site	Character of bed	Channel width (m)	Channel depth (m)	Sediment D_{50} (mm)	Bankfull velocity (cm³/sec)	Bankfull discharge (m³/sec)	Flood height above bankfull (m)	Flood discharge above bankfull (m³/sec)	Incision (m)
1	Bedrock & gravel	5.9	2.2	62	5.5	0.6	0	0	0.22
2	Bedrock & gravel	5.0	1.6	76	10.7	0.89	0.2	8.50	0.33
3	Alluvium	6.6	1.9	42	7.9	0.97	0.4	22.50	0.21
4	Alluvium	5.9	1.5	97	5.9	1.28	0.6	23.71	0
5	Alluvium	25.2	1.6	sand	5.1	2.04	0.8	27.14	0
6	Alluvium	4.0	1.6	sand	1.7	1.09	1.9	20.45	0
7	Estuarine silt	15.0	1.8	sand	1.7	4.48	1.1	26.64	0
8	Estuarine silt	24.4	1.9	silt	1.7	7.97	1.5	114.66	0

Note:
[1] Location of sites is shown in Figure 13.1

This expectation is confirmed by intensive ground survey of the catchment in Figure 13.2 and other areas of the littoral immediately after the flood. This showed the following features. First there was some landsliding on steep slopes throughout the catchment, including some limestone areas, as a result of saturation of the surface soil and subsoil. Second, there was some entrenching of streams in the upper part of the catchment. In all cases entrenching was restricted to small belts with a maximum width and depth of 0.5 m. All entrenching occurred in the area with Carboniferous slate or Triassic marl bedrock and resulted in limited headward erosion or gullying and deepening of existing channels. This did not extend beyond 0.75 km of the channel head, except where major tributary channels entered, and in most cases was restricted to the upper 0.4 km. The volume of material moved was in all cases small. Third, the erosion of channels was highly restricted. The material eroded from headwaters appeared in most cases to have been deposited within 0.1 km of its source. The rest of the channel lengths appeared as before. This was detectable in most cases by the sediment observed to be deposited on the vegetation covering the channel beds. The extensive vegetation in most cases had protected the downstream reaches from channel bed and bank erosion. This was particularly true of turfed areas of channel-in-channel phenomena. But even in the clean sections of the channel only very limited entrenchment and sediment transport appeared to have occurred because of quasi-cementation of much of the bed sediment. In the case of ten dispersed channel sites that it was possible to relate to 1985 surveys, five sites showed no measurable increase in depth and the five other sites showed an average entrenching of 4.5 cm, but only for the channel-in-channel part of the profile and only for very limited reaches of 5–10 m. No measurable change in channel width had occurred at any site. In most cases the floodwaters appear to have fanned out across valley floors, from where much infiltrated or was trapped by vegetation. The erosive power appears to have been extremely limited.

This evidence suggests that even in the case of extreme events the scope for erosion in the southern Portuguese littoral has been switched off almost permanently. Unless a long-term change in precipitation occurs, it is difficult to see how the rivers will be able to transport any significant loads in future. In effect the area has been turned from an ephemeral to a dry-valley environment by the combined forces of lowered water tables allowing high rates of loss to ground aquifers, changed land uses, dammed headwaters and intensification of channel bed and bank vegetation cover. Against these forces the minor changes in precipitation levels and seasonal incidence of rainfall that Devereux observes to have occurred appear insignificant.

SYSTEM INTERFACING

The discussion of this chapter has illustrated the interfacing between geomorphology and human influence that were central to Chorley's concerns. His interpretation of open systems that faced continual disturbances and thresholds opened the way for a fundamental change in thinking in geomorphology from the closed-systems perspective that had pervaded much earlier research. Chorley was particularly unhappy with the evidence that could be used to support monocausal and monodirectional change through studies of landscape evolution based on establishing erosion surfaces and terraces, correlating time with height from highest to lowest. This chapter has shown that emphasis on such closed-systems concepts has 'blinkered' much of the research on the geomorphology of southern Portugal just as it did in Britain in the 1950s and 1960s. As a result, the influence of human interactions with the landscape has been neglected.

The discussion here has sought to demonstrate that understanding current landform processes in this area must start from assessment of system interfacing, as Chorley and others have argued. Examination of recent rainfall, runoff, water table, flood discharge and land-use records has demonstrated that the controlling features of landscape erosion have been radically adjusted over the last 40 or 45 years. Probably most of these changes have occurred in the last 20 years. Water tables have been radically lowered, runoff reduced and retention in the vegetation cover significantly increased. These changes are permanent ones whilst current general precipitation regimes and social uses of water persist.

The limestone base of the area is a fundamental control of the way landscape evolution has occurred. Radiocarbon records of sediments show a steady movement of sediment materials within the catchments over the measured period of 1750 to 520 years BP. In this period there is some evidence to suggest that the predominantly limestone littoral acted as a kind of control system device regulating the flow of sediments: queuing occurred during relatively drier summers or longer term arid periods, flushing out of channel sediments occurred during winter and more pluvial periods when the limestone water table was higher. However, in recent years, the water tables have been reduced to a level where the limestone 'switch' cannot be opened to allow significant channel transport. When a high-magnitude event does occur, the evidence suggests that the water table has been so depleted, and a new quasi-stable vegetation system has now been so fully established, that the floodwaters disperse through seepage. As a result, the erosive capacity of the channels has been all but eliminated. The example of this area is salutary for the study of the effects on geomorphology of social change in other areas. Much of the south European littoral is configured on a similar limestone bedrock (see, e.g., Tyrakowski 1986; Brückner 1986). It may be that much of the Mediterranean littoral has been turned into the 'dry' landscape observed in this area of southern Portugal by the impact of social change. Further research is clearly required.

ACKNOWLEDGEMENTS

Much of the fieldwork for this chapter was undertaken by students of the University of Cambridge whom Dick Chorley had taught. Major field data collection in 1984 and 1985 was undertaken by these students. Updating of the land-use surveys and analysis of the flood discharges was undertaken by students of the London School of Economics in 1990. I am grateful to all those students and colleagues who participated in the fieldwork: Lloyd Martin, Keith Richards, Robin Glasscock, Derek Gregory, Tanya Bowyer-Bower and Chris Board. I am also grateful for critical comments on an earlier draft of this chapter from David Jones and Keith Richards. The usual disclosures apply.

REFERENCES

Bennett, R.J. and Chorley, R.J. (1978) *Environmental Systems: Philosophy, Analysis and Control*, London: Methuen.
Brückner, H. (1986) 'Man's impact on the evolution of the physical environment in the Mediterranean region in historical times', *Geojournal* 13: 7–17.
Brunsden, D. and Thornes, J.B. (1979) 'Geomorphic thresholds: the concept and its application', *Transactions of the Institute of British Geographers* NS 4: 485–515.
Chester, D.K. and James, P.A. (1991) 'Holocene alluviation in the Algarve, southern Portugal: the case for an anthropogenic cause', *Journal of Archaeological Science* 18: 73–87.
Chorley, R.J. (1962) 'Geomorphology and general systems theory', *U.S. Geological Survey Professional Paper* 500-B: 1–10.
——(1967) 'Models in geomorphology', in R.J. Chorley and P. Haggett (eds) *Models in Geography*, London: Methuen.
——(1973) 'Geography as human ecology', in R.J. Chorley (ed.) *Directions in Geography*, London: Methuen.
Chorley, R.J. and Kennedy, B.A. (1972) *Physical Geography: A Systems Approach*, London: Prentice Hall.
Daveau, S. (1973) 'Quelques exemples d'évolution quaternaire des versants au Portugal', *Finisterra: Revista Portuguesa de Geografia* 8: 5–47.
Devereux, C.M. (1983a) 'Recent erosion and sedimentation in southern Portugal', unpublished Ph.D. thesis, University of London.
——(1983b) 'Climate speeds erosion of the Algarve's valleys', *Geographical Magazine*, 10–17.
Feio, M. (1946) 'Os terracos do Guardiana a justante do Ardila', *Comunicações dos Servicos Geológicos de Portugal* 27: 5–83.
——(1949) *Le Bas Alentejo et l'Algarve*, Lisbon: Cong. Internat. Geogr.
——(1951) 'A evoluçao do revalo do Baixo Alentejo e Algarve', *Comunicações dos Servicos Geológicos de Portugal* 32: 1–190.
MHOPT (1982) *Plano, geral de urbanização área territorial do Algarve*, vol. 1: *Sintese de ánalises elaborades*, Lisbon: Minesterio da Habitaçâo, Obras Publicas e Transportes, Direçâo Geral do Planeamento Urbanistico.

Rocha, R.B. da, Ramalho, M., Atunes, M.T. and Coelho, A.V.P. (1983) *Noticia explicativa da folha 52-A: Portimâo*, Lisbon: Carta Geologica de Portugal, Servicos Geológicos de Portugal.

Rocha, R.B. da, Ramalho, M., Manuppella, G. and Zbyszewski, G. (1979) *Carta geológica de Portugal: 51-B, Vila do Bispo*, Lisbon: Servicos Geológicos de Portugal.

Sala, M. (1984) 'The Iberian massif', in C. Embleton (ed.) *Geomorphology of Europe*, London: Macmillan.

Schumm, S.A. (1979) 'Landscape sensitivity and change', *Transactions of the Institute of British Geographers* NS 4: 463–484.

Sweeting, M.M (1972) *Karst Landforms*, London: Macmillan.

Teixeira, C. (1979) 'Plio-Plistocenico de Portugal', *Comunicações dos Servicos Geológicos de Portugal* 65: 35–46.

Tyrakowski, K. (1986) 'The role of tourism in land utilisation conflicts on the Spanish Mediterranean coast', *Geojournal* 13: 19–26.

Vita-Finzi, C. (1969) *The Mediterranean Valleys*, Cambridge: Cambridge University Press.

Zbyszewski, G. (1958) 'Le Quaternaire du Portugal', *Boletim de la Sociedade Geológica de Portugal* 13: 1–227.

14

CHANCE AND NECESSITY IN GEOMORPHOLOGY

Alan Werritty

You believe in a God who plays dice, and I in complete law and order.
(Albert Einstein in a letter to Max Born)

'Whenever anyone mentions theory to a geomorphologist, he instinctively reaches for his soil auger' (Chorley 1978: 1). My own experience of twenty years' teaching and research has amply confirmed the truth of this aphorism. The majority of geomorphologists are both ignorant of and uninterested in theory, or, again in Chorley's own words, they see 'no need to distinguish methodology from techniques' since 'the scientific method is obvious and therefore needs no discussion'. A re-reading of Dick Chorley's papers makes it clear that one of the major threads in his own work has been the task of identifying and developing a sound theoretical basis for geomorphology (Chorley 1962, 1965, 1967, 1978). It thus seemed appropriate to focus on a theoretical issue in selecting a topic to contribute to this Festschrift, the more so since the issue I wish to examine is one which I first explored in my Ph.D. dissertation under Dick Chorley's guidance (Werritty 1976).

The purpose of this chapter is to address a theoretical question which has hovered on the edges of informed debate for more than three decades and which has recently re-emerged as chaos theory has been taken up by geomorphologists. At its simplest the question takes the form: 'In our attempts to explain the origin and development of landforms what should be the role of random processes?' Or alternatively, expressing the question explicitly in terms of the title of this chapter, 'What are the respective roles of "chance" and "necessity" in explaining the shape taken by the earth's surface?'

CONTRASTING MODES OF SCIENTIFIC EXPLANATION

Scientific explanation of phenomena in the natural world has classically taken one of two routes: that of either deterministic or stochastic reasoning.

Initially, following the success of the seventeenth-century Newtonian paradigm of a 'clockwork' or 'mechanistic' universe, deterministic reasoning provided the route to scientific explanation. Associated with this was the development of mathematical theory which could solve differential equations and which was predicated upon nature possessing forms which were smooth and continuous rather than rough and irregular (Mandelbrot 1983). One of the best-known early scientific definitions of the reasoning underpinning the deterministic mode of scientific explanation is that by Laplace (published in 1814):

> Let us image an intelligence who would know at a given instant of time all forces acting in nature and the position of all things of which the world consists; let us assume further that this intelligence would be capable of subjecting all these data to mathematical analysis. Then it could derive a result which would embrace in one and the same formula the motions of the largest bodies in the universe and of the slightest atoms. Nothing would be uncertain for this intelligence. The past and the future would be present to its eyes.
>
> (*Philosophical Essay on Probability*, translated by Wartofsky 1968: 298)

But this, as Bridgman (1959) has noted, is a definition of determinism far too all-embracing and ambitious to serve the needs of modern science. A more realistic and manageable definition based on Caws (1965: 300) is thus offered below:

> A physical system is deterministic if two requirements are met:
> (1) a precise knowledge of the state of the system at some initial time,
> (2) a precise method for predicting future states when the initial state is known.

However, a problem still remains in terms of the precision with which the initial conditions can be known (Smart 1979). Uncertainty in specifying the initial conditions can result in the system becoming indeterminate, thereby leading to a stochastic formulation in which random processes and chance play a vital role in the search for explanation.

This view of the world, which initially emerged from the Gaussian theory of experimental errors, became essential to nineteenth-century physicists as they sought to develop Newtonian dynamics for systems which involved more than two bodies (Stewart 1989). The subsequent development of statistical mechanics by Gibbs enabled the macroscopic properties of bulk matter (be it in a solid, liquid or gaseous state) to be derived from the dynamic laws governing its microscopic constituents and incidentally provided an unexpected theoretical justification for thermodynamics. Statistical mechanics enabled scientists to study the dynamics of systems involving a large number of particles, since the aggregate or group behaviour of these particles could be characterised in terms of probabilities.

In the twentieth century the advent of quantum mechanics enabled very small systems of molecular or atomic size to be investigated for the first time. Again, stochastic methods proved crucial as deterministic laws were replaced by probabilistic ones. One of the developments within quantum theory (the Heisenberg uncertainty principle) also refuted the claim that within macroscopic systems the initial conditions can, in principle, be determined as accurately as is required. In his evaluation of these developments the physicist/geomorphologist Smart observes:

> It is particularly important to recognize that stochastic methods in physics should not be regarded as second-rate substitutes to be employed only until we arrive at exact deterministic models for all physical phenomena. We cannot get rid of them simply by working harder and being smarter. Instead they are embedded in the structure of modern physics.
>
> (Smart 1979: 654)

DETERMINISTIC AND STOCHASTIC REASONING WITHIN FLUVIAL GEOMORPHOLOGY

The respective roles of deterministic and stochastic reasoning within geomorphology have been commented upon by a number of researchers over the last three decades. The majority of geomorphologists take the view that the phenomena they seek to observe and explain can adequately be handled by deterministic reasoning, the route to such explanation involving methods ultimately derived from Newtonian mechanics. Thus geomorphological phenomena typically arise from a shear being generated by a flow across a land surface (potentially by the movement of water, ice or air). Whether or not erosion then occurs and the land surface is deformed depends upon the resistance of the surface materials in relation to the applied shear. Expressed in these terms the links to classical physics are self-evident. Explanations of river channel morphology derived from the hydraulics of open channel flow represent a typical example of this approach (Richards 1982). A particularly lucid expression of this type of explanation is to be found in Schumm's (1977: 63) account of the growth of drainage networks: 'the drainage network develops in response to material erodibility and to the eroding force applied to the surface of the basin in a deterministic manner'.

This acceptance of deterministic reasoning in fluvial geomorphology had in fact been questioned fifteen years earlier in two influential papers by Leopold and Langbein (1962, 1963). In the first of these the authors introduced the concept of entropy as an aid to understanding landscape evolution. Drawing upon statistical mechanics and the principles of thermodynamics (notably the concept of entropy and the most probable distribu-

314

tion of energy levels within a system), they sought to demonstrate that rivers should not be regarded as simple examples of deterministic systems. Having analysed the characterisation of the river long profile, they concluded that it could adequately be modelled by a random walk model. Drawing upon thermodynamic theory, they then concluded that 'the "equilibrium profile" of the graded river is the profile of maximum entropy and the one in which entropy is equally distributed' (Leopold and Langbein 1962: 11). They also demonstrated that Horton's laws of drainage composition could be reproduced from random walk simulations of drainage networks, an idea whose methodological implications were to be more fully examined in their influential follow-up paper (Leopold and Langbein 1963).

In the latter they specifically addressed the nature of the indeterminacy to be found within geomorphic systems. By indeterminacy they mean

> those situations in which the applicable physical laws may be satisfied by a large number of combinations of interdependent variables.... Any individual case ... cannot be forecast or specified except in a statistical sense. The result of an individual case is indeterminate.
>
> (Leopold and Langbein 1963: 189)

As an illustration they offer the example of

> a hill slope of uniform material and constant slope subjected to the same conditions of rainfall.... Assume that the slope, material and precipitation were such that a large number of rills existed on the surface.... Would it be supposed that rills comparable in size and position were absolutely identical? The postulate of indeterminacy would suggest that they would be very similar but not identical.
>
> (Leopold and Langbein 1963: 190)

It is claimed that such a stochastic approach enlarges on the physical relations based upon Newtonian mechanics. Furthermore, probabilistic reasoning proves to be better than deterministic reasoning because the former results in a more specific understanding of the processes involved. In summary, physical laws, whilst necessary, are not sufficient to determine the exact form of the land surface. There always remain unsatisfied conditions (an excess of unknowns over the number of equations) which preclude total explanation. The form of the land surface is thus best described by measures of central tendency around which there is an irreducible variance. Whilst this variance can be expected to decrease with the development of better theory and more precise measurement, it will always be present to some degree. For this reason there is an irreducible indeterminacy in the natural landscape.

But there is an alternative view of the role of probabilistic reasoning within geomorphology. This is given its most cogent expression by Shreve (1975: 529):

Geomorphic systems are descendants of antecedent states that are generally unknown, and are invariably part of larger systems from which they cannot be isolated. Thus, except in very special cases such as laboratory experiments, the initial and boundary conditions needed as input to a deterministic theory are generally unavailable, even in principle. Of much greater importance, however, is the fact that in certain instances geomorphic systems seem to be unstable against small disturbances; their response is often in the direction of a perturbation, rather than opposed to it, as when the slight concentration of flow in a cow track, for example, creates a rill that becomes a gully. Small random events are thus amplified into a large random element in the geomorphic system. An exact deterministic theory of such a system would not be of much use because small errors in the initial and boundary conditions, which are inescapably subject to limitations in physical measurement and numerical representation, would be amplified into large errors in the predicted behaviour. In such cases ... a probabilistic theory that takes account of the apparent randomness is evidently a necessity, because if our theories are to succeed, they must reflect the world as it is, not as we would like it to be.

Leaving aside for the moment the intriguing anticipation of aspects of chaos theory in this quotation, Shreve proposes that geomorphic relationships, although expressed in deterministic terms, are, in essence, probabilistic. The reasons for this are pragmatic in that the initial boundary conditions needed as an input to a deterministic formulation cannot be specified.

Three contrasting approaches to geomorphic explanation can be identified from the above discussion (see also Smart 1979):

(i)   a classical deterministic type of reasoning ultimately owing its origin to Newtonian physics (deterministic approach);
(ii)   a stochastic type of reasoning which is mainly derived from nineteenth-century statistical mechanics and asserts that randomness is an *inherent* property of physical systems (first stochastic approach);
(iii)   a stochastic type of reasoning which identifies an *apparent* randomness in physical systems but is unconcerned about its origins (second stochastic approach).

A CRITIQUE OF THE CONFLICTING CLAIMS

The classical deterministic approach receives a powerful endorsement from Howard (1972), who observes that deterministic reasoning is the universal paradigm of science underpinning the powerful and successful scientific method. He continues with the observation that 'to posit an alternative indeterministic methodology based on the assumption of inherent randomness in nature is ... to assert the impossibility of discovering explanations of ever-increasing generality, accuracy and simplicity, and would abandon the scientific quest as we know it' (Howard 1972: 78). Furthermore, lurking behind a methodology based upon indeterminism is a philosophy of despair which identifies limits beyond which nature is unknowable. It is surely the role of scientific enquiry constantly to be pushing back these limits. Had previous scientists accepted the claim that nature is inherently indeterminate, scientific progress in many areas would have been stultified generations ago (Werritty 1976). It is suggested that the majority of geomorphologists, if pressed, would subscribe to this viewpoint because it sits easily within what is generally perceived to be the scientific method. But what of the two contrasting stochastic approaches?

Leopold and Langbein's advocacy of 'inherent indeterminacy' (the first stochastic approach) has been the subject of repeated rebuttal by engineers and geomorphologists who find the alleged parallelisms between the extremal behaviour of rivers and thermodynamics a metaphysical distraction that is both unnecessary and unconvincing (Davy and Davies 1979; Davies and Sutherland 1983; Griffiths 1984; Ferguson 1986; Phillips 1990). It has also been the subject of an extended philosophical critique. Having noted that no physical law is certain, Watson (1966: 182) develops this critique on the basis of specificity: 'stochastic laws simply give less specific determinations than do so-called deterministic laws, and this ... is based upon the extent of our knowledge of the facts'. He also disputes the claim that indeterminacy constitutes a third factor explaining subtle variations between landforms resulting from virtually identical initial conditions and processes. Such alleged inherent indeterminacy merely reflects current limitations in human comprehension and knowledge rather than being irreducible (as claimed by Leopold and Langbein 1963). Thus, to take the example of a set of sub-parallel rills developing on a hillslope, the reason why the exact configuration and course of a specific rill is indeterminate is because our understanding of soil hydrology and the hydraulics of overland flow at the macro-scale is limited. Could we design a suitable form of instrumentation and develop the theory underpinning rill development, the problem ceases to be inherently insoluble (Werritty 1976). To claim that the problem is insoluble and the final answer indeterminate in the Heisenberg sense is very misleading. Admittedly this latter claim is not advanced by Leopold and Langbein, but other authors (notably Mann (1970) and Krauskopf (1968)) come very close to such a statement. The claim that the universe is indeterminate at the atomic level is of no direct relevance to geomorphology because the scale change involved is so great. Furthermore, a division of nature into determinate and indeterminate parts requires one to subscribe to the view that natural laws exist, but only to explain phenomena above a certain spatial level. Below that level there would be no reason to expect the occurrence of one particular event rather than any other plausible event. But, as Watson (1966: 184) comments, 'to

317

subscribe to this view is ... to deny the basic rationality of the universe'.

One reason for much of this confusion in the geological literature lies in an insufficient distinction between the application of statistical laws about group behaviour and the development of lawlike generalisations about particular events. Even if the outcome of a particular event cannot be predicted by a lawlike statement, it is often possible to predict the average behaviour of the group within which that particular event occurs. In such a situation it is possible to be just as specific about a particular group's behaviour as about a particular event about which an individual lawlike statement is possible. It is important to note that

> statistical laws ... are not certain about given particulars not because they are not just as well evidenced ... but because logically they say nothing about particulars at all. They may be just as certain about groups of particulars as laws may be about particulars.
>
> (Watson 1969: 491)

The claim advanced by Leopold and Langbein in the early 1960s that there is an inherent randomness governing the behaviour of water flowing down hillslopes and within channels has been the subject of extended scrutiny both in terms of a philosophical critique (Watson 1966, 1969) and empirical investigations (see Phillips (1990) for a recent summary). It has not commanded support during three decades of further research and should now be discarded. Nevertheless, Leopold and Langbein should be commended for identifying a set of crucial questions within fluvial geomorphology at a very formative moment within the development of the discipline. Their answers may, in hindsight, have pointed in the wrong direction, but no one can doubt the importance that still attaches to their original question.

The best example of the second stochastic approach is Shreve's claim (1975) that the development of channel networks should, on account of their apparently random behaviour, be based upon a probabilistic theory. The distinction between *inherently* and *apparently* random is the crucial distinction between the two stochastic approaches (see above). In Shreve's view probabilistic theory is inevitable in geomorphology because of the nature and behaviour of the phenomena being analysed. The departures of individual measurements from nominal values predicted, for example by Horton's Laws of Stream Composition, arise not from random measurement errors or extraneous noise, but rather because of geomorphic randomness. Empirical relations, although expressed in deterministic terms, are probabilistic in essence since they characterise general tendencies in large populations rather than exact relations for individual cases (e.g. the reformulation of Horton's Laws in terms of the random model: Shreve 1966). Furthermore, this is to be welcomed since, in the case of the random model, probabilistic reasoning provides results that are 'generally simpler, better, and more practical' (Shreve 1975: 527).

318

Thus, as far as channel networks are concerned, the justification for accepting the random topology model, and with it the 'second stochastic approach', arises from the fact that it is more successful than the earlier, and implicitly deterministic, approach of Horton. But this largely pragmatic justification has been subject to criticism by some geomorphologists who have sought to use the random model empirically (e.g. Werritty 1972). Thus, in his exhaustive review of research derived from Shreve's random model, Abrahams (1984: 185) is rather circumspect in his conclusions that 'any future model that supplants the random model ... will contain stochastic elements, but these elements will have far less influence on the predictions of the model than they do in the present model'. Implicit in this is the anticipation that the initial and boundary conditions governing the initiation and development of channel networks will eventually be better specified and the role of purely stochastic processes thereby reduced. In this statement Abrahams would appear to be speaking on behalf of those scientists for whom deterministic reasoning still remains the ultimate path of scientific enquiry.

But in recent years a very different approach involving both deterministic and stochastic reasoning has developed within science. This is chaos theory, to which I now turn.

CHAOS THEORY

Chaos theory is the collective name given to a number of parallel and interconnected developments in the modelling of dynamic systems that have emerged since the mid-1970s. A useful definition which encompasses many of these developments and makes an explicit reference to the material developed earlier in this chapter is that by Phillips (1992: 178): 'Chaos is complex, apparently random behaviour arising from the non-linear dynamics of (sometimes very simple) deterministic systems.'

As Culling (1987: 57) has noted in his seminal review on chaos theory in geography, one no longer has to choose between the 'Keplerian motion of the planets, and the irregular and unpredictable, as in the Brownian motion'. Chance and necessity are not irrevocably in conflict (Berry 1983). The newly emerging field of chaos theory, particularly in the development of non-linear dynamic theory, provides a fresh approach to what otherwise were deemed to be intractable or mystifying phenomena. For example, analyses of non-linear systems have revealed irregular orbits which nevertheless are completely determined, despite being, in some cases, more intricate than those orbits generated by random processes. A well-known example of this is the sequences of weather patterns which led to Lorenz's discovery of a chaotic non-linear system and its strange attractor (Lorenz 1963). More generally chaotic motion is said to exist if it exhibits the following behaviour (Ott 1981: 656):

(i)   a sensitive dependence upon the initial conditions;
(ii)   a mean correlation function which tends to zero with the time evolution of the system;
(iii) the trajectory is aperiodic.

The evolution of the system appears to be controlled by random processes, but in reality is fully determined and has an immediate future which can be predicted with accuracy. A system which behaves in this manner displays 'deterministic chaos'.

Thus the discovery of chaotic motion has led to some types of irregular behaviour, formerly rejected as random noise (and attributed to either apparently or inherently stochastic processes), as potentially having an internal deterministic origin. But this does not imply that all irregular behaviour can be described as chaotic. In many physical systems chaotic behaviour exists in the presence of noise, which has to be identified and separated out. Not all physical systems at present lend themselves to being investigated in this manner, but for those that do chaos theory provides a powerful new branch of mathematical analysis which has registered many notable successes within the physical and natural sciences (e.g. diffusion limited aggregation: Stewart 1989).

Chaos theory does not provide a philosopher's stone resolving all the perceived conflicts between deterministic and stochastic explanations within science, but it has opened up rich alternative ways of analysing physical systems. Indeed Culling (1987: 69) extols it as having opened a 'magic casement [finding] between chance and necessity, one dimension and the next, a whole new world of chaotic motions, strange attractors and periodic windows'. But is such hyperbole justified? In what sense is the age-old tension between chance and necessity resolved and of what practical significance is this to geomorphologists?

CHAOS, CHANCE AND NECESSITY

It has always been difficult to reconcile chance and necessity in our understanding of the world. Newton's world was essentially a clockwork universe governed by deterministic laws, but this was hard to reconcile with the apparently random pattern of many physical processes (Davies 1987). Maxwell and Boltzmann circumvented this, in part, by introducing statistical mechanics into physics. Thus the aggregate behaviour of the molecules comprising a gas could be determined, even if the behaviour of individual molecules could not. But as Davies (1987: 52) observes: 'it has always been paradoxical how a theory based on Newtonian mechanics can produce chaos merely as the result of including large numbers of particles and making the subjective judgement that their behaviour cannot be observed by humans'. Nevertheless, recent work in chaos theory has succeeded in a partial resolution of this paradox. It now seems possible to build bridges between chance and necessity.

Chaos theory tells us that complex and intricate structures can be derived from equations so simple that some of them can be programmed on pocket calculators. Weather systems, the capsizing of ships, turbulence in fluids, and the seepage of water through rocks all constitute dynamic systems which can be expressed in relatively simple equations and yet yield great complexity in terms of their outcomes. In such systems chaotic behaviour is the norm rather than the exception on account of their extreme sensitivity to initial conditions. The resultant trajectories of these systems are also unpredictable and, for all practical purposes, may be considered to be random.

However, it is important to note that such behaviour does not arise because the system is inherently random (i.e. 'first stochastic approach' in the discussion above). It can be proved mathematically (in support of Laplace's speculation) that the specification of the initial conditions is in fact sufficient to predict the entire future behaviour of the system. The problem, however, arises in seeking to specify those initial conditions with sufficient precision. In practice it is not possible to know the exact initial state of a system since observations, no matter how refined, will inevitably include some error (Davies 1987). In ordinary dynamic systems such a lack of precision is of no practical consequence since the predictions can constantly be updated (e.g. in the prediction of eclipses). However, in chaotic systems such a lack of precision is crucial, since the errors grow exponentially. The randomness of chaotic motions is thus found to be fundamental to the system and not merely an expression of measurement error. The updating and correction necessary for accurate prediction for a chaotic system in fact requires a rate of calculation that cannot keep pace with events. Simulations based upon input information in chaotic systems are thus pointless as the amount of information gained is only equivalent to what has been input (Ford 1983). In this situation even the most powerful computers are reduced to xerox machines (Davies 1987).

Such a conclusion might be thought to be fatally damaging to the quest for scientific explanation, but there is a liberating corollary. Chaotic systems possess great freedom in operating within a vast range of potential trajectories. As the Mandelbrot set has so spectacularly demonstrated, chaos spawns spatial forms and structures potentially infinite in extension and complexity (Peitgen and Richter 1986). Within these structures stability and coherence at the macro-scale go hand in hand with unpredictability at the micro-scale, a particularly good example of this being the behaviour of the red spot within the Jovian atmosphere. It is in this sense that chance and necessity can be reconciled within the framework of chaotic systems.

In the light of chaos theory physicists are increasingly regarding necessity or determinism as a myth or part of the kindergarten of science (Prigogine 1980). This view is given eloquent expression by Ford (1983: 40):

For centuries, randomness has been deemed a useful, but subservient citizen in a deterministic universe. Algorithmic complexity theory and non-linear dynamics together establish the fact that determinism actually reigns only over a quite finite domain; outside this small haven of order lies a largely uncharted, vast wasteland of chaos where determinism has faded into an ephemeral memory of existence theorums and only randomness survives.

Thus an alternative to the linear Newtonian deterministic world is to view the universe as being in some sense open, with hitherto unknown levels of variety and complexity potentially in store.

CHAOS THEORY AND GEOMORPHOLOGY

The application of chaos theory has provided a very fertile area of research within geomorphology in recent years. Indeed, the journal *Geomorphology* has recently published two theme issues (Snow and Mayer 1992; Phillips and Renwick 1992) extensively devoted to research in this field. The two main areas within chaos theory which have proved to be particularly attractive to geomorphologists are non-linear dynamic theory, and the application of fractals and self-similarity to geomorphic forms and processes.

Non-linear dynamic theory has four characteristics (Phillips 1992b). First, it is concerned with examining the trajectories of systems between or in relation to specified equilibria. Second, these systems are generally dissipative in structure (see Huggett (1988) for geomorphic examples). Third, the evolution of the system includes discontinuities known in mathematical terms as bifurcations or catastrophes (see Thornes (1981) for geomorphic examples). Fourth, deterministic chaos may be present, fractals often providing a valuable tool in the detection of such chaos. Focusing on the last of these characteristics, Culling (1987, 1988) has noted that chaos is to be expected in the physical landscape since many of the forcing functions (e.g. Rayleigh-Bernard convection cells in patterned ground, gully growth into a homogeneous plateau, diffusion degradation on soil-covered slopes and turbulence in fluid flows) display chaotic behaviour. Shreve (1975), in a prescient observation, already commented upon above, noted that in some geomorphic systems (such as the evolution of channel networks) small errors in the initial and boundary conditions are amplified into large errors in the predicted behaviour. Such behaviour is typical of deterministic chaos.

A particularly instructive use of non-linear modelling and chaos theory in geomorphology is Phillips's (1992a) analysis of surface runoff and the hydraulic geometry of overland flow. Returning in part to the question addressed by Leopold and Langbein (1963), Phillips speculates as to whether the complexity of the flow system may be due to deterministic chaos. If correct, this would imply that the observed complexity can be attributed to the inherent deterministic dynamics of runoff rather than to environmental heterogeneity. The conclusions derived from the analysis of field data are ambivalent. As a general rule surface runoff is not characterised by chaotic behaviour and saturation–excess overland flow is non-chaotic. But chaos is possible in runoff generation where Hortonian overland flow dominates. These results have intriguing implications. In so far as chaos is present in runoff generation, attempts to reduce uncertainty in rainfall–runoff relationships by more precise measurement are likely to yield only limited improvements (Phillips 1992a). Furthermore, the presence of chaos is encouraging in that some of the complexities of overland flow can be explained in relatively simple deterministic terms and thus short-term prediction and probabilistic prediction are not precluded.

However, more generally within hydrology both deterministic chaos and stochastic complexity are likely to be present, and under these conditions it is often difficult to confirm the presence and extent of chaos, a problem which led Culling (1987) to the conclusion that applying chaos theory within physical geography will prove problematic. Nevertheless, in a series of papers Culling (1987, 1988; and Culling and Datko 1987) remains one of the most stimulating workers in this field, giving substance to Mandelbrot's (1983: 1) assertion that 'Clouds are not spheres, mountains are not cones, coastlines are not circles.' All these objects exhibit irregular rather than smooth boundaries, and often possess the property of being self-similar, i.e. their roughness or fragmentation is scale invariant. The tool developed by Mandelbrot to analyse such phenomena is fractal geometry, in which fractional dimensions become possible. Fractal geometry thus occupies an intermediate middle ground between the excessive geometric order of Euclid (with its integer dimensions and smooth boundaries) and the geometric chaos of general mathematics.

The emergence of 'fractals' as an exciting revitalisation of geometry is well known, as are the haunting and beautiful computer graphics which have accompanied fractals, including the development of randomly generated landscapes (Mandelbrot 1983; Peitgen and Richter 1986). One result of this within physical geography has been an attempt to identify the fractal geometry of different soil-covered landscapes across southern England and the representation of such surfaces by Gaussian random fields (Culling and Datko 1987). Fractals have also entered fluvial geomorphology in the characterisation of channel networks which have variously been attributed dimensions of 1.6–1.7 (Barbera and Rosso 1989) and 2.0 (Tarboton, Bras and Rodriguez-Iturbe 1988). Mandelbrot (1983) has also demonstrated that river lengths can be viewed as a fractal with the dimension of 1.136. This is approximately twice the value of the exponent in the well-known relationship between mainstream length and basin area derived by Hack (1957). Empirical support for Mandelbrot's finding can be found in the reported average fractal dimension of 1.158 for eight rivers in Missouri (Hjelmfelt

1988). Fractals have also been used to analyse the topography of Arizona (Chase 1992) and in characterising the shape of drainage basin perimeters (Breyer and Snow 1992) and sinkholes (Reams 1992). But, except for identifying the scale at which stable diffusive processes yield to unstable channel-forming processes (Tarboton, Bras and Rodriguez-Iturbe 1992) and developing a multifractal spectrum of energy expenditure within the drainage basin (Ijjaz-Vasquez, Rodriguez-Iturbe and Bras 1992), much of this research fails to address truly fundamental questions. The use of fractals in geomorphology may have illuminated some old problems, but as Culling and Datko (1987: 370) wisely caution: 'the problem is not to find fractals in the landscape but to explain them'. That challenge has yet to be seriously taken up.

But what of the broader issues raised by the introduction of chaos theory into geomorphology? If a system possesses complexity that can be explained in terms of deterministic chaos, then predicting future trajectories will not be improved by the reductionist approach of steady improvements in deterministic modelling (Phillips 1992a). Prediction instead must rely upon identifying the strange attractors present and the system's behaviour relative to them. Practical prediction will then require a stochastic approach. This is not far removed from Shreve's (1975) prescription in which apparent randomness in channel networks was accepted and the quest for deterministic explanation abandoned as unnecessary and unprofitable. Furthermore, in rejecting the reductionist approach characteristic of much recent physical science, Phillips (1992c) seeks to promote a geomorphology in which whole systems rather than component parts of systems are investigated. In such a holistic approach the object of study then becomes the system itself.

CONCLUSION

At the end of this review, I return to the original question posed: what are the respective roles of 'chance' and 'necessity' in explaining the shape taken by the earth's surface? It is clear that necessity or determinism in both the ideal formulation of Laplace's 'Omniscient Being' or even Caws's (1965) more restricted version do not reflect the world as found by geomorphologists. The identification of initial conditions with the required precision to apply deterministic reasoning in a formal manner cannot be met in geomorphic research. This does not mean that we should abandon this approach. In many areas, we can proceed as if Newtonian mechanics applied, providing the results obtained are sufficiently accurate to meet our needs. A great deal of traditional geomorphology will undoubtedly continue to be undertaken using such a paradigm. But increasingly chance plays a larger role in our understanding of geomorphic processes – e.g. in the location and development of rills and gully heads, in the entrainment of sediments, in the turbulent structures to be found in fluid flows. Here the findings of Leopold and Langbein (1963), that there is an 'irreducible indeterminacy' at work, seems a counsel of despair which runs counter to the scientific quest. A more positive view is that expressed by Shreve (1975), who identifies an *apparent* rather than an *inherent* randomness within the structure of channel networks. This both encourages and permits further enquiry along the lines recently developed in chaos theory. Indeed, it now seems possible that some of the apparent randomness to be found in geomorphic phenomena can be attributed to deterministic chaos enabling short-term predictions to be made and aggregate behaviour to be characterised. How far this will prove to be successful remains unclear for, with a few notable exceptions (see above), the current application of chaos theory within geomorphology is still largely descriptive and restricted in its range of applications.

Beyond chaos there still exists random behaviour that cannot be reduced to some ordered pattern. Here stochastic modelling will continue to play an important role in characterising, if not initially explaining chance phenomena. The task of the geomorphologist, as in all the physical sciences, is to continue to make forays in this unknown world, constantly searching for order where hitherto none has been found.

REFERENCES

Abrahams, A.D. (1984) 'Channel networks: a geomorphological perspective', *Water Resources Research* 20: 161–168.

Barbera, P.L. and Rosso, R. (1989) 'On the fractal dimension of stream networks', *Water Resources Research* 25: 735–741.

Berry, M. (1983) 'Chance and necessity', *Nature* 305: 456.

Breyer, S.P. and Snow, R.S. (1992) 'Drainage basin perimeters: a fractal significance', *Geomorphology* 5: 143–158.

Bridgman, P.W. (1959) *The Way Things Are*, Cambridge, Mass.: Harvard University Press.

Caws, P.W. (1965) *The Philosophy of Science*, Princeton: Van Nostrand.

Chase, C.G. (1992) 'Fluvial sculpting and the fractal dimension of topography', *Geomorphology* 5: 39–57.

Chorley, R.J. (1962) 'Geomorphology and General Systems Theory', *U.S. Geological Survey Professional Paper* 500B: 1–10.

——(1965) 'A re-evaluation of the geomorphic system of W.M. Davis', in R.J. Chorley and P. Haggett (eds) *Frontiers in Geographical Teaching*, London: Methuen.

——(1967) 'Models, paradigms and the new Geography', in R.J. Chorley and P. Haggett (eds) *Models in Geography*, London: Methuen.

——(1978) 'Bases for theory in geomorphology', in C. Embleton, D. Brunsden and D.K.C. Jones (eds) *Geomorphology: Present Problems and Future Prospects*, Oxford: Oxford University Press.

Culling, W.E.H. (1987) 'Equifinality: modern approaches to dynamical systems and their potential for geographical thought', *Transactions of the Institute of British Geographers* NS 12: 57–72.

——(1988) 'A new view of the landscape', *Transactions of the Institute of British Geographers* NS 13: 345–360.

Culling, W.E.H. and Datko, M. (1987) 'The fractal geometry of the soil-covered landscape', *Earth Surface Processes and Landforms* 12: 369–385.

Davies, P. (1987) *The Cosmic Blueprint*, London: Heinemann.

Davies T.R.H. and Sutherland, A.J. (1983) 'Extremal hypotheses for river behaviour', *Water Resources Research* 19: 141–149.

Davy, B.W. and Davies, T.R.H. (1979) 'Entropy concepts in geomorphology: a re-evaluation', *Water Resources Research* 15: 103–106.

Ferguson, R.I. (1986) 'Hydraulics and hydraulic geometry', *Progress in Physical Geography* 10: 1–31.

Ford, J. (1983) 'How random is a coin toss?' *Physics Today* April 1983, p. 4.

Griffiths, G.A. (1984) 'Extremal hypotheses for river regime: an illusion of progress', *Water Resources Research* 20: 113–118.

Hack, J.T. (1957) 'Studies of longitudinal stream profiles in Virginia and Maryland', *U.S. Geological Survey Professional Paper* 294B.

Hjelmfelt, A.T. (1988) 'Fractals and the river-length catchment-area ratio', *Water Resources Bulletin* 24: 455–459.

Howard, A.D. (1972) 'Problems of interpretation of simulation models of geologic processes', in M. Morisawa (ed.) *Quantitative Geomorphology: Some Aspects and Applications*, Binghamton, N.Y.: Publications in Geomorphology.

Huggett, R.J. (1988) 'Dissipative systems: implications for geomorphology', *Earth Surface Processes and Landforms* 13: 45–49.

Ijjasz-Vasquez, E.J., Rodriguez-Iturbe, I. and Bras, R.L. (1992) 'On the multifractal characterization of river basins', *Geomorphology* 5: 297–310.

Krauskopf, K.B. (1968) 'A tale of ten plutons', *Bulletin of the Geological Society of America* 79: 1–18.

Leopold, L.B. and Langbein, W.B. (1962) 'The concept of entropy in landscape evolution', *U.S. Geological Survey Professional Paper* 500-A.

——(1963) 'Association and indeterminacy in geomorphology', in C.C. Albritton (ed.) *The Fabric of Geology*, Reading, Mass.: Addison-Wesley.

Lorenz, E.N. (1963) 'Deterministic non-periodic flows', *Journal of Atmospheric Science* 20: 130–141.

Mandelbrot, B. (1983) *The Fractal Geometry of Nature*, New York: W.H. Freeman & Co.

Mann, C.J. (1970) 'Randomness in nature', *Bulletin of the Geological Society of America* 81: 95–104.

Ott, E. (1981) 'Strange attractors and chaotic motions of dynamic systems', *Review of Modern Physics* 53: 655–671.

Peitgen, H.-O. and Richter, P.H. (1986) *The Beauty of Fractals: Images of Complex Dynamical Systems*, Berlin: Springer-Verlag.

Phillips, J.D. (1990) 'The instability of hydraulic geometry', *Water Resources Research* 26: 739–744.

——(1992a) 'Deterministic chaos in surface runoff', in A.J. Parsons and A.D. Abrahams (eds) *Overland Flow: Hydraulics and Erosion Mechanics*, London: University College Press.

——(1992b) 'Nonlinear dynamic systems in geomorphology: revolution of evolution?' *Geomorphology* 5: 219–229.

——(1992c) 'The end of equilibrium?', *Geomorphology* 5: 195–201.

Phillips, J.D. and Renwick, W.H. (eds) (1992) 'Geomorphological systems', Proceedings of the 23rd Binghamton Symposium in Geomorphology, September 1992. *Geomorphology* 5: 195–488.

Prigogine, I. (1980) *From Beginning to Becoming: Time and Complexity in the Physical Sciences*, San Francisco: W.H. Freeman & Co.

Reams, M.W. (1992) 'Fractal dimensions of sinkholes', *Geomorphology* 5: 159–165.

Richards, K.S. (1982) *Rivers: Form and Process in Alluvial Channels*, London: Methuen.

Schumm, S.A. (1977) *The Fluvial System*, New York: John Wiley.

Shreve, R.L. (1966) 'Statistical law of stream numbers', *Journal of Geology* 74: 17–37.

——(1975) 'The probabilistic-topologic approach to drainage-basin geomorphology', *Geology* 3: 527–529.

Smart, J.S. (1979) 'Determinism and randomness in geomorphology', *EOS Transactions of the American Geophysical Union* 60: 651–655.

Snow, R.S. and Mayer, L. (eds) (1992) 'Fractals in geomorphology', *Geomorphology* 5: 1–194.

Stewart, I. (1989) *Does God Play Dice?: The Mathematics of Chaos*, London: Penguin Books.

Tarboton, D.G., Bras, R.L. and Rodriguez-Iturbe, I. (1988) 'The fractal nature of river networks', *Water Resources Research* 24: 1317–1322.

——(1992) 'A physical basis for drainage density', *Geomorphology* 5: 59–76.

Thornes, J.B. (1981) 'Structural instability and ephemeral channel behaviour', *Zeitschrift für Geomorphologie* NF 26: 233–244.

Wartofsky, M.W. (1968) *Conceptual Foundations of Scientific Thought*, New York: Macmillan.

Watson, R.A. (1966) 'Discussion: Is geology different?: a critical discussion of "The Fabric of Geology"', *Philosophy of Science* 33: 172–185.

——(1969) 'Explanation and prediction in geology', *Journal of Geology* 77: 488–494.

Werritty, A. (1972) 'The topology of stream networks', in R.J. Chorley (ed.) *Spatial Analysis in Geomorphology*, London: Methuen.

——(1976) 'The topology, geometry and orientation of stream networks in southwest England', unpublished Ph.D. thesis, University of Cambridge.

15

A PLURALIST, PROBLEM-FOCUSED GEOMORPHOLOGY

Olav Slaymaker

Geomorphology at the close of the twentieth century appears still to be dominated by the philosophy of positivism. 'All participants (in the Frankfurt meeting of the International Association of Geomorphologists) are quite clearly working within a single paradigm which derives fundamentally from the European scientific tradition of the eighteenth and nineteenth centuries' (Kennedy 1990). At the same time, geomorphologists are tackling an increasingly wide range of problems and, without a central unifying concept, 'we are in a divergent phase of high plurality' (Barsch 1990). These observations, made at the same scientific event, imply that the unity of the positivist position may be worthy of closer examination.

It was the genius of Chorley (1962), Chorley and Kennedy (1971) and Bennett and Chorley (1978) that recognised the necessity for geomorphology to engage the broader framework of general systems theory in order to tackle problems of relevance to society. Regrettably, as far as geomorphology is concerned, this methodological flexibility was widely interpreted as lacking in philosophical flexibility and as being insensitive to the implicit assumptions of instrumentalism (D. Gregory 1980). My assumption is that geomorphologists will benefit from continuing the agenda, initiated by Chorley, to establish the societal relevance of geomorphology, beyond its strictly positivist and instrumentalist constraints.

I shall argue the case that the social relevance of geomorphology will be increased if philosophical pluralism is engaged, and its programmatic focus will be clarified if a flexible unifying concept can be identified. The single paradigm pursuing many divergent problems is leading geomorphology to intellectual confusion; on the other hand, pluralist philosophies and a unifying concept may well lead to a renewed sense of direction. The alternative approach would seem to be 'that the Golden Rule is that there are no golden rules' (Baker 1993) and that geomorphology is what geomorphologists do. I find little comfort in such an agenda.

A central question which is unresolved is whether geomorphology is a proudly radical backroom field or whether it wishes to engage society, decision making and planning. The former position seems to be espoused by Haines-Young and Petch (1985) in their otherwise admirable exploration of the philosophical underpinnings of geomorphology. A 'real, unified and committed' geomorphology, to use Stoddart's language, must, it would seem to me, embrace a form of philosophical pluralism that extends beyond positivism. Even twenty years ago, Butzer (1973) pleaded for geomorphologists to allow for diversity and to accept pluralism. He was sensitive to the danger of fragmentation and particularism but suggested that 'the Anglo-Saxon prerogative', represented by contemporary process geomorphology, was in danger of becoming as repressive and one-sided as earlier Davisian geomorphology. In 1992, Yatsu wanted to reinforce that narrow vision of geomorphology so as 'to make it more scientific'. He rejected the contributions of Leopold ('a contemporary Kepler'), Chorley ('an exponent of the enlightenment movement') and Brunsden ('a modern Moses') as being 'fanciful'. By contrast with Yatsu's recommendation for geomorphology, this chapter suggests that geomorphology needs more rigorous philosophical debate, a reunifying central concept, and a clearer problem focus.

Geomorphology treats the general configuration of the earth's surface and (a) explores the formative processes of present-day landforms and natural landscapes, (b) determines their relationship to underlying geological structure, (c) investigates long-term landform evolution, (d) establishes societal implications of modifications of the earth's surface, and (e) determines the physical constraints for management of landforms.

There are at least five different groups of geomorphologists in terms of research goals and methods (Slaymaker 1991a): morphographers, who analyse static landform and material parameters (Chorley 1972); historical-genetic geomorphologists, who analyse long-term landform development (Penck 1894); specialists in energy flow and material transfer close to the earth surface (K.J. Gregory 1985); functional geomorphologists, who analyse the dynamic interaction of energy with landforms and materials (Gilbert 1877); and applied geomorphologists, who interpret geomorphology to society and explore the influence of society on landforms (Costa and Fleisher 1984).

Geomorphology shares with geography, ecology and engineering an interest in complex environmental systems which include purposive or inadvertent modification by society; it shares with astronomy, geology and geophysics an interest in vast expanses of time and very large areas. Because of these linked interests with cognate disciplines it might be assumed that philosophical pluralism would prevail at all times. But this has not been the case. Astronomy, engineering, geology, geophysics and, to a lesser extent, ecology, have evolved under broadly positivist assumptions; philosophical pluralism has been little in evidence in these disciplines since the debates surrounding evolution in the nineteenth century.

THE NEED FOR PLURALISM

One of the reasons that a *new* philosophical position can lend excitement to geomorphology lies in the need to relate geomorphology more closely to societal problems; but another one can be illustrated by Figure 15.1, in which a range of ways of knowing is illustrated. A discussion by Hammond (1978) on the characteristics and constraints of various modes of enquiry sets the issue in a broader context. He recognises a continuum of modes of enquiry from 'true experiment' to 'intuitive judgement'; along the continuum, the covertness of the mode of enquiry increases, active manipulation of the variables decreases and the analytical is replaced by the intuitive approach. It is also clear from Figure 15.1 that a positivistic philosophy becomes progressively less relevant in addressing questions that belong in the realm of intuition and passive manipulation of variables. The realm of intuitive

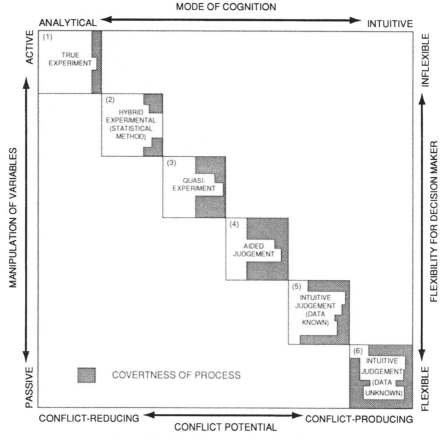

Figure 15.1 Classification of modes of enquiry
Source: After Hammond 1978

judgement is one with which applied geomorphologists and consultants, as well as geomorphologists working on long-term landform development, must engage. The classical experimental method of physical science has to be modified to deal with complex environmental systems, and may be unrecognisable as the same mode of enquiry in principle (Chalmers 1990). Social scientists such as Boruch and Riecken (1975) have discussed the nature of experimentation in their discipline, and Eberhardt (1976), a quantitative ecologist, has recognised that field experiments in ecology cannot be equated with classical physics experiments, especially with respect to the degree of control exerted over the variables. A similar argument holds for geomorphology and relates closely to the fundamental critique of positivism offered by realism (Hesse 1974; Sayer 1984).

Geomorphology, together with other earth sciences, cannot be an exclusively nomothetic science, as Büdel (1982) has explained (Figure 15.2). Geomorphology is multi-tiered, and explanations that hold for one tier do not necessarily hold for another tier. Schumm (1985) has identified seven fundamental problems associated with attempts to extrapolate the findings from functional (climatic dynamic) geomorphology to historical-genetic (climatogenetic) geomorphology as follows: (i) scale (Kennedy 1977), (ii) location (Tricart 1965), (iii) convergence (Albritton 1963), (iv) divergence (Brunsden 1990), (v) singularity (Harvey 1969), (vi) sensitivity (Brunsden and Thornes 1979), and (vii) complexity (Schumm 1979). The implications of his findings are that however complete the explanations of functional geomorphology may be, they are inadequate in themselves to explain long-term relief development (cf. Thornes and Brunsden 1977).

It is clear both that 'many (geomorphologists) operate with a dearth of quantitative rules or laws' (Johnson 1970) and that the fundamental principles of geomorphology are highly variable. Though every geomorphology text contains its list of fundamental 'concepts', ranging from the 'principle' of uniformitarianism to the 'caution' that 'complexity of geomorphic evolution is more common than simplicity' (Thornbury 1969), they are largely qualitative statements which can only with difficulty and extensive reformulation be falsified. Indeed, they are more in the nature of 'hints to young geomorphologists'. In spite of the fact that in the last twenty years 'human geography ... has rushed towards pluralism and relativism at a time when physical geography has bitten the bullet of conducting normal science' (Newson 1992), there are many aspects of geomorphology that do not lend themselves to exploration via normal science. The term 'normal science' refers to a philosophy which deals with 'determination of significant facts, matching of facts with theory and articulation of theory' (Kuhn 1962). Because the bias of normal science is positivistic, and rejects holistic philosophies, important aspects of geomorphology have been neglected. Perhaps more importantly, geomorphology and geomorphologists have been marginalised by decision-makers, not to mention the public at large, who

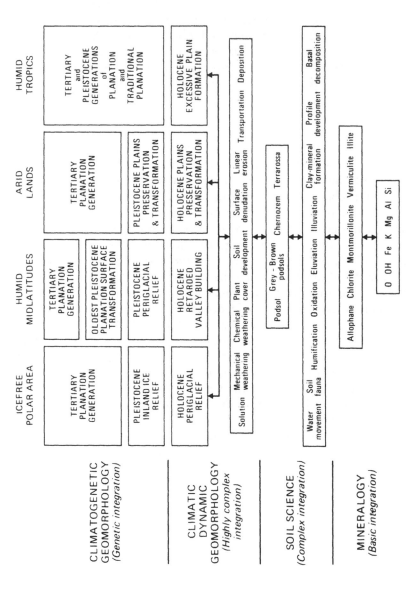

Figure 15.2 Geomorphology in the framework of the natural sciences
Source: Büdel 1982

have concluded that the discipline does not interact well with social structures.

Only two philosophical positions will be identified here and some of their potential value to geomorphology. First, positivism is characterised by empiricism, a unitary scientific method, theories capable of verification and strict functionalism. So powerful is this philosophical position that it has come to be equated with normal science. Realism is an anti-positivist philosophy which distinguishes between what causes something to happen and the number of times and places that it happens. A realist has a multi-tiered conception of reality in contrast to the atomism of positivism. Not only events but also mechanisms and structures are important (Hesse 1974; Bhaskar 1986). Although the value of this framework has been recognised (Kennedy 1979), it is not widely utilised by geomorphologists.

Gould and Olsson (1982) argued the necessity to pursue all ways to knowledge; schools of thought in geomorphology do differ, and while positivists promote social engineering, realists seek to increase understanding. Decision-making under uncertainty and planning require pluralism. Structural explanations of social life and its relationship to environment offer an exciting reconceptualisation through which greater understanding can be achieved in spite of the conventional physical geographers' caveat that the 'biophysical components of these structures remain to be understood both in their own right and in order to open up the possibility of prediction and management' (Clark et al. 1987: 383).

A real, unified and committed geomorphology requires pluralism: holistic geomorphology cannot be achieved through positivism, and integration of pure and applied geomorphology will remain an unachievable goal without pluralism. The contested character of intellectual traditions is surely best reflected through philosophical pluralism, and it can be argued that geomorphology will become a more normal science when it participates more actively in this contest. This is not to succumb to radical relativism and to suggest that any philosophical position will do (Chalmers 1990). We are simply seeking to point out the need for frameworks other than positivism, and identify realism as the most promising candidate (Bhaskar 1986, 1989).

NEED FOR A UNIFYING CONCEPT AND PROBLEM FOCUS

Since the time of W.M. Davis, geomorphology has not had an accepted unifying concept. A flexible unifying concept would, ideally, help to develop distinctive geomorphic understanding. The energy budget in climatology and the hydrologic cycle in hydrology are examples of unifying concepts in various fields of physical geography. In the case of geomorphology, the geographical cycle was central prior to the 1940s and geomorphology grew rapidly in influence under this unifying concept. Since the discrediting of the

Davisian cycle, geomorphology's intellectual energy has been dissipated in a number of directions (Tinkler 1985). These directions do not necessarily add up to a coherent field, and a number of calls for a new focus for the field have appeared recently (e.g. Schumm 1977; Brunsden 1990; Baker and Twidale 1991). As long as the field remains unfocused, geomorphology is vulnerable to internal divisions, and the possibilities of incremental and additive research are reduced.

Moreover, intellectual frameworks which seem to offer cohesion but are intellectually flawed have become common. Timeless, theoretical and utilitarian bandwagons are just three examples. Geomorphology that deals exclusively with functional relationships that ignore actual time sequences, or that considers exclusively theoretical constructs or that defines itself in terms of solutions to applied problems is simply not true to its intellectual roots nor is it able to define a role in the academy that will command respect from other fields. It is my purpose to advance the flexible concept of a sediment budget as one such unifying concept.

The idea of a sediment budget, that of identification of storage sites, transport processes, and linkages among them, and the quantification of storage volumes and rates of transport processes is as old as geomorphology. In Hutton (1795), Gilbert (1877), Darwin (1882) and Davis (1899) there is always implicit the relationship between surface form and variable rates of production, movement and storage of sediment. While Hutton and Davis thought of the problem in terms of erosional forms as residuals, Gilbert and Darwin attempted to measure the sediment movement on slopes and in channels.

Nevertheless, it is relatively recently that the concept has been more precisely defined and its range of application discussed (Swanson et al. 1982; Slaymaker 1988). The sediment budget, defined as the quantitative description of the movement of sediment through a landscape unit, has the potential to embrace the traditional questions in geomorphology, to allow growth in understanding of landforms as part of a dynamically vibrant planet, and to encourage exploration of the societal implications of sediment redistribution patterns.

The sediment budget equation, in its simplest form is

$$I = O + \Delta S \tag{1}$$

where I is input, O is output and ΔS is change of storage. Equation 1 can describe the routing of clastic and dissolved sediments over a specified time increment and with respect to a particular storage reservoir. Such a framework allows the interpretation of form change as the residual difference between input and output at specified spatial and temporal scales. In the unusual case, where

$$I = O \tag{2}$$

local form is unchanged and yet a net lowering of the landform relative to an absolute datum will have occurred.

Examples of application of the sediment budget approach to clastic sediment routing are numerous – Jackli (1957), Rapp (1960), and Swanson *et al.* (1982) are perhaps the best known. They have identified the major sources, pathways and sinks of clastic sediments within relatively small basins or landform units and determined process rates. Oldfield (1977) linked lake basin sediment budget studies with solute routing, nutrient cycling and ecology, and Foster *et al.* (1985) formulated a simple mass balance of sediment transfers in aquatic systems that includes the dissolved solids:

$$\Delta S = (If + Ic + Io + Is + Ib + Ia) - O \qquad (3)$$

where ΔS = net accumulation of lake sediment during a specified time, Δt; I = input of sediment to the lake in time Δt; f = fluvial; c = colluvial; o = organic matter; s = biogenic silica; b = lake bank erosion; a = aeolian dust; and O = output of sediment from the lake in time Δt.

A wide variety of questions related to material removal and storage can be examined using a sediment budget approach. The approach is sufficiently general as to allow the treatment of both human and biophysical effects. Swanson *et al.* (1982) and Roberts and Church (1986) provide examples of how the effects of forestry practices superimposed on natural processes can be evaluated by means of a sediment budget approach. The Lillooet River basin of British Columbia, a basin of over 3,000 km², has been analysed in terms of sediment sources and storages (Jordan and Slaymaker 1991). The spatial variations are huge, levels of uncertainty are high and the residence times of sediments are high. The average turnover time of the sediment reservoir in this very active and steep sub-basin of the Fraser Basin is several hundreds of years; therefore, by implication, the turnover time in the Fraser River basin (over 230,000 km² in area) is of the order of thousands to tens of thousands of years (Slaymaker 1991b). The timescale that needs to be seriously considered in a sediment hazard investigation is thus very different from the standard engineering project timescale (Church and Slaymaker 1989). The value of this approach is that it prompts us to ask: Where and what are the sediment storage reservoirs that must be considered? In what ways are these reservoirs modified over time? What are the appropriate time and spatial scales at which to conduct the sediment budget? What are the societal implications of sediment budgets investigated over very large areas?

The first point to note is that there are significant gaps in our understanding of the degree of interconnectedness of various components of the sediment budget. We understand the workings of most of the diverse phenomena at site scale. But we have considerable difficulty in tracking the pathways followed by sediment to and from the disturbed sites, and consequently we have difficulty in assessing basin-wide effects. At the scale of the Fraser River basin, there is considerable uncertainty about the functioning of the sediment system as a whole. This high level of uncertainty implies both a need to construct alternative scenarios of development and a need for humility in any policy recommendations.

Any policy that influences land use in the basin must be sensitive to local, regional and basin-wide variations in the ability of land to absorb the impacts of human activity. The authority to recommend or to assign land for use in sustainable ways and to change land uses that are not sustainable must be vested in a team that has authority to act over the whole basin. But a further implication is that regulation of sediment impacts exclusively at the basin scale makes little sense. Smaller units within the basin need to be defined, and management options at subregional, local or site-specific scale should be considered. Because all land-use change has primarily local impacts with respect to sediment transfers, all implementation of land-use decisions that impact on erosion and sedimentation should be sensitive to local and subregional communities. This does not negate the necessity for a basin-wide integrated land-use policy.

The geomorphologist's findings from sediment budgets estimated over large areas may involve a heightened and more explicit ethical concern over sustainability of land and community. Words like stewardship, partnership and dependence describe, from my perspective, a community's relationship to the land more adequately than control. Before we prescribe the future use of land, ethical questions and, indeed, the realm of metaphysics which was declared meaningless by positivism, must be respected and addressed.

If it be granted that geomorphology is the study of the earth's surface form and how that form responds to temporally and spatially variable energy inputs, whether biogeochemical or human, then the sediment budget explored in the broader philosophical framework of realism can address both traditional and more societally sensitive questions in geomorphology. As geomorphologists, we have always taken an interest in vast expanses of time and very large areas. Can we now link those interests to questions of environmental sustainability and species survival? I believe we can do so only if we abandon the exclusive attachment to positivism. We have now learned how to do normal science, and our site-by-site analysis and field experiments have brought geomorphology credibility in the earth science community. The next step will be to achieve a similar credibility in environmental science and, even more significantly, to link arms with those who seek not only social engineering but also understanding and change in the direction of a sustainable future.

Geomorphology in the 1990s has encouraged little philosophical debate because normal science is assumed to be adequate to provide explanation in geomorphology. There is no recognisable central concept in geomorphology, and we have no problem focus. In this sense, the appeals by Chorley – that philosophical debate be prized, that an open systems framework be adopted by geomorphology, and that the agendas of functional and historical geomorphology be clearly distinguished (Chorley 1978) – were visionary appeals that the field of geomorphology has neglected to its cost.

ACKNOWLEDGEMENT

The thrust of this argument has been stimulated by team teaching a University of British Columbia geography course with Derek Gregory; several specific comments by him are acknowledged by incorporation in the text. Errors of interpretation that remain are exclusively my responsibility.

REFERENCES

Ahnert, F. (1970) 'Functional relationships between denudation, relief, and uplift in large mid-latitude drainage basins', *American Journal of Science* 268: 243–263.
Albritton, C.C. (ed.) (1963) *The Fabric of Geology*, Addison-Wesley.
Baker, V.R. (1993) 'Tablets of stone: ten commandments or a golden rule', *Proceedings of the Third Conference of the International Association of Geomorphologists* (Abstract).
Baker, V.R. and Twidale, C.R. (1991) 'The reenchantment of geomorphology', *Geomorphology* 4: 73–100.
Barsch, D. (1990) 'Geomorphology and geoecology', *Zeitschrift für Geomorphologie*, Supp. vol. 79: 39–49.
Bennett, R.J. and Chorley, R.J. (1978) *Environmental Systems: Philosophy, Analysis and Control*, London: Methuen.
Bhaskar, R. (1986) *Scientific Realism and Human Emancipation*, London: Verso.
——(1989) *Reclaiming Reality*, London: Verso.
Boruch, R.F. and Riecken, M.W. (1975) *Experimental Testing of Public Policy*, Boulder: Westview Press.
Brunsden, D. (1990) 'Tablets of stone: toward the ten commandments of geomorphology', *Zeitschrift für Geomorphologie*, Supp. vol. 79: 1–37.
Brunsden, D. and Thornes, J.B. (1979) 'Landscape sensitivity and change', *Transactions of the Institute of British Geographers* 4: 463–484.
Büdel, J. (1982) *Climatic Geomorphology*, translated by L. Fischer and D. Busche, Princeton: Princeton University Press.
Butzer, K.W. (1973) 'Pluralism in geomorphology', *Proceedings of the Annual Meeting of the American Association of Geographers*, 39–43.
Chalmers, A. (1990) *Science and Its Fabrication*, University of Minnesota Press.
Chorley, R.J. (1962) 'Geomorphology and general systems theory', *U.S. Geological Survey Professional Paper* 500-B.
——(1972) *Spatial Analysis in Geomorphology*, Harper & Row.
——(1978) 'Bases for theory in geomorphology', in C. Embleton, D. Brunsden and D.K.C. Jones (eds) *Geomorphology: Present Problems and Future Prospects*.
Chorley, R.J. and Kennedy, B.A. (1971) *Physical Geography: A Systems Approach*, London: Prentice-Hall.
Church, M. and Slaymaker, O. (1989) 'Disequilibrium of Holocene sediment yield in glaciated British Columbia', *Nature* 337: 452–454.
Clark, M.J., Gregory, K.J. and Gurnell, A.M. (eds) (1987) *Horizons in Physical Geography*, Macmillan Education.
Costa, J.E. and Fleisher, P.I. (eds) (1984) *Developments and Applications of Geomorphology*, Berlin: Springer-Verlag.

Darwin, C. (1882) *The Formation of Vegetable Mould, through the Action of Worms*, London.

Davis, W.M. (1899) 'The geographical cycle', *Geographical Journal* 14: 481–504.

Eberhardt, L.L. (1976) 'Quantitative ecology and impact assessment', *Journal of Environmental Management* 4: 27–70.

Foster, I.D.L., Dearing, J.A., Simpson, A., Carter, A.D. and Appleby, P.G. (1985) 'Lake catchment based studies of erosion and denudation in the merevale catchment', *Earth Surface Processes and Landforms* 10: 45–68.

Gilbert, G.K. (1877) *Report on the Geology of the Henry Mountains*, Washington, D.C.: Government Printing Office.

Gould, P.R. and Olsson, G. (eds) (1982) *A Search for Common Ground*, London: Pion.

Gregory, D. (1980) 'The ideology of control: systems theory and geography', *Tijdschrift voor Economische en Sociale Geografie* 71: 327–342.

Gregory, K.J. (1985) *The Nature of Physical Geography*, London: Edward Arnold.

Haines-Young, R. and Petch, J.R. (1985) *Physical Geography: Its Nature and Methods*, London: Harper & Row.

Hammond, K.R. (1978) 'Toward increasing competence of thought in public policy formation', in K.R. Hammond (ed.) *Judgment and Decision in Public Policy Formation*, Boulder: Westview.

Harvey, D. (1969) *Explanation in Geography*, London: Edward Arnold.

Hesse, M. (1974) *The Structure of Scientific Inference*, London: Macmillan.

Hutton, J. (1795) *Theory of the Earth with Proofs and Illustrations*, Edinburgh: William Creech.

Jackli, H. (1957) *Gegenwarts Geologie des Bundnerischen Rheingebietes*, Berr: Kummerli and Frey, Beitrage zür Geologie der Schweiz, Geotech, Series no. 36.

Johnson, A.M. (1970) *Physical Processes in Geology*, Freeman.

Jordan, P. and Slaymaker, O. (1991) 'Holocene sediment production in Lillooet River basin: a sediment budget approach', *Géographie Physique et Quaternaire* 45: 45–57.

Kennedy, B.A. (1977) 'A question of scale?', *Progress in Physical Geography* 1: 154–157.

——(1979) 'A naughty world', *Transactions of the Institute of British Geographers* 4: 550–558.

——(1990) 'Paint your boulder', *Zeitschrift für Geomorphologie*, NF Supplement band 79: 213–216.

Kuhn, T.S. (1962) *The Structure of Scientific Revolutions*, Chicago: University of Chicago Press.

Newson, M. (1992) 'Twenty years of systematic physical geography: issues for a "new environmental age"', *Progress in Physical Geography* 16: 209–221.

Oldfield, F. (1977) 'Lakes and their drainage basins as units of sediment based ecological study', *Progress in Physical Geography* 1: 460–504.

Penck, A. (1894) *Morphologie der Erdoberflache*, Stuttgart.

Rapp, A. (1960) 'Recent development of mountain slopes in Kärkevagge and surroundings', *Geografiska Annaler* 42A: 65–200.

Roberts, R.G. and Church, M. (1986) 'The sediment budget in severely disturbed watersheds, Queen Charlotte Ranges', *Canadian Journal of Forest Research* 16: 1092–1106.

Sayer, A. (1984) *Method in Social Science: A Realist Approach*, London: Hutchinson.

Schumm, S.A. (1977) *The Fluvial System*, Chichester: John Wiley.

——(1979) 'Geomorphic thresholds: the concept and its equilibrium', *Transactions of the Institute of British Geographers* NS 4: 485–515.

——(1985) 'Explanation and extrapolation in geomorphology: seven reasons for geologic uncertainty', *Transactions of the Japanese Geomorphological Union* 6: 1–18.

Slaymaker, O. (1988) 'Slope erosion and mass movement in relation to weathering in geochemical cycles', in A. Lerman and M. Meybeck (eds) *Physical and Chemical Weathering in Geochemical Cycles*, Dordrecht: Kluwer.

——(1991a) 'Mountain geomorphology: a theoretical framework for measurement programmes', *Catena* 18: 427–437.

——(1991b) 'Implications of the processes of erosion and sedimentation for sustainable development in the Fraser River basin', in A.H.J. Dorcey (ed.) *Perspectives on Sustainable Development in Water Management*, Vancouver: Westwater Research Centre.

Stoddart, D.R. (1986) *On Geography and Its History*, Oxford: Basil Blackwell.

Swanson, F.J., Janda, R.J., Dunne, T. and Swanston, D. (eds) (1982) *Sediment Budgets and Routing in Forested Drainage Basins*, US Department of Agriculture, Forest Service, General Technical Report PNW-141.

Thornbury, W.D. (1969) *Principles of Geomorphology*, New York: Wiley.

Thornes, J.B. and Brunsden, D. (1977) *Geomorphology and Time*, London: Methuen.

Tinkler, K.J. (1985) *A Short History of Geomorphology*, Barnes & Noble.

Tricart, J. (1965) *Principes et méthodes de géomorphologie*, Paris: Masson & Cie.

Wolman, M.G. and Gerson, R. (1978) 'Relative scales of time and effectiveness of climate in watershed geomorphology', *Earth Surface Processes* 3: 189–208.

Yatsu, E. (1992) 'To make geomorphology more scientific', *Transactions of the Japanese Geomorphological Union* 13: 87–124.

16

CARL SAUER: GEOMORPHOLOGIST

David R. Stoddart

INTRODUCTION

... it is often overlooked that in the earlier period [of the Berkeley School] Sauer worked at length in geomorphology, and that many Berkeley Ph.D.s have done likewise. I enrolled at Berkeley because Sauer was working with topics in physical geography, and I was a member of several of his final seminars in geomorphology.

(Joseph E. Spencer 1976: 8)

Most of those who have written about Carl Sauer from their own memories of him, such as James J. Parsons (1976) and David Hooson (1981), came to know him when his geomorphological interests had waned, the major exception being the extensive recollections of his friend and colleague of over half a century, John Leighly (1963b, 1976, 1978a, 1978b, 1978c, 1979, 1983). There is little concerning Sauer's geomorphic interests in the accounts of the early years of the Berkeley Department of Geography by Speth (1981) and Macpherson (1987). Most commentators on Sauer's early intellectual development concentrate on his monographs on the Upper Illinois valley (Sauer 1916), the Ozarks (Sauer 1920) and Kentucky (Sauer 1927), and on his programmatic statement 'The morphology of landscape' (Sauer 1925), which in spite of its title was not really about geomorphology at all. Only in the magisterial pages of *The History of the Study of Landforms* is there any historical consideration of Carl Sauer as geomorphologist (Chorley, Beckinsale and Dunn 1973: 427–428, 647, 663, 754; Beckinsale and Chorley 1991: 117–119, 246–247, and especially 366–372). It is the purpose of this chapter to elaborate and extend that treatment and to place Sauer's geomorphic period in the social and scientific context of the times.

BEFORE BERKELEY

The years before Sauer's move to Berkeley were especially formative in a number of ways. Born in 1889, in Missouri, he enrolled in 1908 at Northwestern University to read geology, and from that time on, in spite of

340

later changes in interests, always regarded himself as an 'earth scientist' (Sauer to J.R. Smith, 18 November 1948; quoted by Leighly 1987: 405). His coursework centred on petrography and he was also exposed to palaeontology. The following year, however, he transferred to the University of Chicago to study geography in the Department of Geography. Here he was powerfully influenced by the chairman of the Department, Rollin D. Salisbury, whose major textbook, *Physiography*, had recently been published (Salisbury 1907). By temperament Salisbury closely resembled the mature Sauer. He was, wrote Chamberlin (1931: 133),

> the ideal teacher. His flashing personality fascinated the students; frequently outcropping bits of brilliancy kept them continually on the qui vive for thrills; and a never-lapsing autocracy maintained the strictest order. His classroom was no place for the dull or the slow, but the better students profited enormously. He was a man of strange contrasts. He had a reputation for being very gruff and short; some people felt insulted; others developed a strong dislike, but they were those who saw only one side of his nature. The gruffness was but a mask which hid the warm heart, deep human sympathy, and true kindliness which were the real basis of his character.

It was under Salisbury that Sauer learned and adopted the style of fieldwork that stayed with him for the rest of his life. Salisbury

> encouraged learning by independent field observation. At the end of my first year he sent me into the Illinois Valley to make a study for the state geological survey. When I asked how to go about it he answered that that was my business, that he would find out later whether I had seen and learned. He came down twice to visit me in the field, asking me to show him whatever I wished in the way of evidence. He listened to my expositions, asking occasional acute questions but giving me no assistance in interpretation. This was my job and it was up to me to come to my own conclusions.
>
> (Sauer 1966: 69)

In 1914 Sauer began teaching at Salem Normal School in Massachusetts, but in the same year taught a course on physiography at the University of Michigan in Ann Arbor. His thirty-two lectures comprised:

1   Introduction. Place of physiography, aims of study
2   Methods
3   Devices
4   Maps and diagrams
5   Astronomical geography
6   Climate
7   Precipitation

8   Climatic regions
9   Earth sphere [atmosphere, hydrosphere, lithosphere]
10–12 Work of streams

Field trip to Niagara Falls

13   Review of field trip
14   Weathering
15   Work of wind
16   Ground water
17   Waves
18   Coastlines
19   Lakes
20–25 Glaciation
26–29 Vulcanism and diastrophism
30–32 Summary – physiographic types and their origin: mountains, plateaus,
       plains

The course must have been a success, for after completing his doctoral thesis on the Ozark Highland in 1915 he was appointed in January 1916 to the Department of Geography at the University of Michigan. The chairman of the Department was W.H. Hobbs, a physiographer whose *Earth Features and Their Meaning* had appeared in 1912. Hobbs specialised in glaciation, but in spite of the titles of his recent papers in that field (Hobbs 1910a, 1910b) he, like Salisbury, was no Davisian.

It was at Ann Arbor that Sauer met John Leighly and Fred Kniffen, who were later to become his students, and where he also initiated summer field camps. He reconnoitred Kentucky and Tennessee, and in 1920 leased an abandoned mill at Mill Springs, Kentucky. Thus began an interest in the south-eastern United States which continued long after his move to Berkeley (Sauer to W.D. Jones, 9 November 1951). It also led to his monograph on the Pennyroyal area of Kentucky, published in 1927. He was also the driving force behind the establishment of the Michigan Land Economic Survey in 1922, which began with student mapping projects of degraded areas. We see here the beginnings of his concern with land use and resources, as well as developing technical survey issues which he had become interested in at Chicago (Jones and Sauer 1915; Sauer 1921, 1924). These concerns also led to a growing interest in soils (Sauer 1918, 1922). These were significant initiatives in land utilisation studies at that time (Guttenberg 1976; Schmalz 1978). All resonated during the rest of Sauer's career. The very success of the Michigan developments paradoxically proved unsettling: 'Between the University summer camp and the Mich. L. E. Survey I saw myself tied down for an indefinite time into the future and I wanted to get out on my own and roam new areas' (Sauer to W.D. Jones, 9 November 1951).

His experience at the Universities of Chicago and Michigan proved to be an apprenticeship. It provided the background for the turn to geomorphology that he was about to make.

BEYOND THE HIGH SIERRA

In 1923 Sauer accepted appointment as chairman of the Department of Geography of the University of California at Berkeley, a move which Leighly (1976: 339) sees as 'by far the most momentous one of his life'. He came with a substantial background in physical geography and geology, but he also came to a campus where the Department of Geology was a formidable force. He himself recalled that when he arrived in Berkeley he was frightened off the study of landforms, soils and vegetation because there were experts in these fields there already (Sauer in conversation, 11 October 1971). Leighly (1976: 339) goes so far as to say that it was made a condition of Sauer's appointment that introductory physical geography ('physiography') be transferred to Geology. (Sauer did teach Geography 1, Introduction to Geography: Elements, in his first year, stressing climate, and regularly taught Geography 2, Introduction to Geography: Natural and Cultural Regions. He also took charge of Geography 101, the Field Class, and in 1929 introduced a Seminar in Physical Geography. For further details on departmental teaching in the first decade after his arrival, see Speth 1981: 235–237).

Pre-eminent among the experts Sauer encountered were A.C. Lawson and G.D. Louderback. Lawson (1861–1952) had come to Berkeley in 1890 and became chairman of the Geology Department in 1906. Before Sauer's arrival he had started to alternate as chairman with Louderback (Louderback was chair 1923–1924, Lawson 1924–1925, and Louderback again 1925–1927), and although Lawson formally retired in 1928 he lived until the year before Sauer's own retirement. He was ambitious, irascible and dauntingly productive (he fathered his first child at the age of 29 and his last, by a different woman, at the age of 88). He developed Davisian ideas of denudation chronology in the high Sierra Nevada (Lawson 1904), and worked actively on the tectonics of California (Lawson 1893, 1921). He had already contributed a classic paper on the development of desert pediments (Lawson 1915). Louderback made his reputation with studies of the faulted terrain of the Great Basin (Louderback 1904, 1923, 1926). Known as 'Uncle George' in the Geology Department at the time of Sauer's arrival, he 'admired the German tradition of scholarship, which found expression in his insistence on thoroughness and familiarity with the German literature – in the original' (Pettijohn 1984: 105).

There was also N.E.A. Hinds, who taught geomorphology and who twenty years later produced *Geomorphology: The Evolution of Landscape*, a book which 'followed ... the general system of land form evolution developed and so brilliantly advocated by Professor Davis's (Hinds 1943: ix).

In that book Sauer is mentioned but once, in the additional reading for Chapter 5, 'Principles of landform evolution' (which occupied all of 39 lines), and the reference was to 'The morphology of landscape'. At the time of Sauer's arrival Hinds was also beginning his work on fluvial landscapes in the Hawaiian Islands (Hinds 1925, 1931), and he was later to synthesise landform evolution in California (Hinds 1952). He was not the most stimulating of lecturers, as Francis Pettijohn (1984: 107) recalls, though his field classes clearly had their attractions:

> The course by N.E.A. Hinds [on geomorphology] was perhaps the weakest, though the field excursions were very worthwhile. Our first trip was to Yosemite, where we stayed in the floored tents of Camp Curry. At that time the California students combined geologic field-work with a fair amount of carousing. Every evening there were several card games under way accompanied by gin drinking – bootleg gin, of course, since Prohibition was in force. These activities went on long past midnight.

But taken together, the concentration of geomorphology at Berkeley in the hands of Lawson, Louderback and Hinds cannot have been encouraging to a Sauer then still in his early thirties. He saw it as a personal challenge, which he met initially in two ways. The first was to produce a programmatic statement for his subject in 'The morphology of landscape' (Sauer 1925); the second was to launch himself aggressively into geomorphic field research. After some initial reconnaissance in Baja California, he first engaged with the Coast Ranges, on which Lawson (1921) himself was working, and specifically that sector to which Lawson had given the namer 'Peninsular Sierra' (Lawson 1893: 117). Over the next decade Sauer vigorously pursued geomorphic research. It was a period during which, following Lawson's retirement, the Geology Department shifted emphasis, becoming 'a department of petrographers and mineralogists basically, ... [which] requires microscopical technique for everything' (Sauer to J.R. Smith, 2 May 1939). Unfortunately, as the window of opportunity opened for geographers with the changing focus in Geology, Sauer himself had lost his research interest in landforms by the end of the 1930s, though he continued to teach and advise students in physical geography and geomorphology.

To some degree a bridge between the two departments was provided by Richard J. Russell (Kniffen 1973), who in 1923 was working under Lawson and Louderback on basin-range structures in north-east California; his thesis was completed in 1925 (Russell 1927, 1928). He was an assistant in the Geography Department when Sauer arrived, and Sauer retained him. After leaving Berkeley in 1926 he often revisited Sauer's department to teach. He later had the distinction, as had W.M. Davis, of presiding over both the Association of American Geographers and the Geological Society of

America. Russell's relations with Sauer were always cordial, and indeed Sauer came to think that Russell 'would probably be considered the best geomorphologist in the country' (Hewes 1983: 144).

Complexity is added to the interactions between Sauer and his Berkeley colleagues by the figures dominating American and German landform interpretation at that time, W.M. Davis and Albrecht and Walther Penck. Davis was in his seventies when Sauer came to Berkeley, though his life-work was very far from done, while the younger Penck was about to publish his major work *Die morphologische Analyse* (1924). By background and temperament Sauer was attracted to Penck's mode of interpretation, which also had the immense advantage of giving him new intellectual tools with which to assault the Davisian orthodoxy prevailing in the West. There was also the added advantage that few could match the command that Sauer (and later Leighly and Kesseli) had of Penck's tortured prose. Sauer thus leaned heavily on Penck to satisfy both personal and departmental goals as he strove to establish himself in the early years in Berkeley.

By contrast, Sauer's doubts about the Davisian system began when he was a student of Salisbury at Chicago. There

> Davis was not taken too seriously, though he was not held in disrespect. The two were of very different temperaments and physique, and they were likely to meet in combat at the annual meetings, an event always looked forward to. Salisbury big and massive, direct and laconic of speech, never lecturing but the best master of Socratic discourse I have known. Davis almost petite, prim, nimble in movement and discourse and impressive in extemporaneous lecture and debate. I think we derived from Salisbury a distrust of systems, I'd like to say an open horizon; at least that is what he tried for. Davis trained disciples and he trained them well, but it seems to me that they did not leave the path the master had laid out.
>
> (Sauer to R.J. Chorley, 18 October 1961; quoted in Chorley, Beckinsale and Dunn 1971: 428)

Sauer followed Salisbury in distrusting Davis's tidy codification of landforms in the cycle of erosion concept:

> His famous formulation of structure, process, and stage would provide a genetic classification of all land forms. This unifying concept of his was a construct of his mind. Description therefore was illustrative of his thesis instead of empirically informative. What he called explanatory description required identification of forms by terms of his system of explanation, and therefore the introduction of a new vocabulary. A Davisian description did not propose to give the reader data that were independent of his frame of reference. The block diagrams that he drew with art were theoretical models, not graphic descriptions. His thinking

depended on stage and cycle, the metaphors of youth, maturity, and old age, and of rejuvenation by a new cycle.

(Sauer 1966: 255; see also Sauer 1974: 191)

Davis of course had seen this as a virtue, claiming that

the scheme of the cycle is not meant to include any actual examples at all, because it is by intention a scheme of the imagination and not a matter of observation; yet it should be accompanied, tested, and corrected by a collection of actual examples that match just as many of its elements as possible.

(Davis 1905: 152)

For Sauer Davis's system ignored the true complexity and therefore the fascination of landforms:

Davis formulated a theory of recurrent geographical cycles, of uplift, erosion, and wearing down to a peneplain, passing through stages of youth, maturity, and old age.... The cycle might be long or short, its length and position in time were irrelevant. Davis was our first and greatest maker of a system that replaced the complexity of events by a general order. Theory was illustrated by models, the block diagrams which he drew so well to show his concept of how the modelling of the land should pass from stage to stage. Davis continued to develop and expound the cyclic order that he thought he had discovered.

(Sauer 1974: 190)

Echoing what came to be termed Fenneman's Principle (Fenneman 1936), Sauer's 'sense of real, non-duplicated time and place' led him to the belief that

The things with which we are concerned are changing continuously and without end, and they take place with good reason, not anywhere, but somewhere, that is in actual situations or places. That succession of events with which we deal is quite other than the conceptual models that are set up as regular, recurrent, or parallel stages and cycles

and to the conclusion that 'Among geographers, William Morris Davis delayed somewhat our learning about the physical earth by his systems of attractive but unreal cycles of erosion' (Sauer 1952: 2).

Sauer's first student and principal colleague, John Leighly, displayed a more personal animus against Davis and his methods. While still at Central Michigan Normal School in 1919–1920 Leighly had been 'repelled' by Davis's 'stultifying definition' of geography in *Geographical Essays* (Davis 1909), and was 'thoroughly tired' of Davis when required to read the book again by J.P. Buwalda of the Geology Department during his first year at Berkeley (Leighly 1979: 4, 6).

Davis had the reputation of being an autocratic and imperious figure, secure in the knowledge of his own rectitude and infallibility. Sauer recalled that:

> My impression of Davis while I was still in the eastern part of the country was that he was a sharp and sometimes merciless critic of work that departed from his formulations and that young men who were not his followers were turned away from the study of land forms. He had authority and dialectic skill and he would cut up a youngster who had neither. I felt sorry for them and I think physical geography in this country became poorer as Davis exercised his censure and approval.
> (Sauer to R.J. Chorley, 18 October 1961; quoted in Chorley, Dunn and Beckinsale 1973: 428)

Leighly deeply resented being treated in this manner, and this resentment stayed with him all his life. He recalled in 1961 that

> while I was a graduate student here at Berkeley the Department of Geology invited Davis to give a graduate seminar in that department. I had, of course, read much of Davis as a student, and eagerly seized the opportunity to enroll in the seminar. The result was a great disappointment to me. At one of the early meetings of the seminar I brought up a question about some statement Davis had made. Davis pointed a finger at me and said, 'If you had read (some article of his, with which I was familiar) you wouldn't ask that question.' It was evident that he did not want any discussion on the part of the students, and so I kept my mouth shut for the rest of the term. He spent his time expounding ideas he had already, in some instances long before, set forth in his writings. At the end I submitted some kind of paper, which I wrote without any enthusiasm, and wrote off the time spent as lost. Undoubtedly Davis was senile by this time (he was about 75 at the time), and should not have gone about pretending to instruct students.
> (Leighly to R.J. Chorley, 18 October 1961; quoted in Chorley, Beckinsale and Dunn 1973: 569)

And in his review of the volume from which this quotation is taken, Leighly states that

> years have passed since Davis ceased to be the colossus that bestrode the world of geographical learning, but his shadow still stretches indefinitely into the future.... Here he stands, in his pettiness and his greatness: vain, egotistical, contentious, but within a narrow core of mental vision thinking incisively and imaginatively, and wielding a pen capable of translating thought with equal skill into limpid prose and elegant drawings.
> (Leighly 1974: 328)

347

It is thus perhaps not surprising that for reasons both straightforward and more complex both Sauer and Leighly sought alternative modes of thinking in geomorphology, and that they both turned to German sources. Sauer (1925) was certainly familiar with the writings of Von Richthofen, Albrecht Penck and Siegfried Passarge, among others, and must have become directly aware of Hettner's ideas through his fellow student Wellington Jones, who took a semester away from Chicago in 1913 to spend time with Hettner in Heidelberg. Leighly took Hettner's book *Die Oberflächenformen des Festlandes* (1921) to Buwalda's class in Berkeley and read passages from it in translation (Leighly 1979: 6). It was a violently anti-Davisian book and was doubtless brought to Leighly's notice by Davis's denunciation of it in the *Geographical Review* for April 1923, where he objects to Hettner's 'homilies, truisms, hesitations, obstructive misunderstandings, and disputatious objections' to his own scheme of the cycle of erosion (Davis 1923: 318). Sauer also invited Albrecht Penck to teach in Berkeley in 1928 (some recent publications give the date as 1925 and 1927).

But it was Walther Penck's *Die morphologische Analyse*, published in the year following Sauer's move to Berkeley, that provided a viable alternative framework of analysis. Both Sauer and Leighly made Penck's book a feature of their seminars in geomorphology (see Spencer 1979: 48 on this aspect of Sauer's seminar). Leighly (1983: 84) recalled that

> In one of my early efforts I introduced students to Walther Penck's *Die morphologische Analyse* (1924), long before its publication in English translation [1953]. I translated orally the essential parts of the book, paragraph by paragraph, and let the students discuss Penck's ideas.

Leighly waxed enthusiastic:

> We became acquainted with Walther Penck's *Die morphologische Analyse* (1924) soon after it appeared, promptly recognizing the light that his *Massenbewegung* [mass movement] threw on the detailed forms of the Coast Ranges in our part of California, with their highly disturbed structure of mingled weak and resistant rocks. Penck opened our eyes to the meaning of the gently curved lines of fences originally built straight but now sagging downhill, their posts leaning as they were carried with the soil in which they were set; and of the presence in canyons (in California the general term for all narrow valleys) of trees leaning in the direction of slope, their roots dragging behind them as they moved toward the bottoms of the canyons, eventually to fall into the channels of the ephemeral streams. We learned to see the sculpture of the land surface as proceeding more by migration of the cover of weathered rock down slopes than by stream erosion. The work of streams, most of which flow here only in winter, seemed to

consist more in sluicing away the waste fed into them by creep from their tributary slopes than in the erosion of their beds.

(Leighly 1979: 7–8)

J.E. Kesseli, who had been appointed to the Department to teach geomorphology in 1932 and who had studied under Johannes Walther at Halle, also continued the emphasis on Penck, producing a summary of his views for departmental use in 1940. His treatment of Penck's views on slope development under varying conditions (Figure 16.1) was the most detailed available in English until the translation of Penck's book appeared in 1953; compare, for example, the account by Von Engeln (1942), which had a vastly wider circulation. Kesseli's own research, under Sauer's supervision, had been on glaciation in the Sierra Nevada (Kesseli 1938, 1941a), but he also wrote on river dynamics (Kesseli 1941b) and produced an *Outline of Physical Geography* (Kesseli 1939) which also had only local distribution. Leighly described the parallel retreat of slopes in the Badlands of South Dakota and in central Kentucky (Leighly 1940: 227). And Samuel Dicken claimed to have 'applied Walther Penck's principles of slope development' (Dicken 1986: 89) in his work on karst development and soil erosion, carried out under Sauer's direction, in Kentucky (Dicken 1931, 1935a, 1935b, 1935c).

Henry Bruman remembers that on the Saturday field class, in the mid-1930s, Sauer would expatiate after lunch on Penck's ideas,

which he thought useful, and we would ask questions. At times he would discuss the view before us, pointing out accordant summits, remnant terraces, recurring slope changes, evidence of tectonic displacement, the distribution of vegetation associations in relation to slope and exposure, and similar topics.

(Bruman 1987: 130)

These discussions of course followed Salisbury's style. Parsons notes that 'Field techniques as such had no place in Sauer's field instruction' (Parsons 1979: 12). During the extended later fieldwork in Mexico Hewes (1983: 143) found that

Ordinarily, the fieldwork was quite casual, seeing what could be seen.... There was a minimum of mapping or formal note taking. Nonetheless, Sauer made fairly numerous photographs of features or scenes that interested him, using a large camera and tripod, as a form of documentation, I thought.

There was some taking of altimeter readings on some journeys, and checking of valley profiles. It was, however, a method of interrogation of landforms through Penckian eyes, at least from 1923 to 1940, and quite different to the declamatory style of Davis which Sauer experienced at first hand. 'I was never in the field with Davis until after he moved to California when I took

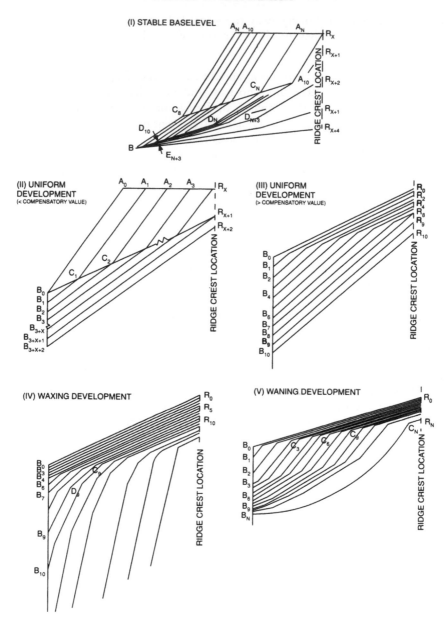

Figure 16.1 Kesseli's (1940) interpretation of five of Penck's slope development models. These figures are original and do not appear in Penck (1924)
Source: Redrawn and reproduced from Beckinsale and Chorley 1991: 359

him out into areas new to him. At any stop he expounded the scene rather than asked questions, which, as I recall the occasion, he was inclined to brush aside' (Sauer to R.J. Chorley, 18 October 1961; quoted by Chorley, Beckinsale and Dunn 1973: 428).

Sauer clearly realised the potential impact of this new direction in geomorphology, which in the United States at that time was virtually confined to the Berkeley Department of Geography. He wrote to Alfred Kroeber from Arizona in 1934:

> the next direction in geography is going to come from the west. Leighly and I and perhaps Kesseli have got some things started that are, with the help of Russell, going to build a couple of bonfires under the physiographic mule, and part of the wood we're using is handiest out there.... The situation has stayed put for a generation in this country and it's due to break.
>
> (Sauer to A.C. Kroeber, n.d.; quoted by Macpherson 1987: 84)

Given this personal and scientific background it is thus not surprising that Sauer's first sustained field research after moving to Berkeley was conceived as 'a trial of the method developed by Penck, to which he gave the name morphologic analysis'; and certainly nowhere other than in his own Department could it then realistically be 'assumed that the terminology and formulation of this method are familiar to the reader' (Sauer 1929: 202).

First, however, it is necessary to set aside one of Sauer's most cited (though not quoted) papers, 'The morphology of landscape' (1925). Suffice it to say that this was not a study in geomorphology but an enquiry into the nature of geography which drew heavily on the ideas of Alfred Hettner (1921). Sauer advocated that geography be concerned with the classificatory non-historical study of the areal differentiation of the Earth's surface, proceeding from the naive classification of surface forms. In the present context it is of interest in revealing the depth of Sauer's immersion in German geomorphology. Thus he cites Peschel (1870, 1879–1880), von Richthofen (1886) and Sapper (1917), but especially Siegfried Passarge (1912, 1920, 1922, 1923a, 1923b) and most notably Volume I (*Beschreibende Landschaftskunde*) of his *Die Grundlagen der Landschaftskunde* (1919). Leighly (1963b: 6) commented that 'The positive effect of this paper was, unfortunately, to stimulate a spate of detailed descriptions of small areas written in the ensuing twenty years, which had little value, either scholarly or practical.' Sauer himself never put its ideas into effect, and indeed came to repudiate its arguments, most notably in his addresses to the Association of American Geographers (Sauer 1941, 1956).

THE PENINSULAR RANGE

Sauer directed his first field energies at Berkeley to the investigation of Baja California. His earliest doctoral students, apart from Leighly, whose work was on Swedish towns, were C. Warren Thornthwaite (who curiously wrote another urban thesis derived from Sauer's interest in Kentucky), Peveril Meigs, Fred Kniffen and Samuel Dicken. Thornthwaite and Meigs went to Baja in 1925. The following year they were joined by Kniffen and Dicken on a six-week expedition for which Sauer bought his first car. Dicken (1988: 34) recalled:

> We observed the change in vegetation as it became drier, and in landforms, such as sand dunes, terraces, granite outcrops and steep-sided arroyos. Professor Sauer took a special interest in evidences of sea level changes and old Indian shell heaps. (He thought they were related.) We argued some about the sea level changes and Professor Sauer was very reluctant to accept an alternative view.

Or, as he put it in the *Itinerant Geographer* in 1959:

> Morphology was then the rage
> We saw each form and guessed the stage
> To these each terrace was related
> The land was sunk, then elevated
> We learned the rise of land and seas
> Was seen by aborigines.

Kniffen, however, was less enthusiastic in his recollections: 'Dicken and I, who had come up through geology, were let down. There was no systematic geomorphology' (personal communication, 9 February 1989).

These Baja forays were, however, for Sauer simply reconnaissance: they were not meant to involve formal research but rather the formulation of problems for further study. In a sense, of course, they encapsulated the field style already discussed. Indeed Dicken (1988: 37), referring to the period 1926–1927, noted that 'It was Professor Sauer's custom on a field trip to take a class to a good vantage point, sit down, look over the landscape, and then talk about it. We did not get much instruction on field methods.' The formal application of the Walther Penck method required somewhat more.

Sauer's decision to work on the landforms of the Peninsular Range was certainly audacious: not only had Lawson given it its name, but he had written on the tectonics of coastal southern California (1893, 1921), and his department had recently published detailed studies of parts of the area (Hudson 1922; Miller 1928). But given Sauer's interest in Penck, the tectonic instability of the area made it highly attractive:

> The surface complex is, almost in its entirety, an expression of the relation of intensity of diastrophism to weathering of rock, denudation of slopes,

352

and accumulation of detritus.... Penck has provided a fundamental method, because it alone recognizes the critical significance of intensity and duration of tectonic movement in surface sculpture.

(Sauer 1929: 202)

Beckinsale and Chorley (1991: 246) term the tectonic environment of the Coast Ranges 'unstable and idiosyncratic'.

The area mainly consists of a granitic batholith and metamorphic rocks, with structures defined by dominant NW–SE-trending faults intersected by other faults at right angles to them. This divides the region into a series of blocks, generally bounded by steep slopes and characterised by smooth summit plateaus, the bounding slopes and the plateaus being linked by convex shoulders. Overall the area is steeply bounded to the east and slopes towards the ocean. Sauer developed the argument that 'the Peninsular Range consists of individual blocks, which are differentiated from each other in surface forms and which have undergone differing morphologic evolution' (Sauer 1929: 210).

The summit plateau, which Sauer terms a 'summit peneplain', he interprets in a Penckian rather than a Davisian sense – 'of the type characterized by Penck as primary peneplain (*Primärrumpf*)' (Sauer 1929: 212), citing an earlier theoretical paper (Penck 1920: 80–81) rather than the 1923 treatise.

We have here not, in so far as is known, a surface once worn to a low level and then uplifted, but an assemblage of forms which, though at summit position, is in process of reduction of relief. The minor elevations are being reduced by denudation, their lower soil-covered flanks are growing at the expense of the upper slopes of discontinuous soil cover, and the dales ['channel-less depressions': 1929: 210] are enlarging headward as well as transversely.

(Sauer 1929: 212)

The area is uninfluenced by any general base-level: the steep bounding slopes have a tectonic origin, while the plateaus are formed by denudation (solifluction). 'The local surfaces were never level though they may be quite continuous.... Unified surfaces here at best have continuous inclination and undulatory continuity, the result of tilt or warp, concurrently modified by increased or decreased intensity of surface removal' (Sauer 1929: 213). He describes one such 'primary peneplain':

The summit of Agua Caliente Mountain is mesa-like, and is the most anciently established surface in this unit area. Its highest parts are bared bedrock surfaces, either domes, heavily exfoliating, or rock monuments.... The greater part of the summit area is occupied by meadows, park-like forest, and brush thickets, mingling on a rolling surface.

(Sauer 1929: 226)

Sauer's explanation of the development of these features thus inverts a traditional Davisian explanation:

> The summit area is a primary peneplain initiated by the detachment of the block from other adjacent surfaces by a period of rapidly increasing uplift during which a girdle of convex slopes became established about the rising block. The great western convex flank presumably started at the western margin of the block, that is, in what is now the Valle San Jose and has receded eastward to its present position. The history of sculpture of the block reads from summit to basal plain. Declining rate of rise of the block allowed thereupon the establishment of a first basal plain, also initiated at the west, followed by a small accelerated rise, during which this first concave surface became bordered at the west by a convex step. An intermittent rising of the block, especially at the east, gradually but not regularly reduced in amount to the present, is adequate to explain the step-like, and as a whole, concave series of surfaces that sweep upward from basin floor to summit shoulder. That is, these surfaces are succcessive from east to west and all of them have been, and are, in process of moving back toward the heart of the mountain mass. They are of course in their present position and elevation purely current features of the landscape, but their arrangement and the relationship of their slopes traces a long history of evolution.
>
> (Sauer 1929: 230)

The sheer novelty of Sauer's interpretation at that time bears quotation of the summary of his conclusions, even though it is readily apparent that being based on belief in Penckian dogma they are fully as a priori as any Davisian interpretation. Sauer makes sixteen points about the Peninsular Range:

1 The Peninsular Range is primarily a batholith, deeply unroofed, with only minor areas of the metamorphic cover left. These older formations are present locally in patches that have been infolded or faulted. The great majority of the exposed surfaces consist of moderately coarse diorite of deep-seated origin and high uniformity.

2 Major faults have developed north-west–south-east, parallel to the early axis of compression. Numerous faults of lesser order intersect the major trend at high, nearly right, angles. The most active faults of the present are of NW–SE orientation.

3 The morphologic evolution of the area is individualized in terms of the individual blocks. Each block has a history of its own, and in general the entire block does not have the same history, since vertical movement does not appear to have been uniform or of equal length in different margins of the same block.

4 No unmodified relief features due to faulting, such as facets, sag ponds, kern buts, rifts, have been definitely recognized in the area. All the relief forms observed appear to owe their present form to denudation, erosion, or aggradation.

5 No graben or other form of dropped block has been identified.

6 The narrow linear depressions are due to long continued erosion in fault zones.

7 The broad intermontane basins are features of denudation primarily, of erosion secondarily. They are developed on the stable or down-tilting margin of a unit block, as indicated by a striking asymmetry of slopes, especially by very long concave slopes ascending in unbroken sweep from basin floor to that mountain crest which is the unreduced margin of the block involved.

8 Convex slopes mark such block margins as are rising more rapidly than they are being reduced by denudation; more quiescent tectonic fronts are marked by convexo-concave and especially by concave slopes, unity of slope assemblage being notably persistent over not inconsiderable linear distances.

9 Summit peneplains are characteristic of all the higher elevations, and form a series of important mesas, on which denudation is nearly balanced by accumulation of weathered material. These primary peneplains are in part growing smoother at the present time, and are most characteristically associated, in agreement with the *Primärrumpf* theory of Penck, with a great convex flank of a rising mass.

10 Basal peneplains, dependent on a local basis of erosion, have been initiated on the quiescent or sinking margins of blocks, adjoined by rising blocks.

11 Anomalous interruptions of stream courses, that is, abrupt changes in gradient unrelated to lithologic changes, occur in identical series along and behind the fronts of several blocks, and indicate interrupted, repeated uplift.

12 The western margin of the area contains deep gravel beds of very resistant pregranitic formations, in positions that leave little doubt as to their marine origin. A rude general level of the western flank of the highland, extending far into Lower California, appears to indicate an Eocene marine terrace, now much dissected and warped (averaging perhaps 2500 feet above sea level).

13 In the mesothermal parts of the area, weathering is largely by chemical disintegration, both the arkose-like debris and the woolsack

forms of the bedrock being produced in the main by the oxidation and hydration of the ferruginous minerals.

14 Hillside and basal swamps and ponds are common features, produced by fissure springs in the areas of greater precipitation, and apparently persistent through the morphological cycle, unless the spring disappears. These became local bases of denudation and give rise to small basins.

15 On the arid margins at the east, pediments appear conspicuously, playas are formed by the partial blocking of surface drainage through interfering fan fronts, and swamp conditions may further develop through the accretions formed by algae in stream courses.

16 In part there is evidence of a recent, slightly greater extension of the desert into the eastern intermontane basins, presumably a local diastrophic feature.

(Sauer 1929: 247–248)

It needs to be noted that these interpretations were highly controversial and continue to be so (Jahns 1954a, 1954b). The gravels interpreted by Sauer as marine have generally been considered fluviatile (Fairbanks 1893, 1901; Ellis and Lee 1919). The summit plateaus are usually viewed as the remnants of a broken and dissected peneplain, on Davisian lines, rather than forming *in situ*. Indeed, in a study of the adjacent San Gabriel Mountains, Miller (1935: 1553) commented that Sauer's interpretation was 'out of the question'. Controversy over the tectonic history of the southern Coast Ranges continues unabated (Crowell 1981; Wood and Elliott 1979; Hill 1990; George and Dokka 1994). Nevertheless, it is impossible not to be impressed by the skill with which Sauer superimposed such an internally consistent view on an area of such confused and confusing topography.

Sauer's Peninsular Range study has the merit of being the first application of Penckian analysis in North America. It thus had a stimulating freshness, as Leighly (though hardly without bias in the matter) pointed out:

I know that if Penck's reasoning is applied to a region of complex structure – we constantly see it exemplified in an area of exceedingly complex structure, the California Coast Ranges – it does at least as much justice to structure as Davis's scheme ever did. It does a great deal more justice, in my opinion, for it enables one to see in one's mind the way in which structure gradually becomes evident in relief. Once on an automobile trip through the local section of the Coast Ranges, the Berkeley Hills, Davis disposed of the complicated series of forms to be observed here by a single phrase, 'maturely dissected.' Penck's reasoning makes rounded hilltops, rock bosses, spatulate canyon heads, and slightly incised stream channels 'come alive' as forms in which the structure of the ranges is gradually becoming more and more evident in their relief.

(Leighly 1940: 226)

What Leighly failed to realise, however, was that his famous criticism of Davis in the same symposium as these laudatory comments on Penck – 'Davis's great mistake was the assumption that we knew the processes involved in the development of landforms. We don't; and until we do we shall be ignorant of the general course of their development' (Leighly 1940: 225) – applied with equal force both to Penck's system and to Sauer's exemplification of it in the Peninsular Range.

Sauer's publication prompted a critical review by Kirk Bryan and Gladys Wickson (1931). Sauer has 'attempted', they say, on the 'basis of a relatively brief study of a small part of the eastern portion' of the Range, to interpret the landforms through 'a thorough knowledge of the methods and terminology of Walther Penck'. Their critique is primarily about the use of the Penck method, though critical observations are made concerning Sauer's views on climate and morphology and on the meaning of the high-level gravels. They thought it

unfortunate that some of the complexities of the region escaped attention, and that detailed work was not done in the western part of the area where the coastal terraces and their extensions inland afford a remarkable series of datum points in the complicated geomorphology of the region.

(Bryan and Wickson 1931: 291)

Sauer was not best pleased. He had originally had a high opinion of Bryan as a geomorphologist: 'I have regarded Bryan's really very honest and sensible observations in the Southwest as some of our best geography of recent years and I was rather jubilant when he went to Harvard' (Sauer to Derwent Whittlesey, 23 March 1929; this passage is quoted by Martin 1987: xiv). But Bryan's criticism was too much. After pointing out that the authors were not familiar with the region, he held that the review was 'in effect an a priori Ablehnung [rejection] of Penck' (Sauer 1932: 246); he commented reasonably on the reviewers' objections. But he was in fact deeply annoyed. He wrote to Richard Russell (24 February 1932):

The more I see of Bryan the more I think he has the bumptiousness of the half literate. He excommunicated me on that Warner Valley [Peninsular Range] study.... Bryan seems to have undertaken to suppress all observing and thinking in this country which does not follow him or his prophets, and believe me he is a long way from having the skill or the knowledge necessary to the success of such censorship.

Here Sauer adds Bryan to the Berkeley demonology of Davis. It was a break

357

which was to have profound consequences in subsequent years and was perhaps the first indication that Sauer was not destined to be a geomorphologist. At the same time it illustrates the degree of commitment Sauer felt towards his early geomorphic work.

The Penckian debate continued. Bowman (1926) reviewed *Die morphologische Analyse* in the *Geographical Review*, and Davis (1932b) offered an extended exegesis, later harshly criticised by Simons (1962). Davis (1930b) incorporated Penckian ideas in a major study of slope evolution. Kesseli (1940), as we have noted, clarified Penck's views for the English-speaking geomorphologist. Tuan (1958) was perhaps the last Berkeley student actively to engage in the debate, the most informative guide to which is undoubtedly presented by Beckinsale and Chorley (1991: 366–372).

THE CHIRICAHUA MOUNTAINS

Sauer was not one to linger on particular projects, and in any case he was becoming interested in quite different issues in Mexico. Political problems aborted a Mexican field trip in 1929, and he diverted to the Chiricahua Mountains in south-east Arizona. His study there (Sauer 1930) he saw as 'primarily a re-examination of some commonly accepted concepts concerning basin-range features'. These were interpreted as being determined by parallel faults in classic papers by Davis (1903, 1913, 1925), as well as by the Berkeley geologists Lawson (1915) and Louderback (1904, 1923, 1926). The most recent research on these features had been the papers which had initially impressed Sauer – by Kirk Bryan (1923, 1925a). In this new study Sauer (1930: 348) took as his text a quotation from *Die morphologische Analyse* (Penck 1924: 207): 'faulting does not have the role in delineating mountains which has been assigned to it.'

Evidence for the ways in which faulting controlled landscape features in the West had been assembled by Davis (1913), Louderback (1926) and Blackwelder (1928), as well as many others. But Sauer (1930: 351), again playing at Daniel in the lion's den, almost inevitably asserted in the Chiricahua study that 'As far as I know, there exists ... no stratigraphic determination of a single fault as forming the front of any of the local ranges.' He saw anticlinal ridges associated with magmatic injection and volcanism, a process he rather disgustingly terms 'intrusive suppuration' (Sauer 1930: 353–354), and found no need to infer post-orogenic faulting. In brief he interpreted the basin-range country of Arizona in terms of Penck's idea of *Grossfaltung* [broad fold] rather than block-faulting.

There is no support for Sauer's interpretation. Drewes (1980, 1981, 1982) found abundant evidence of late Cretaceous thrust faults of 15–35 km, strike-slip faults, and normal faults. He goes so far as to say that the tectonic history of the Chiricahua Mountains is now 'the best known chapter in the development of southeastern Arizona' (Drewes 1981: 90). Sauer's interpretation was again a priori and impressionistic.

It was in the Chiricahua paper that Sauer made his most explicitly Penckian pronunciation: 'Every change in rate and manner of differential movement between range and basin is expressed immediately in change in character of denudation and erosion. Every variation in rate of denudation and erosion is expressed in the fill of the basins' (Sauer 1930: 366). He summarises Penck's views at length (Sauer 1930: 377–380) and offers a diagrammatic interpretation, clearly in the style of Penck, of the development of slope profiles (Figure 16.2).

This was of course the classic territory for Blackwelder's (1931) studies of pediment development, and of Lawson's (1915) and Bryan's (1923, 1925a) interpretations of pediment gaps and passes. Davis's (1930b) major paper on these features was favourably reviewed by Bryan (1932), and was followed by specific studies of the Peacock Range, the Catalina Mountains, the Mojave

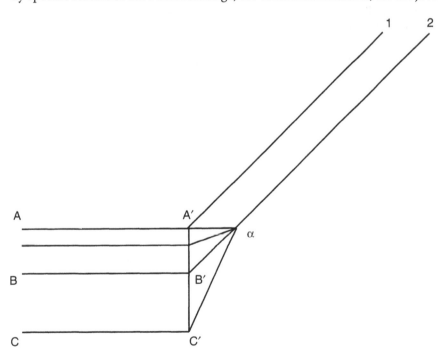

Figure 16.2 Sauer's only graphical essay in the Penckian mode (Sauer 1930: 379). The original caption reads: 'Diagram to show range of conditions under which concave slopes persist. A–A' is initial base level, flanked by denudational slope A'–1. In a given interval of time the latter recedes to position a–2. If in this interval the base is lowered to a position that fails to reach B–B', slope a–2 will be joined to the base by a slope of reduced pitch, or, a concave slope results. If the base is lowered beyond B–B', as to C–C', a steepened slope effects the junction with a–2, or, a convex slope results.'

Desert and the Galiuro Mountains (Davis 1930a, 1931, 1933; Davis and Brooks 1930). Sauer saw the same kinds of slopes as these workers, but through Penckian eyes and without close attention to process. 'McGee's quaint notion of sheet-flood erosion hardly needs to be regarded as to the development of pediments,' he averred (Sauer 1930: 375), a position to be massively rebutted by Davis in a posthumous paper in 1938.

The Chiricahua paper was not one of Sauer's greatest achievements. Indeed, not only was it an unplanned investigation intended to rescue a Mexican field season, but one has the distinct impression that most of his observations were from the railroad car. Certainly he fails to mention the most spectacular geomorphological features of the area, which now form the basis of the Chiricahua National Monument. It was a study characterised by all the faults of deductive reasoning attributed by Sauer and Leighly to W.M. Davis, and it attracted little attention. Sauer continued geomorphic work in Arizona, though with a quite different research focus. He claimed to maintain his original posture, as he told Kroeber while engaged in Arizona fieldwork in the summer of 1934: 'Like this volcanic country fine and can see several good-sized morphologic jobs ahead provided one dared to proceed from assumptions non-Davisian, which is doubtful' (Sauer to A.C. Kroeber, 19 July 1934; quoted by Macpherson 1987: 83). Departmental research interest in Arizona and adjacent California was continued at a later date in Tuan's thesis, supervised by Kesseli (Tuan 1957, 1959, 1962), and by Oberlander (1972, 1974), but in a tradition different from that of Sauer.

GEOMORPHOLOGY AND SOIL CONSERVATION

There is a further body of Sauer's geomorphic work which is much less well-known than the Peninsular Range and Chiricahua studies, but on which he spent much effort in the 1930s. President Roosevelt established a Science Advisory Board by executive order in July 1933, and the geographer Isaiah Bowman, then chairman of the National Research Council as well as Director of the American Geographical Society, was an influential member. The Board established a Land-Use Committee with Bowman as chairman and Sauer as a member. It proved a significant association, though Leighly recalled that Sauer 'despised' Bowman in the 1930s, without giving a reason (though Bowman had given an 'atrocious' address at Berkeley in 1930 or 1931) (Leighly to G.J. Martin, 24 March 1976). The Committee held two conferences in December 1933 and May 1934, of which the second involved nine geographers (one of whom was R.J. Russell) in a total of eleven. The other two members, O.E. Baker and Homer Shantz, both had impressive geographical credentials. A further meeting in June 1934 included H.H. Bennett and W.C. Lowdermilk from the Soil Erosion Service of the Department of the Interior, and C.F. Marbut of the Bureau of Chemistry and Soils and Baker of the Bureau of Agricultural Economics in the Department of Agriculture, as well as Bowman and Aldo Leopold. Ironically Sauer's new involvement in process studies coincided with yet a further paper by Lawson, this time on rain-wash erosion (Lawson 1932). This third meeting discussed a substantial report on 'Land resource and land use in relation to public policy', prepared by Sauer and published in April 1934. It is odd that this report has attracted virtually no interest, in contrast to the continuing attention given to 'The morphology of landscape'.

'Land resource and land use' is a powerful and indeed polemical document. It called for an integrative 'science of the land', applied to economic problems, which would examine 'the different complexes formed by man and nature, which we call regions':

> We cannot get along without this form of synthesis, since we do not deal in relations that are placeless; throughout we are concerned with spatially differentiated phenomena of different categories which form areal combinations. Thus the final questions are reached, whether we have appropriated the land wisely or unwisely to our present needs and whether we have made proper provision for the future.... The organized knowledge of the land is geography.
>
> (Sauer 1934: 134–135)

Succinctly stated these were themes which were to dominate the rest of Sauer's life (Stoddart 1991).

The report consisted of three parts: Land resource, Land use, and 'Land, the geographic totality'. Throughout Sauer emphasised the importance of mapping, and indeed speaks of 'the map as evidence and process of thought': 'Land data are inherently cartographic data and their meaning can be understood only by getting their actual distribution down on a map.... Without this there is no ordered knowledge of the land, no effective approach to land planning' (Sauer 1934: 244). He called for detailed mapping of soils, and saw soils as necessarily congruent with surface morphology. Citing Walther Penck, he argued that diastrophism controls slope profile, hence denudation rate, and therefore soil profile: maps are needed of such soil-slope complexes (Sauer 1934: 184–185). Processes need to be investigated, especially sheet wash, rill wash and gullying, and climatological, hydrological and vegetational studies are needed to understand these. The rest of the report, dealing with such hallmark Sauerian themes as 'the impress of man on the land' and the need for maps of population, migration, agriculture, industry and urbanisation, does not concern us here.

Sauer's paper led to specific recommendations relating to 'Soil erosion and critical land margins' (Sauer et al. 1934). This included specific programmes in climatology put forward by Leighly, together with a detailed 'Field project for slope-soil inquiry' which was presumably written by Sauer (Sauer et al. 1934: 150–151). Given the fact that Sauer's lack of interest in field procedures as opposed to field understanding has been repeatedly remarked, this is an astonishingly detailed operational design for field investigation and took Sauer far beyond his earlier purely morphological studies. The 'Field project for slope-soil inquiry' is designed for implementation in Soil Erosion Administration Unit A within Climatic Region X:

1 Inspection of area A for selection of locality to be observed as outlined below. The site is to have the following qualities, or the nearest approach to them possible:
   a. Climate, slopes and mantle rock are to be typical of the conditions encountered in administrative unit A.
   b. Vegetation cover to be undisturbed by destructive exploitation and, if possible, to be in virgin condition.
   c. Location if possible to be near surface experiencing destruction to facilitate comparisons ...
2 Arrange for protection of tract from disturbance during period of observation.
3 Prepare accurate topographic and cover map of site selected (area will normally not exceed forty to eighty acres).
4 Establish control points and set out markers at measured distances along surveyed lines entered on map.
   a. Control points must be on bed rock or anchored to such foundation that they remain unmoved.
   b. Markers (tin discs with staples) to penetrate soil not to exceed two inches.
5 Read displacement of markers at seasonal intervals, as at end of dry season and again at end of rainy season, or before and after frost period, to get rate of mass movement of surface.
6 Prepare test pits for determining decrease of movement with increase of depth. Experimentation with several devices. One useful procedure has been to place layers of colored chalk in the test pit, which is then covered and left for several years.
7 Classify slope types in erosion administration unit sufficiently to be familiar with the characteristic slope assemblage or assemblages. Having thus determined by inspection dominant slope forms, select an undisturbed typical tract (may include site of observations 1–6) for making a profile on a large scale (say 1:1000). Along this slope profile soil sections are taken at short intervals (sufficient to characterize changes of slope).
8 A number of these soil sections are to be obtained by trenching so that the actual soil structure can be drawn or photographed.
9 Physical, organic and bacteriologic analyses of such samples.
10 Construct from 7, 8 and 9 a slope-soil profile in reduced and simplified form which will serve as symbol for that characteristic slope-soil complex.
11 By use of non-soluble stains applied to soil, determine whether different

types of vegetation have different rates of gravity movement of soil through root mesh.

These recommendations and proposals were put into effect when the Soil Erosion Service directed by H.H. Bennett became the Soil Conservation Service in 1935. An Office of Research was established, first under R.V. Allison and later W.C. Lowdermilk, and it included a Division for Climatic and Physiographic Research, which became the vehicle for the execution of Sauer's proposals. Sauer can thus be credited with originating soil-geomorphology studies in the US Department of Agriculture (Effland and Effland 1992: 190).

Sauer began work for the Soil Erosion Service as Consulting Geographer in 1934. He was appointed Senior Soil Conservationist in the Soil Conservation Service for the summer period 1 May–31 July (actually ending 14 August) 1936 at $4600 per annum. He served as Collaborator (without pay), 15 August 1936–30 June 1937, a position then made permanent.

Fieldwork began in the summer of 1934 in Polacca Wash, a 30-mile long channel (arroyo) in the Navajo–Hopi Reservation in north-east Arizona. It was distinguished by the fact that Sauer was accompanied by John Leighly, who, taking his inspiration from a paper by Kirk Bryan (1925b), made a special study of arroyo meanders in north-west New Mexico. This was characterised not only by meticulous field mapping of a number of arroyo stretches but by the ingenuity of the mathematical analysis applied to the results. Leighly (1936: 282) showed by plotting meander radius of curvature against distance along the channel that the meanders described logarithmic spirals. The analysis, reminiscent of Buchanan's (1907) paper on the same problem (and equally subsequently overlooked), was virtually unique in fluvial geomorphology at that time; to my knowledge the use of the logarithmic spiral did not re-enter geomorphology until the work of Yasso (1964) on coastal spits.

The Polacca Wash work was taken up in 1935 by the new Division for Climatic and Physiographic Studies under its Director, C. Warren Thornthwaite. Thornthwaite, now remembered chiefly as a climatologist, had gained his doctorate at Berkeley under Sauer in 1930 (somewhat oddly with a dissertation on the urban geography of Louisville, Kentucky), and Leighly attests (1978a: 41) that it was Sauer who got him the job. The Polacca project was scarcely a success (though the results were published by Thornthwaite, Sharpe and Dorch in 1942) and in fact caused grave dissension, especially with Bryan, who doubtless resented Sauer's intrusion into his own preserve. As Leighly put it (1978a: 41): 'Mr Sauer's elaborate plan came to naught because of dissension in the party after he returned to Berkeley.' The leader of the field party was a student named Richard Norman, and it included two geologists, Frank Johnson and Parry Reiche, both of whom left after a year. It was, however, the intervention of Bryan, by now deeply unsympathetic to Sauer, which exacerbated the situation.

He detailed his objections directly to Bowman on 28 July 1936. First, 'There are too many of these Washington research units each under a zealous and jealous chief': the local scientists disliked them. Second:

> The original plans of Sauer for the Polacca project were ill-advised at best and definitely stupid at the worst. It is my considered judgment that the man knows very little about erosive processes and nothing about the genesis of land forms. He issued definite instructions that Johnson and Reiche should not look at the rocks as 'Geology' has nothing to do with 'Geomorphology'. Such a statement is almost beyond belief.

(The charge seems inherently unlikely since a neglect of geology was one of Sauer's chief complaints against Davis: Sauer 1966: 255; also Sauer to R.J. Chorley, 18 October 1961, quoted in Chorley, Beckinsale and Dunn 1973: 428.)

Bryan then stated that in the first year the survey ran a line of levels through the system, and in the second mapped geomorphic units in the channel at 2 inches to the mile: this 'represents an incredible amount of work. But I cannot see just what value it has to anyone for any purpose.' Of course it was characteristic of Sauer that he had told his assistant that

> I am not planning on the assembling of any instrumental equipment. The S.C.S. should have a sufficient number of engineers, levels, and plane tables so that if and as we pick spots for detailed observation we should be able to borrow men or equipment or both as needed.
>   (Sauer to H.A. Ireland, 13 April 1936)

Instrumental survey was hardly Sauer's chief priority in fieldwork.

> At present [Bryan continued] the orders are to study 'The Dynamics of Stream Erosion by Geomorphic Methods'. Rain gages are being set up and observation quadrats adjacent to them. No stream gaging or measurement of run-off is to be used. Now just what does this mean? Either I am very old-fashioned and even out of date [he was one year older than Sauer] or it means nothing and cannot accomplish anything.
>   (K. Bryan to I. Bowman, 28 July 1936)

Bryan advised Bowman that in spite of the difficulties Johnson had had some success, and that he had himself advised him (doubtless without Sauer's knowledge) about working out the geomorphic history. He recommended that Sauer's influence should be removed. Bryan asked Bowman to intervene 'before too many young scientists have been ruined by stupid programs laid out by men of inadequate scientific knowledge, training, and ethics'. It was an intemperate and unworthy letter from someone with Bryan's reputation.

Thornthwaite explained to Bowman (13 August 1936) that the work

Johnson was doing had been planned and executed by Johnson himself, and that neither Sauer nor Thornthwaite had anything to do with it, though both clearly had responsibility for the project. 'I think you will agree that Sauer has more ideas in a minute than Bryan has in an hour.... Our aim is to substitute brains for equipment.' He called Bryan's letter 'vicious'. Bryan told Bowman (26 August 1936) that the trouble with Thornthwaite was that 'He learned about geomorphology from Sauer and consequently is thus in error.' And he renewed the attack:

> Now as to Sauer: He is a fellow able with words and his report to the Land-Use Committee sounds well. It is something like his paper on 'The Morphology of Landscape' which sounds as if it meant something but is nonsense. Have you read Sauer's paper on the Peninsular Range in Southern California, or his paper on the Chiricahua Mountains in Arizona? Both are full of egregious errors of observation and analysis. It would be easy to forgive Sauer for a little carelessness or for ignorance but he is wrong-headed and has completely erroneous notions about geomorphology.... He is a man who goes desperately wrong in Science.... You can judge whether my worm's-eye view has not detected the feet of clay supporting some glorious-looking Gods.

Bowman annotated the letter: 'What an ass! I shall not reply' – and sent it on to Thornthwaite. Sauer himself did not see these letters for nearly a quarter of a century, when Thornthwaite sent them to him (Thornthwaite to Sauer, 15 December 1959). Sauer's reaction is not recorded and Bryan himself was long since dead.

It is difficult to understand the depth of animosity shown by Bryan to Sauer, especially considering the similarity of their interests (Whittlesey 1951). They were virtually the same age, though Bryan was born in New Mexico and had published his major studies of the Papago country in Arizona before Sauer had started his own fieldwork in the west: he doubtless considered Sauer an intruder. He certainly did not share the Berkeley Department's antipathy to Davis (Bryan 1935c) but at the same time developed an interest in Penck's ideas (Bryan 1940). Sharp (1993: 189) says that Bryan's 'forte was relationships with people' and that 'he exuded human warmth', qualities not apparent in his relationships with Sauer. Sharp (1993: 191) also says 'He had strong feelings on most subjects and about most people,' and this was also true of Sauer. On soil erosion and arroyo development there was a genuine disagreement between the two, with Bryan emphasising climatic change and Sauer land-use practices.

> Bryan was no man to refuse combat in support of conclusions he had drawn from his scientific studies, and on occasion he took sharp issue

with those who put the whole blame for deterioration of the habitat upon unwise use of the land.

(Whittlesey 1951: 91)

The scientific problems raised by Polacca Wash and similar arroyos continue to be debated (Cooke and Reeves 1976), not least by Berkeley geographers of diverse persuasions (Tuan 1966; Denevan 1967). The upheaval probably did Thornthwaite himself no good. In 1944 Sauer urged Bowman as president of Johns Hopkins to appoint Thornthwaite chairman of the Department of Geography there, but Bowman declined to take the advice (Sauer to Bowman, 21 May 1944; Bowman to Sauer, 27 May 1944).

In spite of these difficulties the Soil Conservation Service programme continued (Trimble 1985). In 1936 Sauer initiated gully erosion studies in the Piedmont of South Carolina near Spartanburg. This study involved C.F. Stewart Sharpe, whose later book (1938) became something of a classic. The results were published by Ireland, Sharpe and Eargle (1939) and by Eargle (1941) and are discussed by Hall (1940, 1948). A study of erosion in the loess of south-western Louisiana, directed by R.J. Russell, began in 1936, but it lasted only two months because the principal field scientist Stewart Sharpe was needed in South Carolina. In the same year erosion studies were conducted in the karst lands of Kentucky by Samuel Dicken; the results were published by Dicken and Brown (1938). Finally, work on mass movements in the Muskingum watershed in the Appalachians of eastern Ohio, again under Sharpe, began in 1938 (Sharpe and Dorch 1943). The programme was suspended during the Second World War, and was revived on a different basis under Robert Ruhe in the 1950s (Effland and Effland 1992). The questions addressed during these several field projects remain of urgent concern (Trimble 1974, 1985).

The choice of both sites and of personnel during the years of operation of this Soil Conservation Service programme indicate the dominant influence of Sauer and his students. Sauer himself did not contribute to the published results of the surveys, nor did he conduct any further geomorphological research. His interests moved squarely into the areas on which his reputation rests. A brief note on soil conservation, written in 1936 and omitted from the standard bibliography of his writings, is given in Appendix 1.

There was, however, one final contribution from Sauer's geomorphological phase.

HANDBOOK FOR GEOMORPHOLOGISTS

As Director of the Division of Climatic and Physiographic Studies, Thornthwaite was concerned to systematise the Division's investigations. Thus a series of handbooks were prepared: of emergency conservation work (1937), wildlife management (1937), farm woodland management (1941), and climatology (5 volumes, 1937–1939). The first to appear was the *Handbook for Geomorphologists* in 1936. Though authored by Carl O. Sauer, it does not appear in the bibliographies of his works (Leighly 1963a; Sauer 1981). I have been unable to locate any copy in Berkeley, where it appears to be unknown. There is a copy in the US Department of Agriculture files in the National Archives in Washington, D.C., but the box in question cannot be located. Two copies are to be found, however, in the National Agricultural Library in Bethesda, Maryland. It (Sauer 1936a) ranks as his final formal contribution as a geomorphologist.

It seems clear that the initiative for its preparation came from Thornthwaite. He sent part of it to Bowman on 13 August 1936, saying it was 'comprised entirely of excerpts from the various letters and memoranda which Sauer wrote while engaged in initiating the fieldwork in the Southwest'. In the Sauer Papers there is a memorandum from Thornthwaite giving a tentative list of contents and referring to a completion date of 1 June 1937.

Part I is entitled 'The inauguration of geomorphological research in the Southwest' and consists entirely of communications from Sauer to H.H. Bennett, W.C. Lowdermilk and W.A.F. Stephenson. Part II is entitled 'The expansion of geomorphological research to the piedmont – plans and progress', and likewise comprises communications to Bennett and to Thornthwaite. Appendix I reprints the 'Preliminary recommendations of the Land-Use Committee of the Science Advisory Board relating to Soil Erosion and Critical Land Margins' (Sauer *et al.* 1934), including J.B. Leighly's 'Statement on dynamic climatology'. Appendix II gives 'Excerpts from land resource and land use in relation to public policy' (Sauer 1934). Appendix III and Appendix IV deal at length with 'The dynamics of soil erosion. Research Project no. C-1'. To these are annexed a 'Preliminary outline of the climatic groupings of destructional land forms' and a paper by Thornthwaite on 'The land forms of the San Francisco Peninsula: an exercise in classification' which does not appear to have been otherwise published. The illustrations include views of various types of erosion in the states of California, Idaho, Illinois, Iowa, Louisiana, Missouri, New Mexico, Oklahoma, Pennsylvania, South Carolina, Texas, Washington and Wyoming. Tantalisingly the extensive bibliography includes a reference to an unpublished paper by Sauer on 'Mechanics of slope-denudation under natural conditions' which cannot now be traced.

Most of the communications in Parts I and II are addressed directly to the head of the Soil Erosion Service, H.H. Bennett, rather than to Thornthwaite as head of the Division for Climatic and Physiographic Research. Bennett – 'Big Hugh' (Brink 1951) – was termed by Sauer 'the field marshal of soil erosion' (Sauer to Bennett, 14 July 1934); he had opened the decade with a geographical survey of soil erosion in the United States (Bennett 1931) and he closed it with his massive book on *Soil Conservation* (Bennett 1939). In this same letter to Bennett, written from the field in New Mexico, Sauer saw soil erosion as a disease. 'Badly infected surfaces' needed 'land physiologists and pathologists' in order to 'diagnose normal and abnormal conditions of the land surface' and especially the 'malignancy' of gullying. He set out detailed procedures for the study of mass movement, ablation and deflation, and linear erosion through measurement of the rates of operation of these processes and their local and regional variation in response to climate, vegetation and geomorphic environment. Prefiguring his later dominant interests he stated that the most important problem he had identified was 'the reconstruction of the erosion history as shown by the deep sections of the active washes [gullies], both as to historic and prehistoric occupation and as to prehuman cutting and filling'. To Lowdermilk he wrote (9 October 1934) that the Soil Erosion Service had 'an extraordinary opportunity to develop a newer science of geomorphology'.

He gave detailed analyses of problems and procedures for the study of processes and erosion history, first at Polacca Wash (Sauer to Bennett, 5 August 1934) and then in the South Carolina Piedmont (Sauer to Bennett, 31 December 1934). In a memorandum dated June 1936 (Sauer 1936a, Part II: 13–15) he specified the topics with which the field parties should deal. The book concludes with a climatic classification of landforms, outlining regional variations in weathering, erosion, landforms and soil groups for seven major climatic types defined in terms of the then novel classification scheme of Thornthwaite (1931). These were Warm humid climates, Cold season humid climates, Seasonal rainfall (Mediterranean) climates, Subhumid climates, Steppe climates, Desert climates, and Polar climates. While developing this theme from Albrecht Penck (1905) and others such as Passarge (1926), Sauer's analysis long predates the well-known scheme of Peltier (1950), and in terms of the development of the discipline it is unfortunate that it was not more formally published at the time.

EPILOGUE

Thus ended Mr Sauer's formal commitment as a research geomorphologist. I think he became impatient with the Soil Conservation Service projects for the same reason that he became impatient in Michigan. He was not a research administrator, nor was he an experimental field scientist. He wished to roam free, identifying problems in specific places and speculating about answers to them. His contribution to the Service was to identify places where problems needed to be addressed, and then to identify people (usually colleagues and students at Berkeley) who could do the work. That done, as Salisbury had taught him, it was up to them, and time for him to move on. The discord over Polacca Wash could not have enamoured him of the ways of government science.

He did retain an interest in geomorphology for the rest of his active career. He had supervised Dicken's doctoral research on the Kentucky karst at an early stage (Dicken 1931). He also supervised Kesseli's work on glaciation in the eastern Sierra Nevada (Kesseli 1938). Isaiah Bowman's son Robert (of whom Sauer had at the time a higher regard than he apparently did for his father) worked under him on soil erosion in Puerto Rico (R. Bowman 1941). Sauer's physical and cultural interests were brought together in Edwin Doran's (1953) research on Grand Cayman Island. And Sauer's last student in geomorphology, Brigham Arnold, somewhat fittingly returned to Sauer's earliest explorations in Baja California, with a thesis in 1954 and a major paper on landforms, climate and archaeology in the year of Sauer's retirement. And of course, throughout, there was the collaboration and support given by his colleague John Leighly, without question one of the most incisive and innovative intellects in the Berkeley Department, though he chose to spread his talents over a wide diversity of fields and was somewhat hampered (though not in print) by a degree of personal reticence (Miller 1988). We have already noted the innovative mathematics he employed in his work on New Mexico arroyos. Both Leighly and Kesseli had a profound interest in how rivers function. Kesseli (1941b) published a widely quoted paper on the concept of the graded river. Leighly (1932, 1934, 1937) applied his knowledge of atmospheric turbulence to the question of turbulence and suspended sediment transportation in flowing water using calculations of the Austausch Coefficient. In the English-speaking world no other physical geographer displayed such technical versatility at the time (see Stoddart 1987). But the potential for Berkeley to become known as a centre for geomorphic research was eclipsed by Sauer's achievements in other fields.

Sauer's work in geomorphology had little lasting impact. It is not mentioned in the mid-century evaluation of American geography edited by James and Jones (1954). Even his work for the Soil Conservation Service receives no reference by Bennett (1939). His pioneer advocacy of the views of Walther Penck retains a certain historical interest, but his substantive contributions have not survived the passage of time. I believe that Sauer undertook his geomorphic work for political rather than scientific reasons. Coming fresh to the west coast he was faced with a formidable array of talent and experience in the Department of Geology – a department much larger and more prestigious than his own. He needed to carve out a territory and his strategy was to meet them on their home ground, the study of landforms in the West. His secret weapon in this assault was his unparalleled knowledge of German geomorphology and the timely ammunition provided by Walther Penck. By the mid-1930s the need for self-justification was past, and Sauer had moved on to other concerns. Leighly and Kesseli continued the geomorphic interest but lacked the panache to make a general impact on the science. For Sauer and the Berkeley Department geomorphology had served its purpose: other sirens called.

Shortly before he died, one of Sauer's oldest disciples, Fred Kniffen,

DAVID R. STODDART

reavowed his debt to Mr Sauer as the greatest geographer of his time (letter, 9 February 1989). But he also wrote an epitaph for Sauer's geomorphic work, which while seemingly somewhat harsh was almost certainly just:

> Geomorphology was not his forte. The imagination that served him so well in one direction did not in the other.

ACKNOWLEDGEMENTS

I gratefully acknowledge my indebtedness through correspondence and conversation to Henry Bruman, Clarence Glacken, David Hooson, Fred Kniffen, John Leighly, Jim Parsons and Jesse Walker, and of course Mr Sauer himself. Mr Myron M. Weinstein, formerly of the Library of Congress, searched the National Archives for me for the *Handbook for Geomorphologists* and finally located it in the National Agricultural Library: he has my best thanks. The record of *60 Years of Berkeley Geography 1923–1983* (Parsons and Vonnegut 1983) has been an invaluable source. My excursion to the Peninsular Range was in the company of Francis J. Murphy and Michelle Goman, and that to the Chiricahua Mountains with Francis Murphy, Sir T. Adams, Eric Grossman, and Aldabra and Michael Stoddart; on the latter by a miracle we managed not to freeze to death at night. My debt to the historical insights of Richard J. Chorley and his collaborators and to *The History of the Study of Landforms* will be readily apparent. The letters quoted are from the Sauer Papers, Bancroft Library, University of California at Berkeley.

APPENDIX: SOIL CONSERVATION, BY CARL O. SAUER (Sauer 1936b)

The wastage of soil by man has won attention tardily though it is perhaps to date the most destructive form of exploitation of which man is guilty. European geographers and other social scientists have given little heed to this occurrence and the German term *Raubbau* is not necessarily concerned with the destruction of the soil profile at all. The science of soil erosion is principally American; and though it is getting public attention only lately from us, attentive observers have been concerned about it for a long time. Thornthwaite's division in the Soil Conservation Service is expecting soon to publish a study that will utilize largely the observations of Thomas Jefferson and his contemporaries. Hilgard and McGee were concerned with this problem. Hugh Bennett in a modest way carried on field and experimental erosion studies for many years before the present national administration launched the Soil Erosion Service, now recognized and enlarged as the Soil Conservation Service. Bennett chiefly has made the American people

370

'erosion conscious' and has developed a national bureau which is attacking the problem on all fronts.

It may not be amiss to point out the connection between soil erosion and commercial exploitation of the land. Soil erosion is far less a problem of the old areas of the world than of the new ones. It is an abnormal feature of man's expansion over the earth, based primarily on specialized economies directed toward maximum cash returns. Where men have settled with a primary regard for permanent living there has rarely been such trouble. Neither agriculture nor grazing per se leads to soil erosion. It is principally where men have considered land in terms of quick and large monetary returns that this ill has arisen. Land as a speculation and soil erosion are closely related. Hence cash crop systems and year to year tenant contracts play an important role in the incidence of soil erosion. All of this means that soil erosion cannot be studied alone as a physical condition – though the physical conditions thereof must be studied – but that it must be regarded as an economic maladjustment. Good farming does not lead to soil erosion, bad farming does. Perhaps the anthropologists would speak of a pathologic acculturation.

Hence, any scientific study of the processes and expressions of soil erosion must regard both the physical and cultural factors and forms. The conditioning may lie in nature, the cause is man operating unnaturally, and in the long run unsocially and uneconomically.

This summer I had occasion to participate in the work of the Soil Conservation Service in the Piedmont, where the pathogenic role of man is most marked. This is one of the great endemic hearths of soil erosion, the disease dating in the north well back into colonial times. The summer was spent as a field seminar in which various members of the Division of Climatic and Physiographic Research participated. The principal base was South Carolina, but the party ranged from Virginia to Alabama. Naturally the work fell into the two parts, erosion morphology (physical) and erosion history (cultural). The individual students are specialized, but their observations and field association is interlocking. The erosion historian must know his way about in soil profiles and erosion forms and the erosion morphologist must be aware of the history of misuse that has loosed the agencies of destruction which he is studying. The observations and recommendations for study made this summer and during my two previous periods as consultant in the field are being issued in mimeographed form by Thornthwaite's office and indicate somewhat the manner in which the problem of soil erosion may be broken into manageable themes by a geographic approach. Forms, rate, causes, and effects of soil destruction constitute a melancholy major theme in the United States for the geographer who is concerned with seeing 'man's relation to the earth'. I know of no theme that better illustrates the unity of physical and cultural geography, the necessity of competence and curiosity in both, and the impossibility of doing economic geography without

371

historical geography. If the major end of geography is to find the realization or failure of symbiosis of man and nature, the conservative use of natural resources is at the heart of all human geography and the use of the soil is one of the major topics with which we must deal.

REFERENCES

Arnold, B.A. (1954) 'Land forms and early human occupation of the Laguna Seca Chapala area, Baja California, Mexico', Ph.D. thesis, University of California at Berkeley.

——(1957) 'Late Pleistocene and Recent changes in land forms, climate and archeology in central Baja California', *University of California Publications in Geography* 10: 201–318.

Beckinsale, R.P. and Chorley, R.J. (1991) *The History of the Study of Landforms or the Development of Geomorphology*, vol. 3: *Historical and Regional Geomorphology 1890–1950*, London: Routledge.

Bennett, H.H. (1931) 'Problems of soil erosion in the United States', *Annals of the Association of American Geographers* 21: 147–170.

——(1939) *Soil Conservation*, New York: McGraw-Hill.

Blackwelder, E. (1928) 'The recognition of fault scarps', *Journal of Geology* 36: 289–311.

——(1931) 'Rock-cut surfaces in the desert ranges', *Journal of Geology* 20: 442–450.

Bowman, I. (1926) 'The analysis of landforms: Walther Penck on the topographic cycle', *Geographical Review* 16: 122–132.

Bowman, R.G. (1941) 'Soil erosion in Puerto Rico', unpublished Ph.D. thesis, University of California at Berkeley.

Brink, W. (1951) *Big Hugh: The Father of Soil Conservation*, New York: Macmillan.

Bruman, H.J. (1987) 'Carl Sauer in midcareer: a personal view by one of his students', in M.S. Kenzer (ed.) *Carl Ortwin Sauer: A Tribute*, Corvallis: Oregon State University Press.

Bryan, K. (1923) 'Erosion and sedimentation in the Papago country, Arizona', *Bulletin of the United States Geological Survey* 730B: 19–90.

——(1925a) 'The Papago country, Arizona', *United States Geological Survey Water Supply Paper* 499: 1–436.

——(1925b) 'Date of channel trenching in the arid Southwest', *Science* 62: 338–344.

——(1932) 'Geographic similarities in arid and humid regions', *Zeitschrift für Geomorphologie* 7: 250–253.

——(1935a) 'Processes of formation of pediments at Granite Gap', *Zeitschrift für Geomorphologie* 9: 125–135.

——(1935b) 'The formation of pediments', *Reports of the 16th International Geographical Congress* 2: 765–775.

——(1935c) 'William Morris Davis', *Annals of the Association of American Geographers* 25: 23–31.

——(1940) 'The retreat of slopes', *Annals of the Association of American Geographers* 30: 254–268.

Bryan, K. and Wickson, G.G. (1931) 'The W. Penck method of analysis in southern California', *Zeitschrift für Geomorphologie* 6: 287–291.

Buchanan, J.Y. (1907) 'The windings of rivers', *Nature* 77: 100–102.

Chamberlin, R.T. (1931) 'Memorial of Rollin D. Salisbury', *Bulletin of the Geological Society of America* 42: 126–138.

Chorley, R.J., Beckinsale, R.P. and Dunn, A.J. (1973) *The History of the Study of*

Landforms or the Development of Geomorphology, vol. 2: *The Life and Work of William Morris Davis*, London: Methuen.

Cooke, R.U. and Reeves, R.W. (1976) *Arroyos and Environmental Change in the American Southwest*, Oxford: Oxford University Press.

Crowell, J.C. (1981) 'An outline of the tectonic history of southeastern California', in W.G. Ernst (ed.) *The Geotectonic Development of California*, Englewood Cliffs: Prentice Hall.

Davis, W.M. (1903) 'The mountain ranges of the Great Basin', *Bulletin of the Museum of Comparative Zoology at Harvard College* 42: 127–178.

———(1905) 'Complications of the geographical cycle', *Report of the 8th International Geographical Congress, Washington 1904*: 150–163.

———(1909) *Geographical Essays*, Boston: Ginn & Co.

———(1913) 'Nomenclature of surface forms on faulted structures', *Bulletin of the Geological Society of America* 24: 187–216.

———(1923) 'The explanatory description of land forms', *Geographical Review* 13: 318–321.

———(1925) 'The basin range problem', *Proceedings of the National Academy of Sciences* 11: 387–392.

———(1930a) 'The Peacock Range, Arizona', *Bulletin of the Geological Society of America* 41: 293–313.

———(1930b) 'Rock floors in arid and humid regions', *Journal of Geology* 38: 1–27, 136–158.

———(1931) 'The Santa Catalina Mountains, Arizona', *American Journal of Science* (5) 22: 289–317.

———(1932a) 'Basin range types', *Science* 76: 241–245.

———(1932b) 'Piedmont benchlands and Primärrumpfe', *Bulletin of the Geological Society of America* 43: 399–440.

———(1933) 'Granitic domes of the Mojave desert, California', *Transactions of the San Diego Society for Natural History* 7: 211–258.

———(1936) 'Geomorphology of mountainous deserts', *Reports of the 16th International Geological Congress* (1933) 2: 703–714.

———(1938) 'Sheetfloods and streamfloods', *Bulletin of the Geological Society of America* 49: 1337–1416.

Davis, W.M. and Brooks, B. (1930) 'The Galiuro Mountains, Arizona', *American Journal of Science* (5) 19: 89–115.

Denevan, W.M. (1967) 'Livestock numbers in nineteenth-century New Mexico, and the problem of gullying in the Southwest', *Annals of the Association of American Geographers* 57: 691–703.

Dicken, S.N. (1931) 'The Big Barrens: a morphologic study in the Kentucky karst', unpublished Ph.D. thesis, University of California at Berkeley.

———(1935a) 'The Kentucky Barrens', *Bulletin of the Philadelphia Geographical Society* 33: 42–51.

———(1935b) 'A Kentucky solution cuesta', *Journal of Geology* 43: 539–544.

———(1935c) 'Kentucky karst landscapes', *Journal of Geology* 43: 708–728.

———(1959) 'Baja California, 1926', *Itinerant Geographer* 1959, unpaginated.

———(1986) *The Education of a Hillbilly: Sixty Years in Six Colleges. The Memoirs of Samuel Newton Dicken*, Eugene, Oreg.: privately published.

Dicken, S.N. and Brown, H.B., Jr (1938) 'Soil erosion in the karst lands of Kentucky', *U.S. Department of Agriculture, Soil Conservation Service Circular* 490: 1–61.

Doran, E. (1953) 'A physical and cultural geography of the Cayman Islands', unpublished Ph.D. dissertation, University of California at Berkeley.

Drewes, H. (1980) 'Tectonics of southeastern Arizona', *U.S. Geological Survey Professional Paper* 1144: 1–96.

——(1981) 'Geologic map and structure sections of the Bowie Mountains, south quadrangle, southeastern Arizona', *U.S. Geological Survey Miscellaneous Investigations Series* Map I-1363.

——(1982) 'Geological map and geologic sections of the Cochise Head quadrangle and adjacent area, southeastern Arizona', *U.S. Geological Survey Miscellaneous Investigations Series* Map I-1312.

Eargle, D.H. (1941) 'The relation of soils and surface in the South Carolina Piedmont', *Science* 91: 337–338.

Effland, A.B.W. and Effland, W.R. (1992) 'Soil geomorphology studies in the U.S. Soil Survey program', *Agricultural History* 66: 189–212.

Ellis, A.J. and Lee, C.H. (1919) 'Geology and ground water resources of western San Diego County, California', *U.S. Geological Survey Water Supply Paper* 446: 20–49.

Fairbanks, H.W. (1893) 'Geology of San Diego; also of portions of Orange and San Bernardino Counties', *Reports of the California Bureau of Mines* 11: 76–120.

——(1901) 'The physiography of California', *Bulletin of the American Bureau of Geography* 2: 232–252, 329–350.

Fenneman, N.B. (1936) 'Cyclic and non-cyclic aspects of erosion', *Science* 83: 87–94; *Bulletin of the Geological Society of America* 47: 173–186.

George, P.G. and Dokka, R.K. (1994) 'Major late Cretaceous cooling events in the eastern Peninsular Ranges, California, and their implications for Cordilleran tectonics', *Geological Society of America Bulletin* 106: 903–914.

Gilbert, G.K. (1928) 'Studies of basin range structure', *U.S. Geological Survey Professional Paper* 153: 1–92.

Guttenberg, A.Z. (1976) 'The land utilization movement of the 1920s', *Agricultural History* 50: 477–490.

Hall, A.R. (1940) 'The story of soil conservation in the South Carolina Piedmont 1800–1860', *U.S. Department of Agriculture Miscellaneous Publication* 407.

——(1948) 'Soil erosion and agriculture in the southern Piedmont: a history', unpublished Ph.D. thesis, Duke University.

Hettner, A. (1921) *Die Oberflächenformen des Festlandes*, Leipzig and Berlin: Teubner. Translated by P. Tilley as *The Surface Features of the Earth*, London: Macmillan (1972).

Hewes, L. (1983) 'Carl Sauer: a personal view', *Journal of Geography* 82: 140–147.

Hill, M.L. (1990) 'Transverse Ranges and neotectonics of southern California', *Geology* 18: 23–25.

Hinds, N.E.A. (1925) 'Amphitheatre valley heads', *Journal of Geology* 33: 816–818.

——(1931) 'The relative ages of the Hawaiian landscapes', *University of California Publications, Bulletin of the Department of Geological Sciences* 20: 143–260.

——(1943) *Geomorphology: The Evolution of Landscape*, New York: Prentice-Hall.

——(1952) 'Evolution of the California landscape', *California Department of Natural Resources, Division of Mines, Bulletin* 158: 1–240.

Hobbs, W.H. (1910a) 'The cycle of mountain glaciation', *Geographical Journal* 35: 146–163, 268–284.

——(1910b) 'Studies of the cycle of glaciation', *Journal of Geology* 29: 370–386.

——(1912) *Earth Features and Their Meaning*, New York: Macmillan.

Hooson, D. (1981) 'Carl O. Sauer', in B.W. Blouet (ed.) *The Origins of Academic Geography in the United States*, Hamden: Archon Books.

Hudson, F.S. (1922) 'Geology of the Cuyamaca region of California with special reference to the origin of the nickeliferous pyrrhotite', *University of California Publications, Bulletin of the Department of Geological Sciences* 13: 175–252.

Ireland, H.A., Sharpe, C.F.S. and Eargle, D.H. (1939) 'Principles of gully erosion in the Piedmont of South Carolina', *U.S. Department of Agriculture Technical Bulletin* 633.

Jahns, R.H. (1954a) 'Geology of the Peninsular Range province, southern California and Baja California', *California Department of Natural Resources, Division of Mines, Bulletin* 170: 29–52.

——(1954b) 'Geology of Southern California. Geologic guide no. 5. Northern part of the Peninsular Range province', *California Department of Natural Resources, Division of Mines, Bulletin* 170, guide no. 5, 1–59.

James, P.E. and Jones, C.F. (1954) *American Geography: Inventory and Prospect*, Syracuse: Syracuse University Press.

Jones, W.D. and Sauer, C.O. (1915) 'Outline for field work in geography', *Bulletin of the American Geographical Society* 47: 520–525.

Kenzer, M.S. (1985) 'Milieu and the "intellectual landscape": Carl O. Sauer's undergraduate heritage', *Annals of the Association of American Geographers* 75: 258–270.

Kesseli, J.B. (1938) 'Pleistocene glaciation in the valleys between Lundy Canyon and Rock Creek, eastern slope of the Sierra Nevada', unpublished Ph.D. thesis, University of California at Berkeley.

——(1939) *Outline of Physical Geography*, Berkeley: California Book Company.

——(1940) 'Summary: the development of slopes: studies by Walther Penck and Sieghart Morawitz', Berkeley: University of California, Department of Geography (mimeo).

——(1941a) 'Studies in the Pleistocene glaciation of the Sierra Nevada, California', *University of California Publications in Geography* 6: 315–362.

——(1941b) 'The concept of the graded river', *Journal of Geology* 49: 561–588.

Kniffen, F.B. (1973) 'Richard Joel Russell, 1895–1971', *Annals of the Association of American Geographers* 63: 241–249.

Lawson, A.C. (1893) 'The post-Pliocene diastrophism of the coast of Southern California', *University of California Publications, Bulletin of the Department of Geological Sciences* 1: 115–120.

——(1904) 'Geomorphology of the upper Kern basin', *University of California Publications, Bulletin of the Department of Geological Sciences* 3: 291–376.

——(1915) 'The epigene profile of the desert', *University of California Publications, Bulletin of the Department of Geological Sciences* 9: 23–48.

——(1921) 'The mobility of the coast ranges of California: an exploitation of the elastic rebound theory', *University of California Publications, Bulletin of the Department of Geological Sciences* 12: 431–473.

——(1932) 'Rain-wash erosion in humid regions', *Bulletin of the Geological Society of America* 43: 703–724.

Leighly, J.B. (1932) 'Toward a theory of the morphologic significance of turbulence in the flow of water in streams', *University of California Publications in Geography* 6: 1–22.

——(1934) 'Turbulence and the transportation of rock debris by streams', *Geographical Review* 24: 453–464.

——(1936) 'Meandering arroyos in the dry Southwest', *Geographical Review* 26: 270–282.

——(1937) 'Steps toward a physical theory of stream action', *Geographical Review* 27: 331–333.

——(1940) 'Comments', *Annals of the Association of American Geographers* 30: 223–228.

——(1955) 'What has happened to physical geography?', *Annals of the Association of American Geographers* 45: 309–318.

——(ed.) (1963a) *Land and Life: A Selection from the Writings of Carl Ortwin Sauer*, Berkeley: University of California Press.

———(1963b) 'Introduction', in J.B. Leighly (ed.) *Land and Life: A Selection from the Writings of Carl Ortwin Sauer*, Berkeley: University of California Press.

———(1974) [Review of Chorley, Beckinsale and Dunn 1973], *Annals of the Association of American Geographers* 64: 326–328.

———(1976) 'Carl Ortwin Sauer, 1889–1975', *Annals of the Association of American Geographers* 66: 337–348.

———(1978a) *Notes for the Author of my Obituary*, Berkeley: University of California, Department of Geography.

———(1978b) 'Scholar and colleague: homage to Carl Sauer', *Yearbook of the Association of Pacific Coast Geographers* 40: 117–133.

———(1978c) 'Carl Ortwin Sauer, 1889–1975', *Geographers: Biobibliographical Studies* 2: 99–108.

———(1979) 'Drifting into geography in the twenties', *Annals of the Association of American Geographers* 69: 4–9.

———(1983) 'Memory as mirror', in A. Buttimer (ed.) *The Practice of Geography*, London: Longman.

———(1987) 'Ecology as metaphor: Carl Sauer and human ecology', *Professional Geographer* 39: 405–412.

Louderback, G.D. (1904) 'Basin range structure of the Humboldt region', *Bulletin of the Geological Society of America* 15: 289–349.

———(1923) 'Basin range structure in the Great Basin', *University of California Publications, Bulletin of the Department of Geological Sciences* 14: 329–376.

———(1926) 'Morphologic features of the basin range displacements in the Great Basin', *University of California Publications, Bulletin of the Department of Geological Sciences* 16: 1–42.

Macpherson, A. (1987) 'Preparing for the national stage: Carl Sauer's first ten years at Berkeley', in M.S. Kenzer (ed.) *Carl Ortwin Sauer: A Tribute*, Corvallis: Oregon State University Press.

Martin, G.J. (1987) 'Foreword', in M.S. Kenzer (ed.) *Carl Ortwin Sauer: A Tribute*, Corvallis: Oregon State University Press.

Miller, D.H. (1988) 'John B. Leighly, 1895–1986', *Annals of the Association of American Geographers* 78: 347–357.

Miller, W.J. (1928) 'Geomorphology of the southwestern San Gabriel Mountains of California', *University of California Publications, Bulletin of the Department of Geological Sciences* 17: 193–240.

———(1935) 'Geomorphology of the southern Peninsular Range of California', *Bulletin of the Geological Society of America* 46: 1535–1562.

Nolan, T.B. (1943) 'The basin and range province in Utah, Nevada and California', *U.S. Geological Survey Professional Paper* 197-D: 141–196.

Oberlander, T. (1972) 'Morphogenesis of granitic boulder slopes in the Mojave Desert, California', *Journal of Geology* 80: 1–20.

———(1974) 'Landscape inheritance and the pediment problem in the Mojave Desert of southern California', *American Journal of Science* 274: 849–875.

Parsons, J.J. (1976) 'Carl Ortwin Sauer, 1889–1975', *Geographical Review* 66: 83–89.

———(1979) 'The later Sauer years', *Annals of the Association of American Geographers* 69: 9–15.

———(1986) 'John Leighly 1895–1986', *Itinerant Geographer* 1986: 3–5.

Parsons, J.J. and Vonnegut, N. (1983) *60 Years of Berkeley Geography 1923–1983. Bio-bibliographies of 159 Ph.D.s Granted by the University of California, Berkeley, since the Establishment of a Doctoral Program in Geography in 1923*, Berkeley: University of California, Department of Geography.

Passarge, S. (1912) 'Physiologische Morphologie', *Mitteilungen der geographischen Gesellschaft in Hamburg* 26: 133–337.

———(1919) *Die Grundlagen der Landschaftskunde*, vol. I: *Beschreibende Landschaftskunde*, Hamburg: L. Friedrichsen.

———(1920) *Die Grundlagen der Landschaftskunde*, vol. II: *Klima, Meer, Pflanzen- und Tierwelt in der Landschaft*, Hamburg: L. Friedrichsen.

———(1922) *Die Grundlagen der Landschaftskunde*, vol. III: *Die Oberflächengestaltung der Erde*, Hamburg: L. Friedrichsen.

———(1923a) *Vergleichende Landschaftskunde*, Berlin: L. Friederichsen.

———(1923b) *Die Landschaftsgürtel der Erde*, Breslau: Natur und Kultur.

———(1926) 'Morphologie der Klimazonen öder Morphologie der Landschaftsgürtel?', *Petermanns Geographische Mitteilungen* 72: 173–175.

Peltier, L.C. (1950) 'The geographic cycle in periglacial regions as it is related to climatic geomorphology', *Annals of the Association of American Geographers* 40: 214–236.

Penck, A. (1894) *Morphologie der Erdoberfläche*, Stuttgart: J. Engelhorns Buchhandlung. 2 volumes.

———(1905) 'Climatic features of the land surface', *American Journal of Science* (4) 19: 165–174.

Penck, W. (1920) 'Wesen und Grundlagen der morphologischen Analyse', *Bericht der Mathematischen-Physikalischen Klasse der Sächsischen Akademie der Wissenschaften (Leipzig)* 72: 65–102.

———(1924) *Die morphologische Analyse. Ein Kapitel der physikalischen Geographie*, Stuttgart: J. Engelhorns Nachforschung. Translated by H. Czech and K.C. Boswell as *Morphological Analysis of Land Forms: A Contribution to Physical Geography*, London: Macmillan (1953).

Peschel, O. (1870) *Neue Probleme der vergleichende Erdkunde als Versuch einer Morphologie der Erdoberfläche*, Leipzig: Duncker & Humblot.

———(1879–1880) *Physische Erdkunde*, Leipzig: Duncker & Humblot.

Pettijohn, F.J. (1984) *Memoirs of an Unrepentant Field Geologist: A Candid Profile of Some Geologists and Their Science, 1921–1981*, Chicago: University of Chicago Press.

Richthofen, F. von (1886) *Führer für Forschungsreisende*, Berlin: R. Oppenheim.

Russell, R. J. (1927) 'The land forms of Surprise Valley, northwestern Great Basin', *University of California Publications in Geography* 2: 323–358.

———(1928) 'Basin range structure and stratigraphy of the Warner Range, northeastern California', *University of California Publications in Geology* 17: 387–496.

Salisbury, R.D. (1907) *Physiography*, New York: Henry Holt.

Sapper, K. (1917) *Geologischer Bau und Landschaftsbild*, Braunschweig: F. Vieweg.

Sauer, C.O. (1916) 'Geography of the upper Illinois valley and history of development', *Bulletin of the Illinois Geological Survey* 27: 1–208.

———(1918) 'A soil classification for Michigan', *20th Annual Report of the Michigan Academy of Sciences*: 83–91.

———(1920) 'The geography of the Ozark Highland of Missouri', *Bulletin of the Geographical Society of Chicago* 7: 1–245.

———(1921) 'The problem of land classification', *Annals of the Association of American Geographers* 21: 3–16.

———(1922) 'Notes on the geographic significance of soils – I', *Journal of Geography* 21: 187–190.

———(1924) 'The survey method in geography and its objectives', *Annals of the Association of American Geographers* 14: 17–33.

———(1925) 'The morphology of landscape', *University of California Publications in Geography* 2: 19–53.

———(1927) 'Geography of the Pennyroyal', *Kentucky Geological Survey* (6) 25: 1–303.

——(1929) 'Land forms in the Peninsular Range of California as developed about Warner's Hot Springs and Mesa Grande', *University of California Publications in Geography* 3: 199–290.

——(1930) 'Basin and range forms in the Chiricahua area', *University of California Publications in Geography* 3: 339–414.

——(1932) 'Land forms in the Peninsular Range', *Zeitschrift für Geomorphologie* 7: 246–248.

——(1934) 'Preliminary report to the Land-Use Committee on Land Resource and Land Use in relation to public policy', *Report of the Science Advisory Board 1933–1934*: 165–260.

——(1936a) *Handbook for Geomorphologists*, Washington, D.C.: Department of Agriculture, Soil Conservation Service, Division of Climatic and Physiographic Research.

——(1936b) 'Soil conservation', *Geographical Error* 2: 1–2.

——(1941) 'Foreword to historical geography', *Annals of the Association of American Geographers* 31: 1–24.

——(1952) *Agricultural Origins and Dispersals*, New York: American Geographical Society.

——(1956) 'The education of a geographer', *Annals of the Association of American Geographers* 46: 287–299.

——(1966) 'On the background of geography in the United States', *Heidelberger Geographische Arbeiten* 15: 59–71.

——(1974) 'The fourth dimension of geography', *Annals of the Association of American Geographers* 64: 189–195.

——(1976) 'Casual remarks', *Historical Geography Newsletter* 6: 70–76.

——(1981) *Selected Essays 1963–1975*, Berkeley: Turtle Island Foundation.

——(n.d.) 'Mechanics of slope-denudation under natural conditions', unpublished manuscript.

Sauer, C.O., Leith, C.K., Merriam, J.C. and Bowman, I. (1934) 'Preliminary recommendations of the Land-Use Committee ... relating to soil erosion and critical land margins', *Report of the Science Advisory Board 1933–1934*: 137–161.

Schmalz, N.J. (1978) 'Michigan's Land Economic Survey', *Agricultural History* 52: 229–246.

Sharp, R.P. (1993) 'Recollections of Kirk Bryan: a biographical sketch', *Geomorphology* 6: 189–205.

Sharpe, C.F.S. (1938) *Landslides and Related Phenomena*, New York: Columbia University Press.

Sharpe, C.E.S. and Dorch, E.F. (1943) 'Relation of soil-creep to earthflow in the Appalachian plateaus', *Journal of Geomorphology*: 5: 312–324.

Simons, M. (1962) 'The morphological significance of landforms: a new review of the work of Walther Penck', *Transactions of the Institute of British Geographers* 31: 1–14.

Spencer, J.E. (1976) 'What's in a name? – "the Berkeley School"', *Historical Geography Newsletter* 6: 7–11.

——(1979) 'A geographer west of the Sierra Nevada', *Annals of the Association of American Geographers* 69: 46–63.

Speth, W.W. (1981) 'Berkeley geography, 1923–33', in B.W. Blouet (ed.) *The Origins of Academic Geography in the United States*, Hamden: Archon Books.

Stoddart, D.R. (1987) 'Geographers and geomorphology in Britain between the wars', in R.W. Steel (ed.) *British Geography 1918–1945*, Cambridge: Cambridge University Press.

——(1991) 'Carl Sauer: the man and his work', *American Association for the Advancement of Science, Pacific Division, Newsletter* 16: 17–20.

Thornthwaite, C.W. (1931) 'The climates of North America according to a new classification', *Geographical Review* 21: 633–655.

Thornthwaite, C.W., Sharpe, C.F.S. and Dorch, E.F. (1942) 'Climate and accelerated erosion in the arid and semi-arid Southwest with special reference to the Polacca Wash drainage basin', *U.S. Department of Agriculture Technical Bulletin* 808: 1–134.

Trimble, S.W. (1974) *Man-Induced Soil Erosion in the Southern Piedmont 1700–1970*, Soil Conservation Society of America.

——(1985) 'Perspectives on the history of soil erosion control in the eastern United States', *Agricultural History* 59: 162–180.

Tuan, Yi-Fu (1957) 'Pediments in southeastern Arizona', Ph.D. thesis, University of California at Berkeley.

——(1958) 'The misleading antithesis of Penckian and Davisian concepts of slope retreat in waning development', *Transactions of the Indiana Academy of Sciences* 67: 212–214.

——(1959) 'Pediments in southeastern Arizona', *University of California Publications in Geography* 13: 1–140.

——(1962) 'Structure, climate and basin land forms in Arizona and New Mexico', *Annals of the Association of American Geographers* 52: 51–68.

——(1966) 'New Mexican gullies: a critical review and some recent observations', *Annals of the Association of American Geographers* 56: 573–597.

Von Engeln, O.D. (1942) 'The Walther Penck geomorphic system', in *Geomorphology: Systematic and Regional*, New York: Macmillan.

Whittlesey, D. (1951) 'Kirk Bryan, 1888–1950', *Annals of the Association of American Geographers* 41: 88–94.

Wood, S.H. and Elliott, M.R. (1979) 'Early 20th-century uplift of the northern Peninsular Ranges province of southern California', *Tectonophysics* 52: 249–265.

Yasso, W.E. (1964) 'Geometry and development of spit-bar shorelines at Horseshoe Cove, Sandy Hook, New Jersey', *Columbia University, Department of Geology, Technical Report* 5: 1–166.

EPILOGUE

17

RICHARD J. CHORLEY AND MODERN GEOMORPHOLOGY

David R. Stoddart

It cannot be too often that one witnesses a revolution in the science of one's choice and can see how it occurs. Indeed, I have long been sceptical of Kuhn's whole notion of paradigm change (Stoddart 1981), without any adequate recognition of the fact that it had happened before my eyes.

Dick Chorley graduated at Oxford in 1951, and in this book Robert Beckinsale has evocatively recalled those early years. Dick came to the Cambridge Department of Geography in the Michaelmas Term of 1958. Peter Haggett had understandably abandoned Darby's department at University College London and returned a year before: the results were momentous. I was a third-year undergraduate specialising in geomorphology when Dick arrived and thus saw it from the start.

The Cambridge to which he came was decorous, respectable and really intellectually somewhat dull, though with people like B.H. Farmer, E.A. Wrigley and Peter Haggett it was far ahead of any other Department of Geography in Britain. Indeed, the general climate of opinion in the country at that time was far from receptive to any kind of change (Stoddart 1987b). Sir Winston Churchill had only just resigned as prime minister at the age of 81. The failure of his successor Sir Anthony Eden during the Suez escapade was the first sign of the winds of change ahead in British life, though the antics of Miss Christine Keeler and Miss Mandy Rice-Davies with one of the Ministers of Defence and others in Mr Macmillan's government (which so diverted Mr Chorley in the early 1960s) were yet to be divulged. (Lord Montagu's proclivities with boy scouts occurred somewhat earlier but also caused us great amusement.) Professor Dudley Stamp (1898–1966), soon to be knighted, had announced in his presidential address to the Institute of British Geographers in 1957 that he was 'alarmed by the view that the geographer must add to his training a considerable knowledge of statistics and statistical methods' (Stamp 1957: 2). His agenda for future research included work on the distribution of public house names (these had already been mapped for Cambridge before the war by the future Sir Vivian Fuchs) and the international distribution of what with his broad experience he termed 'Coco-Cola' (this was reminiscent of Lord Curzon's attempt to join the hoi polloi when he referred to the children's comic – which he had obviously never read and which was a solace of my childhood – as the 'Bé-ahno'). There was no awareness of any impending conceptual break-through. Later, as President of the Royal Geographical Society he went even further after Peter Haggett had given a seminal paper on Brazil: 'he has introduced some of us to the modern quantitative approach,' he opined. 'Some of us may not like the look of it. We may even feel that here is an enormous steam hammer! How is it working in the cracking of nuts?' (in Haggett 1964: 380). Sadly, he never found out: he died in an elevator in Mexico City in 1966: at least the event was announced on the BBC One-o'clock News. In his last posthumous paper he confessed that a decade before he had been 'blissfully unaware of the tsunami of quantification about to break' (Stamp 1966: 11), and claimed to identify a state of 'civil war' between the quantifiers and the more traditional geographers such as himself. Describing Haggett's pioneering *Locational Analysis in Human Geography* (Haggett 1965) as an 'introduction to the new Theology', he made the bizarre statement that 'Quantification has many points in common with Communism: it has become a religion to its devotees, its golden calf is the computer.'

Stamp could not of course claim to be a geomorphologist, in spite of some of his early writings. The leader of that field was S.W. Wooldridge (1900–1963), a figure equally given to *ex cathedra* pronouncements from on high and equally out of touch with the cutting edge of his own discipline (Figure 17.1). In a review mistitled 'The progress of geomorphology' in 1951 he had grudgingly managed to admit that 'the general subject of slope evolution is a fit one for active research', but nevertheless saw geomorphology as 'a tool in the elucidation of earth-history' and 'the complement of stratigraphical geology' (Wooldridge 1951: 170, 176). He made a clarion call for the reprinting of W.M. Davis's *Geographical Essays* (1909), a wish that was granted in 1954. Wooldridge reviewed the new edition in the *Geographical Journal*, and took the opportunity to accuse Strahler and even Sauer of introducing into their criticisms of Davisian cyclic geomorphology 'a distinctly bitter or rancorous note which many of us ... must resent and, indeed, flatly repudiate' (Wooldridge 1955: 90).

In 1958 he gave his views on 'The trend of geomorphology', using the occasion to denounce Strahler's (1950a) criticisms of Davis, Peltier's (1950) climatic geomorphology, and King's (1953) 'Canons of landscape evolution'. In a remarkably out-of-touch statement he said that 'I regard it as quite fundamental that Geomorphology is primarily concerned with the inter-pretation of forms, not the study of processes' (Wooldridge 1958: 31), failing to see that the former was out of the question without the latter. On Walther Penck's views on slope evolution, 'we can leave the geologists to deal with the matter and cease to bother our heads about it' (Wooldridge 1958: 32);

Figure 17.1 Professor Wooldridge in the field. His walking stick was a trademark characteristic.

Source: Photograph courtesy of Professor Denys Brunsden

Penck in any case was guilty of 'a good deal of muddled and simply erroneous thinking' (Wooldridge 1958: 34). Penck's treatise (1924) had been published a quarter of a century before: it could scarcely be called a 'trend'. Inevitably he concluded that the 'rather trivial scraps of algebra, geometry and calculus' which constituted 'mathematical methods would seem to offer very limited chances of success' (Wooldridge 1958: 32). After further denunciation of what he termed 'the periglacial extremists' and 'the morpho-metric squad' he simply footnoted a number of Strahler's papers but did not discuss their contents (Wooldridge 1958: 35). It is extraordinary that the Institute of British Geographers should have published such papers by Stamp and Wooldridge, though symptomatic of those times when reputation and status meant everything (I have been shocked to find that the editors of the *Transactions* in those years were friends whom I hold in the highest regard). The year after this lamentable effort (though hopefully not because of it) Wooldridge was elected a Fellow of the Royal Society: it is perhaps not surprising that no geographer has been so recognised since. It is true, of course, that after a stroke in 1954 he was no longer the man who had collaborated with David Linton on *Structure, Surface and Drainage in South-East England* (Wooldridge and Linton 1939, 1955). When he died, *The Times* noted that while 'to him is due the modern flowering of geomorphol-ogy in Britain', he was 'inclined to bluster.... His tone of voice suggested that all was vanity, and also, in his later years, that all was vexation of spirit too' (Anon. 1963). Martin (1963) recalled that his views were 'expressed in his characteristically intemperate manner and with his own idiosyncratic vocabulary' and years later Balchin (1984: 143) remarked on Wooldridge's penchant for 'difficult disputation' and noted that in his later years he became 'increasingly dogmatic'. Poor Wooldridge: he spent a term in Cambridge in 1960, apparently to write a book on the geography of south-eastern England. But as he was at that time a Congregationalist he spent all his time at Cheshunt College (a theological not a university institution) and when he did come to the Department of Geography for tea I do not recall that anyone ever spoke to him or he to them. In fact his presence was an embarrassment to all. His time had gone, I felt unable to approach him (he would have dismissed me, as he had done before), and I never saw him again.

There were two other dominant figures in British geomorphology in the 1950s. One was Wooldridge's collaborator David Linton (1906–1971). Linton was not one given to philosophical pronouncements, and in his prime had moved away from geomorphology to interests in urbanisation, popula-tion, food supply and the tropical world (Anon. 1972). His work in denudation chronology had led him to the advocacy of morphological mapping (Linton 1951), and his central role in the early years of the British Geomorphological Research Group led too many in that organisation into that particular intellectual cul-de-sac – from which I suffered on Dartmoor as a schoolboy, in the presence of the apoplectic Professor Wooldridge. (It says something for the organisation of British geography at the time that when I showed up for that particular Geographical Association field trip, having coaxed large sums out of my father, there were no beds available in the meeting hotel; I was instead offered three pillows in the tub in the public bathroom; fortunately, given the state of British personal hygiene at that time, no one turned the taps on during the night throughout the entire week.)

The other major figure of course was Alfred Steers (1899–1987), head of the Cambridge Department of Geography and the man who launched Chorley on his university career (Figure 17.2). I suspect he always felt about this unlikely initiative as FitzRoy did after having given Darwin volume 1 of

Figure 17.2 Professor Steers as a gentleman fieldworker
Source: Photograph courtesy of the Department of Geography, Cambridge University

Lyell's *Principles of Geology* in 1830 and then witnessing his biblical world collapse around him with *The Origin of Species*; it would of course never have entered Alfred's mind to take FitzRoy's option and cut his throat, though he did from time to time get profoundly agitated at what he had done. Dick had a rather rough ride with Alfred in the early 1960s (as indeed did we all – Haggett alone had the political finesse to ride the storm), but he and Rosemary were in fact deeply attached to him: I was with Alfred at his deathbed at Addenbrooke's Hospital in Cambridge when they both came to say farewell.

Now Alfred was not a theoretician. The anonymous author of his obituary in *The Times* (Anon. 1987) – in fact his long-time colleague A.A.L. Caesar, who owed his career to Alfred and who should have known better (and I told him that he ought to be ashamed of himself on the day it was published) – said of him: 'No theories will carry his name and no students will argue on his views' (which coming from Caesar, a man with neither theories nor views, only prejudices, was pretty rich). More appreciative assessments of Steers are possible (Stoddart 1987a, 1988). When his time came Caesar was more generously treated in the same newspaper (Anon. 1995); no prizes are offered for identifying the author of that particular mythology. Dick was particularly incensed in those early years that Caesar (who was never ever there, in spite of drawing a university salary) should allow his departmental office to be used for prayer meetings of the more rabid undergraduates where the Almighty was publicly implored to redeem Chorley's soul: Dick viewed it as an invasion of privacy, but perhaps it worked. He has since religiously followed in Alfred Steers's footsteps at St Botolph's Church (but has notably followed an equally individual commitment to the Red Lion at Grantchester).

Steers gave the presidential address to the Institute of British Geographers on 30 December 1959, under the title 'Physiography: some reflections and trends' (Steers 1960). Alas, it was no better than the contemporary announcements of both Stamp and Wooldridge. Saying that he wished to 'try to indicate some of the ways in which physiographical research is developing, or has already developed, in the last few decades' (Steers 1960: 3), Steers praised the ancient work of G.K. Gilbert (1877, 1890) and François Matthes (1930), but then fell back on logistics: the sterling work of the Fenland Research Committee (whose chairman, the coral reef specialist J. Stanley Gardiner, cleaned out his pipe and burned down their chief reserve, Wicken Fen) and the Weald Research Committee, advocated 'the full development of the aqualung' and radioactive tracers for beach studies, and noted that (in those days) you could hire a light aeroplane for two hours for ten pounds. He did in fact quote Strahler's 'Dynamic basis of geomorphology' (1952a) paper and remarked on 'modern techniques of analysing rivers, slopes, forms of drumlins' (the last a direct allusion to the paper Chorley (1959) had just published in which he had drawn particular attention to the forces shaping the avian egg as it was extruded from the hen's interior). 'But it may be asked, how far do these techniques and measurements help us in understanding the land forms analysed?' (Steers 1960: 11): answer came there none. He took another swipe at Chorley's (1957) paper on climate and morphometry:

> That we may represent ... the order, number, and length of streams in given regions as straight lines or curves on some form of graph paper, and that we may find an index between climate and vegetation or some other pair of phenomena may well be an advance in technique; but it is not in itself an explanation.
>
> (Steers 1960: 12)

He did not explain what better approach there might be. Chorley had been in post at Cambridge for only fifteen months, and can scarcely have been encouraged by these public observations on his achievements.

The Cambridge Department to which he had come included three other geomorphologists. Pre-eminent was Vaughan Lewis (1907–1961). Tressilian Nicholas, one of the powers in the University and who recently died at the age of 100, remembered his

> warm and optimistic temperament. He was an enthusiast in everything that he did and, for a don, remarkably unreserved in his personal relations, rich in friendship and ever ready with offers of help. In the field no hardships got him down and any who shared his company on British beaches, or mountains in the Alps, or glacier-caps on Austerdalsbreen, will recall his friendly relations with every member of the party.
>
> (Nicholas 1961: 38)

In his first undergraduate year at the university Lewis had read mathematics, and this 'allowed him to appreciate the value of quantitative and theoretical analysis to guide the analysis of the field observations' (King 1980: 117). Dick knew him for so short a time before Vaughan was tragically killed in a car accident in the United States in June 1961: at the time he was in line to become President of the Institute of British Geographers, and would undoubtedly have given an address of quite a different calibre from those to which I have referred. His major collaborative field study on 'Norwegian cirque glaciers' appeared the year before he died (Lewis 1960) and he had also carried out many investigations in coastal geomorphology in a tradition quite different from that of Alfred Steers. Dick and he were in many respects fellow spirits, and Dick (Chorley 1993) has recently evocatively and affectionately recalled him as 'vital, enthusiastic, mercurial, emotional, vulnerable, and approachable' – he could of course have been speaking about himself. It was Vaughan's position to which Dick was thus able to succeed.

Bruce Sparks (1923–1988) had been appointed to the Department in 1949 (Stoddart 1989). He had been a student of Wooldridge and his original research on the denudation chronology of the South Downs had been squarely in the Wooldridgean tradition (Sparks 1949). He was very much a geological and historical geomorphologist and uninterested in process studies: his lectures emphasised Jukes-Brown and the Clay-with-flints rather than Horton or Leopold. His text on *Geomorphology* appeared in 1960 – to my distress he completely forgot to include deltas – and it was a huge success: Bruce drove a succession of expensive cars and was constantly at the Covent Garden opera for the rest of his life (before *Geomorphology* he rode a motorbike – a lesson I believe Dick took on board). His true interests were revealed when he published *Rocks and Relief* in 1971. There was no quantitative geomorphology in either book, and by the time the third edition of *Geomorphology* appeared in 1986 it had become an anachronism. With increasing ill-health he became somewhat disillusioned and misanthropic, and never produced any philosophical speculations on the future of the discipline. But in the early years he and Dick were constantly quaffing ale at the Bun Shop (now replaced by a hideous Holiday Inn), and with his respectable geological interests he served as an interlocutor and negotiator with Alfred. Dick Grove, happily and most productively still with us, seemed always to be in (or around) Lake Chad, and his sterling work in Africa on Holocene lake levels, dune formations and trepanation in the Tibesti hardly intersected with Dick's concerns (though there are doubtless many Dick would wish to have trepanned).

This was the Cambridge at which Chorley arrived: my wife has reminded me on reading this reminiscence of the degree of indignation (a word she stresses) that students felt on what they were offered. The standard book for students was Thornbury's *Principles of Geomorphology* (1954), a thoroughly Davisian treatment, or indeed Wooldridge and Morgan's text of 1937 – as old as the students it was recommended to. (It took Wooldridge until 1959 to include a discussion of Penck's views on slope development, thirty-five years after they were generally published.) Vaughan Lewis lectured on Horton's (1945) classic paper, but though he had all of Strahler's offprints I do not recall him talking about them. Bakker and Le Heux (1952) (though see Wooldridge 1958: 135 for a calamitous rendition of that reference) was all the rage, and in process terms we all read Hjulström (1935) from more than twenty years before. In my final undergraduate year Steers gave such a disastrous course on the Massif Central (the great authority being Baulig 1928) that Sparks was obliged to post a notice saying he would repeat it in the following term to save the students from nervous breakdowns. King's (1953) 'canons of landscape evolution', in spite of Twidale's (1992) homage, simply fizzled: nobody (rightly) was interested in that kind of declamatory declaration.

Chorley had the great good fortune to work under Arthur Strahler at Columbia University at the latter's most productive time, before he gave up science and simply started writing textbooks. After earlier more traditional work Strahler had been deeply impressed by Horton's landmark paper 'Erosional development of streams and their drainage basins: hydrophysical approach to quantitative morphology' (1945), which unusually began with a quotation from Playfair. Strahler has called that paper a

gold mine of fluvial process concepts based on a lifetime of field studies by a hydraulic engineer. This was a remarkable interdisciplinary transfer of information from hydrology, a geophysical area of knowledge, to a geomorphology largely rooted in geological concepts.

He records (Strahler 1992: 69) that Horton had inscribed the copy of the paper which he sent to him: 'He who runs may read.' Strahler had published the first fruits of his conversion in his papers on slopes (1950b), on the 'dynamic basis of geomorphology' (1952a), and on the hypsometric integral, in the last of which he introduced what came to be known as 'Strahler ordering' for streams (1952b).

Strahler assembled a remarkable collection of graduate students at Columbia (though Dick initially came to study under Lobeck). I will not name them all, but those most important to Dick (and to his own graduate students) were Stanley Schumm (a lifelong colleague, collaborator and friend), Mark Melton and Marie Morisawa, all of whom became icons for his undergraduates. Schumm had just produced his landmark papers on gully erosion in New Jersey and South Dakota (1956a, 1956b), was working on diastrophism and rates of erosion (1963a) and meanders in the Great Plains (1963b), and was shortly to publish his truly epoch-making paper on 'Time, space and causality in geomorphology' (Schumm and Lichty 1965). Mark Melton was publishing papers (1958a, 1958b) which when they appeared seemed somewhat less accessible but had a similar effect on how one went about things.

Dick's other inspiration was Luna Leopold (who celebrated his eightieth birthday on the day I wrote this paragraph), then with the US Geological Survey, and with whom he worked after being with Strahler. Leopold had produced a number of seminal papers on stream dynamics (Leopold and Maddock 1953; Leopold and Miller 1956; Leopold and Wolman 1957). He and his collaborators were also entering more theoretical territory. He published on floodplains with M.G. Wolman (following the latter's classic Brandywine Creek paper in 1955) and then turned to entropy, indeterminacy and equilibrium (Leopold and Langbein 1962, 1963; Langbein and Leopold 1964). Langbein and Schumm's paper on the climatic control of sediment yield, which truly excited Chorley and electrified his student audiences, appeared in 1958. Wolman and Miller's brief but central paper on the magnitude and frequency of operation of geomorphic processes came in 1960. This period of extraordinary productivity culminated in Leopold, Wolman and Miller's text on *Fluvial Processes in Geomorphology* in 1964. (Scheidegger's *Theoretical Geomorphology* (1961) had appeared somewhat earlier but had nothing like the same impact.) Chorley was also much influenced by W.C. Krumbein's work on sampling and multivariate analysis (Krumbein 1955, 1959) and in his teaching he became somewhat obsessed not only with the frequency of the chirping of crickets (from his bible at the time, Croxton and Cowden 1955: 452) but also with dropping cannon balls on beaches; though I might add that on the few occasions I have been with him on a beach I have never seen him drop any balls at all.

Chorley's first years at Cambridge (Figure 17.3) rode the crest of this wave of innovation and were extraordinarily productive, at least in part because of the synergism between him and Peter Haggett (Chorley 1995a).

Figure 17.3 Richard Chorley in 1962, shortly after he came to Cambridge

Doubtless from Strahler he brought the insights of Ludwig von Bertalanffy's General Systems Theory into geomorphology (Chorley 1962). Strahler (1992: 72–73) has recalled how on reading von Bertalanffy in 1950 'It was as if a closed door had opened before me, revealing an entirely new and powerful epistemology of science – a paradigm capable of unifying all dynamic processes and forms that can be observed in the universe': indeed a powerful claim. Chorley's paper on analogue theory came in 1964, and much later a widely quoted consideration of 'Bases for theory in geomorphology' (1978). A major paper on statistical analysis was published in 1966.

Organising the Madingley Lectures at Cambridge with Peter Haggett, beginning in the summer of 1963, led first to *Frontiers in Geographical Teaching* (Chorley and Haggett 1965) and then to *Models in Geography* (Chorley and Haggett 1967). Both volumes had enormous impact at all levels of British geography, and the first included Chorley's celebrated attack on the Davisian geomorphic system (Chorley 1965): had Wooldridge still been alive it would certainly have finished him off. *The Times Literary Supplement* somewhat sourly commented on the first that 'One man's frontier is another man's base camp,' prompting the thought that at that time the latter must have been deserted by all but the geriatric. At the time of the planning meeting for the *Models* conference I recall going to the Cambridge post office, with Dick's connivance, and sending two telegrams to the organisers. One read: 'Wish I were with you – Halford Mackinder', and the other: 'Drop dead – L. Dudley Stamp'. Both were expeditiously delivered soon after the meeting began and read to general acclamation. Haggett and Chorley (1989) have recalled how members of the meeting became so ecstatic at the thought of the new frontier (or someone) that they punched each other on the nose somewhere between Cambridge and Bristol. It was appropriately enough in the Madingley walled garden during the *Models* meeting that *Progress in Geography* (affectionately known as PIG) was conceived. This began publication as an annual review in 1969 and became the present journals *Progress in Physical Geography* and *Progress in Human Geography* (with which Dick is still associated) in 1977.

This is not the place to go through all of Chorley's voluminous publications, a list of which concludes this book. But mention must be made of two sets of contributions. The first are pedagogic. *Atmosphere, Weather and Climate* (Barry and Chorley 1968) has proved the basic English textbook on the subject and is now in its sixth edition. *Water, Earth and Man* (Chorley 1969) soon followed. The book he wrote with Barbara Kennedy, *Physical Geography: A Systems Approach* (Chorley and Kennedy 1971) was perhaps too far ahead of its time for general adoption as a text, but it was a work, like Haggett's *Locational Analysis,* that shed illumination simply through its contents list. *Network Analysis in Geography* (Haggett and Chorley 1969) was followed by *Environmental Systems: Philosophy, Analysis and Control* (Bennett and Chorley 1978). And in his spare time he also managed to edit

Spatial Analysis in Geomorphology (Chorley 1972) and *Directions in Geography* (Chorley 1973); the royalties from the former transformed the finances of the British Geomorphological Research Group. Finally, with Stanley Schumm and David Sugden, he produced the now-standard text on *Geomorphology* in 1984. Authoring any one of these would have satisfied a lesser man. Taken together and with Haggett's books they introduced a new and distinctive style into geographical writing (Figure 17.4).

Apart from the role he played in the transformation of the discipline, Dick's longest-lasting contribution will surely be *The History of the Study of Landforms*, which began publication in 1964 and has now reached three volumes and over two thousand pages (Chorley, Dunn and Beckinsale 1964; Chorley, Beckinsale and Dunn 1973; Beckinsale and Chorley 1991). The first of these was perhaps the easiest to write, dealing with the earlier history of geomorphology. It also provided the material for some of Dick's best lectures, notably with the slide he used of G.K. Gilbert sitting motionless on horseback, illustrating dynamic equilibrium. But as Dorothy Sack (1992) has shown, Chorley's style of history is far from merely narrative but develops an argument to sustain his own methodological position. The biography of Davis and exegesis of his cyclic geomorphology in volume 2 will certainly never be surpassed. The mastery of an increasingly complex terrain in volume 3 takes the story up to 1950, and at least one more volume is planned: perhaps it should be called *The Age of Chorley*, though that title might be misinterpreted and indeed I have a paper in preparation on that subject.

It is perhaps difficult to recognise the scale and the scope of the changes that Dick's work brought about and how rapidly they were achieved. We have noted how, when he started his Cambridge career, there was neither appreciation of nor sympathy for the new geomorphology in the presidential addresses and other pronouncements of the leaders of the discipline. But by the time that Ronald Peel (another Cambridge man) came to give his presidential address to Section E of the British Association for the Advancement of Science in 1967 Wooldridge and Stamp had mercifully departed and all that had changed. While quietly lamenting the demise of Davisian geomorphology as a teaching device, Peel (1967: 207) saw the reaction against Davis as 'one of the most notable features of geomorphological thought in the last decade or two'. Suddenly and for the first time the names of Horton, Leopold, Langbein, Miller, Wolman, Maddock, Hack, Schumm, Strahler, Melton and of course Chorley himself appear in such a review. Wooldridge's opposition was castigated as 'ill-judged' and even at that early date Chorley was recognised as 'the leading champion of this [new] approach in Britain' (Peel 1967: 212).

And that of course was only the beginning.

Figure 17.4 Richard Chorley constructing one of his many works in the early 1960s, as viewed by a local wag. The artist also distributed a review (of quite another book by a different person) which stated that 'it was richly illustrated with diagrams, symbols, directive arrows, charts of the night sky, bits of algebra, plus and minus and equal signs, and an array of small boxes shaded with hatchings and cross-hatchings'. The cognoscenti will recognise the style at once

Source: Courtesy of an anonymous contributor

395

REFERENCES

Anon. (1963) 'Prof. S.W. Wooldridge: geographer of distinction', *The Times*, 27 April 1963.

——(1987) 'Professor J.A. Steers: mapping the coastline', *The Times*, 14 March 1987.

——(1995) 'Gus Caesar', *The Times*, 30 September 1995.

Bakker, J.P. and Le Heux, J.W.N. (1952) 'A remarkable new geomorphological law', *Proceedings of the Koninklijke Nederlandse Akademie van Wetenschappen, Amsterdam* (B) 55: 399–570.

Balchin, W.G.V. (1984) 'Sidney William Wooldridge 1900–1963', *Geographers Bibliographical Studies* 8: 141–149.

Barry, R.G. and Chorley, R.J. (1968) *Atmosphere, Weather and Climate*, London: Methuen (6th edition, London: Routledge, 1992).

Baulig, H. (1928) *Le Plateau Central de la France et sa bordure Méditerranée*, Paris: Armand Colin.

Beckinsale, R.P. and Chorley, R.J. (1991) *The History of the Study of Landforms or The Development of Geomorphology*, vol. 3: *Historical and Regional Geomorphology 1890–1950*, London: Routledge.

Bennett, R.J. and Chorley, R.J. (1978) *Environmental Systems: Philosophy, Analysis and Control*, London: Methuen.

Chorley, R.J. (1957) 'Climate and morphometry', *Journal of Geology* 65: 630–638.

——(1959) 'The shape of drumlins', *Journal of Glaciology* 3: 339–344.

——(1962) 'Geomorphology and General Systems Theory', *U.S. Geological Survey Professional Paper* 500-B: 1–10.

——(1964) 'Geography and analogue theory', *Annals of the Association of American Geographers* 54: 127–137.

——(1965) 'A re-evaluation of the geomorphic system of W.M. Davis', in R.J. Chorley and P. Haggett (eds) *Frontiers in Geographical Teaching*, London: Methuen.

——(1966) 'The application of statistical methods to geomorphology', in G.H. Dury (ed.) *Essays in Geomorphology*, London: Heinemann.

——(1972) *Spatial Analysis in Geomorphology*, London: Methuen.

——(1978) 'Bases for theory in geomorphology', in C. Embleton, D. Brunsden and D.K.C. Jones (eds) *Geomorphology: Present Problems and Future Prospects*, Oxford: Oxford University Press.

——(1993) 'William Vaughan Lewis', *University of Cambridge, Department of Geography, The William Vaughan Lewis Seminars* 1: i–ii.

——(1995a) 'Haggett's Cambridge: 1957–1966', in A.D. Cliff, P.R. Gould, A.G. Hoare and N.J. Thrift (eds), *Diffusing Geography*, Oxford: Blackwell.

——(1995b) 'Classics of physical geography revisited: Horton, 1945', *Progress in Physical Geography* 19: 533–554.

Chorley, R.J., Beckinsale, R.P. and Dunn, A.J. (1973) *The History of the Study of Landforms or The Development of Geomorphology*, vol. 2: *The Life and Work of William Morris Davis*, London: Methuen.

Chorley, R.J., Dunn, A.J. and Beckinsale, R.P. (1964) *The History of the Study of Landforms or The Development of Geomorphology*, vol. 1: *Geomorphology before Davis*, London: Methuen.

Chorley, R.J. and Haggett, P. (eds) (1965) *Frontiers in Geographical Teaching*, London: Methuen.

——(eds) (1967) *Models in Geography*, London: Methuen.

Chorley, R.J. and Kennedy, B.A. (1971) *Physical Geography: A Systems Approach*, London: Prentice-Hall.

Chorley, R.J., Schumm, S.A. and Sugden, D.E. (1984) *Geomorphology*, London: Methuen.

Croxton, F.E. and Cowden, D.J. (1955) *Applied General Statistics*, Englewood Cliffs: Prentice Hall and London: Pitman.

Gilbert, G.K. (1877) *Report on the Geology of the Henry Mountains*, Washington: US Department of the Interior, Geographical and Geological Survey of the Rocky Mountain Region.

——(1890) 'Lake Bonneville', *Monographs of the U.S. Geological Survey* 1: 1–438.

Haggett, P. (1964) 'Regional and local components in the distribution of forested areas in south east Brazil: a multivariate approach', *Geographical Journal* 130: 365–378, discussion 378–380.

——(1965) *Locational Analysis in Human Geography*, London: Arnold.

Haggett, P. and Chorley, R.J. (1989) 'From Madingley to Oxford: a foreword to *Remodelling Geography*', in W. Macmillan (ed.) *Remodelling Geography*, Oxford: Blackwell.

Hjulström, F. (1935) 'Studies of the morphological activity of rivers as illustrated by the River Fyris', *Bulletin of the Geological Institute of the University of Upsala* 25: 221–527.

Horton, R.E. (1945) 'Erosional development of streams and their drainage basins: hydrophysical approach to quantitative morphology', *Bulletin of the Geological Society of America* 56: 275–370.

King, C.A.M. (1980) 'William Vaughan Lewis 1907–1961', *Geographers: Bio-bibliographies* 4: 113–120.

King, L.C. (1953) 'Canons of landscape evolution', *Bulletin of the Geological Society of America* 64: 721–762.

Krumbein, W.C. (1955) 'Experimental design in the earth sciences', *Transactions of the American Geophysical Union* 36: 1–11.

——(1959) 'The "sorting out" of geological variables illustrated by regression analysis of factors controlling beach firmness', *Journal of Sedimentary Petrology* 29: 575–587.

Kuhn, T.S. (1962) *The Structure of Scientific Revolutions*, Chicago: University of Chicago Press.

Langbein, W.B. and Leopold, L.B. (1964) 'Quasi-equilibrium states in channel morphology', *American Journal of Science* 262: 782–794.

Langbein, W.B. and Schumm, S.A. (1958) 'Yield of sediment in relation to mean annual precipitation', *Transactions of the American Geophysical Union* 39: 1076–1084.

Leopold, L.B. and Langbein, W.B. (1962) 'The concept of entropy in landscape evolution', *U.S. Geological Survey Professional Paper* 500-A: 1–20.

——(1963) 'Association and indeterminacy in geomorphology', in C.C. Albritton (ed.) *The Fabric of Geology*, Reading: Addison-Wesley.

Leopold, L.B. and Maddock, T. (1953) 'The hydraulic geometry of stream channels and some physiographic implications', *U.S. Geological Survey Professional Paper* 252: 1–57.

Leopold, L.B. and Miller, J.P. (1956) 'Ephemeral streams – hydraulic factors and their relation to the drainage net', *U.S. Geological Survey Professional Paper* 282-A: 1–37.

Leopold, L.B. and Wolman, M.G. (1957) 'River channel patterns – braided, meandering and straight', *U.S. Geological Survey Professional Paper* 282-B: 39–85.

Leopold, L.B., Wolman, M.G. and Miller, J.P. (1964) *Fluvial Processes in Geomorphology*, San Francisco: W.H. Freeman.

Lewis, W.V. (ed.) (1960) 'Investigations on Norwegian cirque glaciers', *Royal Geographical Society Research Series* 4: 1–104.

Linton, D.L. (1951) 'The delimitation of morphological regions', in L.D. Stamp and

S.W. Wooldridge (eds) *London Essays in Geography*, London: G. Philip.

Martin, A.F. (1963) 'Professor S.W. Wooldridge', *The Times*, 8 May 1963.

Matthes, F.E. (1930) 'Geologic history of the Yosemite valley', *U.S. Geological Survey Professional Paper* 160: 1–137.

Melton, M.A. (1958a) 'Geometric properties of mature drainage and their representation in an E-4 phase space', *Journal of Geology* 66: 35–54.

——(1958b) 'Correlation structure of morphometric properties of drainage systems and their controlling agents', *Journal of Geology* 66: 442–460.

Nicholas, T.C. (1961) 'W.V. Lewis', *Trinity College, Cambridge, Annual Record 1961*, 37–39.

Peel, R.F. (1967) 'Geomorphology: trends and problems', *Advancement of Science* 24: 205–216.

Peltier, L.C. (1950) 'The geographic cycle in periglacial regions as it is related to climatic geomorphology', *Annals of the Association of American Geographers* 40: 214–236.

Penck, W. (1924) *Die morphologische Analyse. Ein Kapitel der physikalischen Geographie*. Stuttgart: J. Engelhorns Nachforschung.

Sack, D. (1992) 'New wine in old bottles: the historiography of a paradigm change', *Geomorphology* 5: 251–263.

Scheidegger, A.E. (1961) *Theoretical Geomorphology*, Berlin: Springer-Verlag.

Scheidegger, A.E. and Langbein, W.B. (1966) 'Probability concepts in geomorphology', *U.S. Geological Survey Professional Paper* 500-C: 1–14.

Schumm, S.A. (1956a) 'Evolution of drainage systems and slopes in badlands at Perth Amboy, New Jersey', *Bulletin of the Geological Society of America* 67: 597–646.

——(1956b) 'The role of creep and rainwash in the retreat of badland slopes', *American Journal of Science* 254: 693–706.

——(1963a) 'The disparity between present rates of denudation and orogeny', *U.S. Geological Survey Professional Paper* 454-H: 1–13.

——(1963b) 'Sinuosity of rivers on the Great Plains', *Geological Society of America Bulletin* 74: 1089-1100.

Schumm, S.A. and Lichty, R.W. (1965) 'Time, space and causality in geomorphology', *American Journal of Science* 263: 110–119.

Sparks, B.W. (1949) 'The denudation chronology of the dip-slope of the South Downs', *Proceedings of the Geologists' Association* 60: 165–207.

——(1960) *Geomorphology*, London: Longmans (2nd edition 1972, 3rd edition 1986).

Stamp, L.D. (1957) 'Geographical agenda: a review of some tasks awaiting geographical attention', *Transactions and Papers of the Institute of British Geographers* 23: 1–17.

——(1966) 'Ten years on', *Transactions of the Institute of British Geographers* 40: 11–20.

Steers, J.A. (1960) 'Physiography: some reflections and trends', *Geography* 45: 1–15.

Stoddart, D.R. (1981) 'The paradigm concept and the history of geography', in D.R. Stoddart (ed.) *Geography, Ideology and Social Concern*, Oxford: Blackwell.

——(1987a) *Alfred Steers: 1899–1987. A Personal and Departmental Memoir*, Cambridge: Department of Geography.

——(1987b) 'Geographers and geomorphology in Britain between the wars', in R.W. Steel (ed.) *British Geography 1918–1945*, Cambridge: Cambridge University Press.

——(1988) 'James Alfred Steers 1899–1987', *Transactions of the Institute of British Geographers* NS 13: 109–115.

——(1989) 'Obituary: Bruce Wilfred Sparks 1923–1988', *Transactions of the Institute of British Geographers* NS 14: 492–495.

Strahler, A.H. (1950a) 'Davis' concept of slope development viewed in the light of recent quantitative investigations', *Annals of the Association of American Geographers* 40: 209–213.

——(1950b) 'Equilibrium theory of erosional slopes, approached by frequency distribution analysis', *American Journal of Science* 248: 673–696, 800–814.

——(1952a) 'Dynamic basis of geomorphology', *Bulletin of the Geological Society of America* 63: 923–938.

——(1952b) 'Hypsometric (area–altitude) analysis of erosional topography', *Bulletin of the Geological Society of America* 63: 1117–1142.

——(1980) 'Systems theory in physical geography', *Physical Geography* 1: 1–27.

——(1992) 'Quantitative/dynamic geomorphology at Columbia 1945–60: a retrospective', *Progress in Physical Geography* 16: 65–84.

Thornbury, W.D. (1954) *Principles of Geomorphology*, New York: John Wiley.

Twidale, C.R. (1992) 'King of the plains: Lester King's contributions to geomorphology', *Geomorphology* 5: 491–509.

von Bertalanffy, L. (1950) 'The theory of open systems in physics and biology', *Science* 111: 23–28.

Wolman, M.G. (1955) 'The natural channel of Brandywine Creek, Pennsylvania', *U.S. Geological Survey Professional Paper* 271: 1–56.

Wolman, M.G. and Leopold, L.B. (1956) 'River flood plains: some observations on their formation', *U.S. Geological Survey Professional Paper* 282-C: 87–109.

Wolman, M.G. and Miller, J.P. (1960) 'Magnitude and frequency of forces in geomorphic processes', *Journal of Geology* 68: 54–74.

Wooldridge, S.W. (1951) 'The progress of geomorphology', in G. Taylor (ed.) *Geography in the Twentieth Century*, London: Methuen.

——(1955) 'The study of geomorphology [review]', *Geographical Journal* 121: 89–90.

——(1958) 'The trend of geomorphology', *Transactions and Papers of the Institute of British Geographers* 25: 29–35.

Wooldridge, S.W. and Linton, D.L. (1939) 'Structure, surface and drainage in south-east England', *Transactions and Papers of the Institute of British Geographers* 10: 1–124.

——(1955) *Structure, Surface and Drainage in South-East England*, London: G. Philip.

Wooldridge, S.W. and Morgan, R.S. (1937) *The Physical Basis of Geography: An Outline of Geomorphology*, London: Longmans Green.

——(1959) *An Outline of Geomorphology: The Physical Basis of Geography*, London: Longmans.

PUBLICATIONS OF
RICHARD J. CHORLEY

1956

'Some neglected source material in quantitative geomorphology', *Journal of Geology*
64: 422–423.
'The relationships between angle of land slope and soil profile characteristics in the
U.S.A.', *First Report of the Commission for the Study of Slopes, International
Geographical Union*: 42–43.

1957

'Illustrating the laws of morphometry', *Geological Magazine* 94: 140–150.
'Climate and morphometry', *Journal of Geology* 65: 630–638.
'A new standard for estimating drainage basin shape', *American Journal of Science*
255: 138–141 [with D.E.G. Malm and H.A. Pogorzelski].

1958

'Group operator variance in morphometric work with maps', *American Journal of
Science* 256: 208–218.
'Aspects of the morphometry of a "poly-cyclic" drainage basin', *Geographical
Journal* 124: 370–374.

1959

'The geomorphic significance of some Oxford soils', *American Journal of Science* 257:
503–515.
'A simplified approximation for the hypsometric integral', *Journal of Geology* 67:
566–571 [with L.S.D. Morley].
'The shape of drumlins', *Journal of Glaciology* 3: 339–344.

1961

'Early slope development in an expanding stream system', *Geological Magazine* 98:
117–130 [with C.S. Carter].

RICHARD J. CHORLEY: A BIBLIOGRAPHY

1962

'Comparison of morphometric features, Unaka Mountains, Tennessee and North Carolina and Dartmoor, England', *Bulletin of the Geological Society of America* 73: 17–34 [with M.A. Morgan].
'Geomorphology and General Systems Theory', *U.S. Geological Survey Professional Paper* 500-B: 1–10.

1963

'Diastrophic background to twentieth-century geomorphological thought', *Bulletin of the Geological Society of America* 74: 953–970.

1964

The History of the Study of Landforms or The Development of Geomorphology, vol. 1: *Geomorphology before Davis*, London: Methuen, xvi, 678 pp. [with A.J. Dunn and R.P. Beckinsale].
'The fall of Threatening Rock', *American Journal of Science* 262: 1041–1054 [with S.A. Schumm].
'The Vigil Network system', *Journal of Hydrology* 2: 19–24 [with H.O. Slaymaker].
'Geography and analogue theory', *Annals of the Association of American Geographers* 54: 127–137.
'An analysis of the areal distribution of soil size facies on the Lower Greensand rocks of east-central England by the use of trend surface analysis', *Geological Magazine* 101: 314–321.
'Geomorphic evaluation of factors controlling the shearing resistance of surface soils in sandstone', *Journal of Geophysical Research* 69: 1507–1516.
'The nodal position and anomalous character of slope studies in geographical research', *Geographical Journal* 130: 70–73.

1965

Frontiers in Geographical Teaching, London: Methuen, 379 pp. [edited with P. Haggett].
'A re-evaluation of the geomorphic system of W.M. Davis', in R.J. Chorley and P. Haggett (eds) *Frontiers in Geographical Teaching*, London: Methuen, Chapter 2, 21–38.
'The application of quantitative methods to geomorphology', in R.J. Chorley and P. Haggett (eds) *Frontiers in Geographical Teaching*, London: Methuen, Chapter 8, 147–163.
'Frontier movements and the geographical tradition', in R.J. Chorley and P. Haggett (eds) *Frontiers in Geographical Teaching*, London: Methuen, Chapter 18, 358–378 [with P. Haggett].
'Trend-surface mapping in geographical research', *Transactions of the Institute of British Geographers* 37: 47–67 [with P. Haggett].
'Scale standards in geographical research: a new measure of areal magnitude', *Nature* 205: 844–847 [with P. Haggett and D.R. Stoddart].

1966

'The application of statistical methods to geomorphology', in G.H. Dury (ed.) *Essays in Geomorphology*, London: Heinemann, 275–387.

'Regional and local components in the areal distribution of surface sand facies in the Breckland, eastern England', *Journal of Sedimentary Petrology* 36: 209–220 [with D.R. Stoddart, P. Haggett and H.O. Slaymaker].

'Talus weathering and scarp recession in the Colorado Plateaus', *Zeitschrift für Geomorphologie* NF 10: 11–36 [with S.A. Schumm].

1967

Models in Geography: The Second Madingley Lectures, London: Methuen, 816 pp. [edited with P. Haggett].

'Models, paradigms and the new geography', in R.J. Chorley and P. Haggett (eds) *Models in Geography*, London: Methuen, Chapter 1, 19–41.

'Models in geomorphology', in R.J. Chorley and P. Haggett (eds) *Models in Geography*, London: Methuen, Chapter 3, 59–96.

'Throughflow, overland flow and erosion', *Bulletin of the International Association of Scientific Hydrology* 12: 5–21 [with M.J. Kirkby].

'Trend surface mapping of raised shorelines', *Nature* 215: 611–612 [with S.B. McCann].

'Application of computer techniques in geology and geography' (abstract), *Proceedings of the Geological Society of London* 1642: 183–186.

1968

Atmosphere, Weather and Climate, London: Methuen, 319 pp. [with R.G. Barry].

Socio-economic Models in Geography, London: Methuen [edited with P. Haggett].

'Base level', in R.W. Fairbridge (ed.) *The Encyclopedia of Geomorphology*, New York: Reinhold, 58–60.

'History of geomorphology', in R.W. Fairbridge (ed.) *The Encyclopedia of Geomorphology*, New York: Reinhold, 410–416.

1969

Water, Earth and Man: A Synthesis of Hydrology, Geomorphology and Socio-economic Geography, London: Methuen, xix, 588 pp. [editor].

'The drainage basin as the fundamental geomorphic unit', in R.J. Chorley (ed.) *Water, Earth and Man*, London: Methuen, Chapter 2.ii, 77–99.

'The role of water in rock disintegration', in R.J. Chorley (ed.) *Water, Earth and Man*, London: Methuen, Chapter 3.ii, 135–155.

Physical and Information Models in Geography, London: Methuen [edited with P. Haggett].

Integrated Models in Geography, London: Methuen [edited with P. Haggett].

'The elevation of the Lower Greensand ridge, south-east England', *Geological Magazine* 106: 231–248.

'The Standing Committee on the Role of Models and Quantitative Techniques in Geographical Teaching', *Geography* 54: 1–4.

Network Analysis in Geography: An Exploration in Spatial Structure, London: Edward Arnold, xii, 348 pp. [with P. Haggett].

1970

Frontiers in Geographical Teaching, 2nd edition, London: Methuen, 385 pp. [edited with P. Haggett].
Atmosphere, Weather and Climate, 1st American edition, New York: Holt, Rinehart & Winston, 320 pp. [with R.G. Barry].

1971

Atmosphere, Weather and Climate, 2nd edition, London: Methuen, 379 pp. [with R.G. Barry].
'Forecasting in the earth sciences', in M. Chisholm, A.E. Frey and P. Haggett (eds) *Regional Forecasting*, London: Butterworth, 121–137.
'Gabriel-Auguste Daubrée (1814–1896)', in C.C. Gillispie (ed.) *Dictionary of Scientific Biography*, New York: Charles Scribner's Sons, vol. 3: 586–587.
'The role and relations of physical geography', *Progress in Geography* 3: 89–109.
'An experiment in terrain filtering', *Area* 3: 78–91 [with K. Bassett].
Introduction to Physical Hydrology, London: Methuen, 211 pp. [editor].
Introduction to Fluvial Processes, London: Methuen, 218 pp. [editor].
Introduction to Geographical Hydrology, London: Methuen, 206 pp. [editor].
Physical Geography: A Systems Approach, London: Prentice-Hall, xiii, 370 pp. [with B.A. Kennedy].

1972

Network Analysis in Geography: An Exploration in Spatial Structure, 2nd edition, London: Edward Arnold, 348 pp. [with P. Haggett].
Spatial Analysis in Geomorphology, London: Methuen, vii, 393 pp. [editor].
'Spatial analysis in geomorphology', in R.J. Chorley (ed.) *Spatial Analysis in Geomorphology*, London: Methuen, Chapter 1, 3–16.
'Cartographic problems in stream channel delineation', *Cartography* 7: 150–162 [with P.F. Dale].
'Albert Heim (1849–1937)', in C.C. Gillispie (ed.) *Dictionary of Scientific Biography*, New York: Charles Scribner's Sons, vol. 6: 227–228.

1973

Comments on 'Systems modelling and analysis in resource management' by J.N.R. Jeffers, *Journal of Environmental Management* 1: 29–31.
The History of the Study of Landforms or The Development of Geomorphology, vol. 2: *The Life and Work of William Morris Davis*, London: Methuen, xxii, 874 pp. [with R.P. Beckinsale and A.J. Dunn].
Directions in Geography, London: Methuen, xii, 331 pp. [editor].
'Geography as human ecology', in R.J. Chorley (ed.), *Directions in Geography*, London: Methuen, Chapter 7, 155–169.
'Douglas Wilson Johnson (1878–1944)', in C.C. Gillispie (ed.) *Dictionary of Scientific Biography*, New York: Charles Scribner's Sons, vol. 7: 143–145.

'Willard Drake Johnson (1859–1917)', in C.C. Gillispie (ed.) *Dictionary of Scientific Biography*, New York: Charles Scribner's Sons, vol. 7: 148–150.

1974

'Walther Penck (1888–1923)', in C.C. Gillispie (ed.) *Dictionary of Scientific Biography*, New York: Charles Scribner's Sons, vol. 10: 506–509.

1976

Atmosphere, Weather and Climate, 3rd edition, London: Methuen, 432 pp. [with R.G. Barry].
'Some thoughts on the development of geography from 1950 to 1975', *Oxford Polytechnic Discussion Papers in Geography* 3: 29–35.

1978

'The hillslope hydrological cycle', in M.J. Kirkby (ed.) *Hillslope Hydrology*, Chichester: John Wiley, Chapter 1, 1–42.
'Glossary of terms', in M.J. Kirkby (ed.) *Hillslope Hydrology*, Chichester: John Wiley, 365–375.
'Bases for theory in geomorphology', in C. Embleton, D. Brunsden and D.K.C. Jones (eds) *Geomorphology: Present Problems and Future Prospects*, Oxford: Oxford University Press, 1–13.
Environmental Systems: Philosophy, Analysis and Control, London: Methuen and Princeton: Princeton University Press, xii, 624 pp. [with R.J. Bennett].

1980

'G.K. Gilbert's geomorphology', in E.L. Yochelson (ed.) *The Scientific Ideas of G.K. Gilbert*, *U.S. Geological Survey Special Paper* 183: 129–142 [with R.P. Beckinsale].

1981

'Optimization: control models', in N. Wrigley and R.J. Bennett (eds) *Quantitative Geography: A British View*, London: Routledge & Kegan Paul, 219–224 [with R.J. Bennett].
'William Morris Davis 1850–1934', *Geographers: Biobibliographical Studies* 5: 27–33.

1982

Atmosphere, Weather and Climate, 4th edition, London: Methuen, 407 pp. [with R.G. Barry].

1983

Geomorphic Controls on the Management of Nuclear Waste, Nuclear Regulatory Commission Report NUREG/CR-3276: 1–137 [with S.A. Schumm].

RICHARD J. CHORLEY: A BIBLIOGRAPHY

1984

Geomorphology, London: Methuen, xxiii, 607 pp. [with S.A. Schumm and D.E. Sugden].

1987

Atmosphere, Weather and Climate, 5th edition, London: Methuen, 460 pp. [with R.G. Barry].
'Perspectives on the hydrosphere', in M.J. Clark, K.J. Gregory and A.M. Gurnell (eds) *Horizons in Physical Geography*, London: Macmillan Education, 378–381.

1989

'From Madingley to Oxford: a foreword to *Remodelling Geography*', in W. Macmillan (ed.) *Remodelling Geography*, Oxford: Blackwell, xv–xx [with P. Haggett].

1991

The History of the Study of Landforms or The Development of Geomorphology, vol. 3: *Historical and Regional Geomorphology 1890–1950*, London: Routledge, xxiii, 496 pp. [with R.P. Beckinsale].

1992

Atmosphere, Weather and Climate, 6th edition, London: Routledge, 400 pp. [with R.G. Barry].

1993

'Spatial and temporal mapping of water in soil by magnetic resonance imaging', *Hydrological Processes* 7: 279–286 [with M.H.G. Amin *et al.*].
'William Vaughan Lewis', Cambridge University, Department of Geography, *The William Vaughan Lewis Seminars* 1: i–ii.

1994

'Magnetic resonance imaging of soil–water phenomena', *Magnetic Resonance Imaging* 12: 319–321 [with M.H.G. Amin *et al.*].

1995

'Haggett's Cambridge: 1957–1966', in A.D. Cliff, P.R. Gould, A.G. Hoare and N.J. Thrift (eds) *Diffusing Geography*, Oxford: Blackwell, 355–374.
'Classics of physical geography revisited: Horton, 1945', *Progress in Physical Geography* 19: 533–554.
'Studies of soil–water transport by MRI', *Magnetic Resonance Imaging* (in press) [with M.H.G. Amin *et al.*].

INDEX

Page numbers in italics refer to figures and tables.

a/s index 98; water concentration, and piping occurrence 107, *108–9*
acid rain, and acid runoff 96–8
active-channel drainage density 25, 26; maximum, under driest conditions 27; more related to weathered soil mantle characteristics 28–9
afforestation, negative effects of 184
agricultural practices, and soil erosion 371
Allerod climatic phase 48
aluminium concentrations, Maesnant Basin 96–7, *97, 98*
amino-acid geochronology 49, 52
Ancaster Gap, Lincoln Edge 57
Anglian and Wolstonian cold phases, West Somerset 231–2, *233*
arroyos 365–6, *see also* Polacca Wash
Australia, underfit streams and former lakes 55–6
avalanche chutes 150–1
Avon River, Warwickshire 51; average interval for Terraces 55; difficulties of Pleistocene geochronology 53; Terrace sequences revised 52

bajada formation 226–7
Balchin, W.G.V., erosion surfaces, West Somerset 228–9
bankside seeps 84
basal sliding 148, 150, 160
base flow 184
basin infilling, Mesozoic sediments, West Somerset 225–7
basin and range tectonics 135–8

basin stretching 137, *139*
basins: formed by tectonic activity 223–5; intermontane 355; secondary, West Somerset 223–7; subglacial 154
beaches, mobile, evolution of 236
bedrock blocks, entrainment of 154
Bennett, H.H., in communication with Sauer 367–8
bergschrund hypothesis: limitations of 160; modified 164
bergschrunds 162
Berkeley, University of California, and Carl Sauer 343–60
block–fault system 223
Blue Lias 227, 235
bottom water, cool and warm episodes 196
Bowman, Isaiah: and the Land–Use Committee 360; in receipt of Bryan's objections to Sauer 364
breccias, Vale of Porlock 226
Breislak, Scipione, rejection of Hutton's views 63–4
British Columbia: calibration of cirques and troughs 157, *158*, 159; calibration of fjords 157
Bruman, Henry 349
Bryan, Kirk: further attacks on Sauer 365; invoked Sauer's wrath 357–8; objected to Sauer's Polacca Wash work 363–4
Bryan, Kirk and Wickson, Gladys, critical of Sauer's use of Penckian method 357
Büdel, Julius 188

Budleigh Salterton Pebble Beds 227
Buffon, J.M.L. 64, *65*
Burbage Brook, Derbyshire, perennial
  pipes 90, 92, *94*
bypass flow 88
bypass infiltration 85

Cambridge students, indignant at the
  geomorphological offerings 390
Carson, M.A., on river meander
  planform 283–5
case studies 278; realist 274–7, 278–80
catastrophic processes 191
catchment orientation, and piping
  frequency *112*, 112
Central Somerset basin 225
centripetal ordering 243, 247;
  geometrical series laws (Horton) 258
chance 324; and necessity 320–2
channel density 68–9
channel geometry, above and below
  limestone junction, southern
  Portugal *299*, 299
channel pattern change, South Platte
  River, differing accounts 280
channel–in–channel phenomenon 305–6
chaos theory 319–20; chance and
  necessity 320–2; and geomorphology
  322–4
characteristic forms 128–9, 143; as first
  order diffusion model 133–5
Chester, D.K. and James, P.A., human
  intervention and geological processes,
  southern Portugal 298, 302
Chiricahua Mountains, Sauer's
  fieldwork in 358–60
chloride aerosols 200
Chorley, Richard J. 215–16, 383–99;
  early life and education 3–5; in
  America 6–7; in Cambridge 8–10,
  383–94; recent years 10–11;
  application of disequilibrium and
  threshold concepts 300–8; with
  Beckinsale and Dunn *The Life and
  Work of William Morris Davis* 10;
  climate and morphometry paper
  attacked by Steers 389; and
  interfacing 293; Madingley Lectures
  with Haggett 393; open systems as
  disequilibrium systems 293–5;
  Progress in Geography journals 393;
  publications 393–4, 400–5; *The*

History of the Study of Landforms
  60–1, 394; work under Strahler at
  Columbia University 390–1
cirques 169; erosion by rotational flow
  159–60; frost shattering 162, 164–5;
  high–alpine (van) 161–2, *163*; rock
  avalanches from headwalls *168*; size
  of 164–5; wearing down vs wearing
  back 160–1
Clan Alpine Range, Nevada,
  back–analysis of landforms 136–8, *139*
climate: 'equable' 207; implications of
  Quaternary shifts 187–8; lithology
  and relief, affecting drainage density
  18–33, 41; role of in shaping
  landscapes 187
climate system: almost intransitive
  behaviour 189; climate models 192–5;
  components of 189–90; parameters
  relevant to geomorphology 191–2;
  timescales 190–1
climatic change, western Algarve 301–2
climatic conditions, past, influence on
  resistant lithologies 42
climatic states, mean, long–term 190
cloud–radiation feedback 190
coastal processes, W Somerset 235–6
complex response 189
conifer planting, no–go areas 98
contour–crenulation 42
coupled ocean–atmosphere climate
  models 194
Cretaceous: climatic conditions 196–7,
  197; sea–floor spreading in 195
critical rationalism 273; objections to
  hypothesis–experiment–test–falsify
  procedure of 266
crustal thinning 143
cultivation, abandonment of 303
Cuvier, G. 63
cycles: regressive and transgressive 220;
  theme of 252, 258–9

Dana, J.D. 67
dating problems, underfit meanders 49
Dave Johnston Mine *30*, 33, 34
Davis, W.M. 121, 345; and Chiricahua
  mountains 358, 359–60; river
  network 259; seen as overcritical 347
Davisian cycle 128, 259, 333–4, 343–4;
  attacked by Chorley 9; Sauer's
  doubts concerning 345–6

demarcation criteria 273–7
denudation: and orogeny 121; relative, and steady state downcutting 129, *130*
denudation rates 134; chemical and mechanical 191
desiccation cracking 83, 89, 112
determinism: definitions 313, *see also* necessity
deterministic chaos 322; in surface runoff and overland flow 322–3
deterministic reasoning 312–13, 314, 316–17; in fluvial geomorphology, questioned 314–15
Devereux, C.M.: changes in channel morphometry, southern Portugal 301–2; valley fills in southern Portugal 296, 298
Devonian 237; effects of facies variation on scenery 221–3; lithological variations within 218–21
Dicken, S.N. 352; erosion studies 366
Dietrich, W.E., tried to link non-fluvial with fluvial 68–9
discharge: bankfull 48; channel-forming 47; flood, southern Portugal 306, *307*; mid-latitude fluctuations, correlation of 55–6
discharge maximum, latest, and its aftermath 47–9
domino block faulting *see* faults/faulting, extensional
Dorn stream (Cotswolds) 47
downcutting: constant 128, 129, *131*, 132; steady state 129, *130*
drainage, subglacial 146
drainage basins: developed on shale, Texas 20, *21*, *24*, 25; on granite, California *22–3*, *24*, 25–9; Mancos Shale *31*, 32; Mesa Verde Group 30, *31*
drainage density 15–45, 69, 134; effects of climate, lithology and relief on 18–33, 41; human influences on 17–18; site specific studies *30*, 33–6; total 29
drainage density–relief ratio relationship 34–5, *35*, *36*
drainage networks 15, 314; stability of, and changes in drainage density (experiment) 36–41; unmodified cf. modified 39–41

drainage patterns 37; stable, drainage density of 42
Drewes, H., on Chiricahua Mountains 358–9

earthquakes, triggering rock avalanches 166, 167
El Niño–Southern Oscillation (ENSO), climate model studies 201
electromagnetic current meters (EMCMs), distorting 'turbulent structures' 272
entropy, concept of 314–15
envelope profiles 132–3, 143
environmental degradation, and piping 95
ephemeral pipes 83, 85; acidic discharge 96, *97*; lag times 79, *81*; and nutrient concentrations 99, *100*
equifinality (convergence) 188, 269–70
erosion: adjustment of controlling features 309; by rotational flow 159–60; fluvial, on/off model 299–308; glacial, rates of 146–9; of glacial troughs 151–9; headward, valley heads 161; hillslope 138–9; pipe-gully 91; spatially restricted 91–3; Tertiary 227
Evenlode, River 57
explanations, reinforcing 274–7
extension 135, *137*

faults/faulting 223, 227; controlling landscape 358; extensional 135–6, *137*, *139*; pattern round Minehead 223–5; Peninsular Range 354; step 225
feedback mechanisms 190
field area, choice of critical 278–9
fieldwork 277–85; for spatially distributed simulation models 286–8
fjords: British Columbia, calibration of 157; and ribbon lakes 155–6
flash floods 191
flexural uplift 142
flood events: Lynmouth 234–5; southern Portugal, effects of 306, *307*, 308, *see also* flash floods
flow convergence, effect of 133
fluvial geomorphology: deterministic and stochastic reasoning within 314–16; fractals in 323–4

fluvial processes, present, West Somerset 234–5
forcing functions, and chaotic behaviour 322
forest cover, effects of reduction 183–4
fractal geometry 323
fractals 260, 322; in fluvial geomorphology 323–4
fractures, conjugate 166–7
Frey, A.E., on West Somerset 229–31
frost shattering 162, 164–5; closed- and open-system 162, 164

Gangetic Plain, fieldwork, control of long-term river behaviour 278–9
gelifluction 149
general circulation models (GCMs): applied to palaeoclimatic questions 196, *197*; atmospheric *194*, 194–5
General Systems Theory 259, 293, 328, 392–3
generational ordering system 243, 245, 256
geomorphic explanation, contrasting approaches to 316
geomorphic systems, concept of intrinsic thresholds 188–9
geomorphology: and chaos theory 322–4; climatic 16–17; and geomorphologists, marginalised by decision-makers 331, 333; groupings of research goals and methods 329; linked to social problems 328, 330; in a natural sciences framework 331, *332*; and positivism 328, *see also* fluvial geomorphology
Gilbert River, Queensland, dating of deltaic sediments 56
glacial channels, and ice discharge 152
glacial hydrology study: realist experiment 274–7, 378; spatially distributed simulation model 286–8
glacial quarrying 154, 164
glacial troughs: in ablation zones 157; catenary profile 153; cross profile 151–3; fjords and ribbon lakes 155–6; intersecting 159; long profile: steps and basins 153–5; in three dimensions: calibration 156–9
glacier bed lowering, rates of 146
glacier outburst floods 148
glaciers: erosion rates 146–9; surging 148;

temperate/'mixed' 149; tidewater 157; upward flow towards lower end 155, 156; valley and wet-based cirque 148–9
global climate models 192–5, 201; palaeoclimatic reconstructions of geomorphologically pertinent variables 207–8; palaeoclimatic simulations 196
Goosenecks reach, San Juan River, Utah 51, 53
grand design, vs local topographic interpretation, problems of 67
granite, California, drainage density on *22–3*, *24*, 25–9, 41–2
gravel-bed stream (lowland), relationships study 277
Gregory, David 253–4
groundwater 85, *87*, 88, 101
Grove, Dick 390
gullies 90; erosion studies, Sauer 366; gully development 55, 90–1

Haggett, P. 384, 392; Madingley lectures with Chorley 393
Hales, Stephen, use of experiments 249–50
Hangman Grits 218, 220, 221, 223
Harvey, William 253
haystack hills 89
head 234
headwall retreat 167
headwalls: cirque, and pressure release 166; and subsurface erosion 91
Heisenberg uncertainty principle 314
HEP reservoirs, importance of year-long water supply 182–3
hillslope drainage processes, comparative hydrology of 75–88
hillslope evolution models: Davis and Penck 121; one-dimensional 122–32, 139–41; two-dimensional 132, *see also* envelope profiles; stream/slope profile
hillslope length 134
hillslope stability 127–8
hillslopes, climatic influence on characteristics 19
Hinds, N.E.A. 343–4
Hodder/Holford stream, Quantocks, course of 232, *233*
hog's-back cliffs 223, 234, 235, 236
Hooke, J.M. 154; on changes in river

meanders 281–3; update of meltwater hypothesis 164
Horton, R.E. 67; geometrical series laws 258; impressed Strahler 390–1
Hortonian overland flow 75, *76*, 78, 80, 82, 95; deterministic chaos in 323
human intervention and geological processes *see* Portugal, southern
Hutton, James 61–3, 66–7; cyclic ideas 258–9
hydrology: comparative, hillslope processes 75–88; deterministic chaos and stochastic complexity 323; glacial 274–7, 286–8, 378; tropical stream, and land use changes 175–86; use of models 285, 286

iatromechanism, and Newtonian ideas 253–6
ice sheets, erosional selectivity 149
ice streams, East Antarctica 148
ice-albedo feedback 190
Ilfracombe Slates 220
incision, and sinuosity 52–5
incision rate–passage of time models 49–53
infiltration capacity 30, 33, 83
infiltration rates 16, 20, 184; southern Portugal 303–4, *305*
inherent indeterminacy 315, 317–18, 325
interdisciplinarity 289
interfluves 132–3
interglacial conditions 200
irrigation 303
isostatic effects 141, *142*
isostatic flexure 138, *142*
isostatic uplift 143

Jameson, Robert 60, 67; attack on Hutton and Playfair 62
Jedburgh unconformity 62–3
jet streams 206; displacement and splits 201, *202*
Jim Bridger Mine *30*, 33, 34–5

Keill, James 243–64; attractive force between molecules 255–6; concept of infinite regression in size in fractal tree 258; geometric series scaling laws 244–52; geometrical progression laws 256–7
Keill, John 255

Kesseli, J.E. 369; and Penckian principles 349, *350*
Kniffen, F.B. 352, 369–70
Krumbein, W.C., influenced Chorley 392

La Métherie, J.-C. de 63
lake levels, high and low, tropics/subtropics 203–5
land use, and sustainability 336
land use changes: Mahaweli basin 178, *181*, 184; southern Portugal 302–4, *304*, *305*; trade–off, evapotranspiration changes and infiltration rates 184; and tropical stream hydrology 175–86
landforms 208; interpretation of and process geomorphology 190–1; and modelling 285–6; subsurface, longevity of 94–5; surface, and subsurface processes 89–93, *94*
landscape evolution: characteristic forms method 134–5; diffusion model for 121–2, *see also* hillslope evolution models
landscapes 208; polygenetic origin of 188
Late Glacial, mean annual discharges 48
'law of adjusted cross–sections' 156
Laws of Stream Composition, Horton 315, 318
Lawson, A.C. 343, 361
lee cavities, water pressure fluctuations in 154
Leighly, John 346, 369; on application of Penckian analysis to Peninsular Range 356–7; enthusiasm for *Die morphologische Analyse* 348–9; resented treatment by Davis 347; work on arroyo meanders 363
Leopold, Luna, inspired Chorley 391
Lewis, W.V. 8, 389, 390
Lillooet River basin, British Columbia, sediment sources and storage analysis 335–6
limestone: depressing influence on flood events 306, 308; and erosional development of southern Portugal 298–309
liming, to reduce acid rain impact 98, 289
limit cycles 189
Linton, David, collaboration with Wooldridge 386–7

lithology: climate and relief, effects on drainage density 18–33, 41; and glacial steps 154–5
logarithmic spiral 363
Louderbeck, G.D. 343
Lower Devonian 220
Lyell, Charles 63; on valley formation 67, 68–9
Lynton Slates 218

McKinley Mine 30, 33, 34, 35
Maesnant basin: environmental impact of piping 96, 97, 98–104, 105, 106; intrabasin distribution of piping 107, 108–9; peak lag times 79–80, 81; perennial pipes and groundwater bogs 89; runoff coefficients 77; runoff sources 85–8; sources of pipeflow: old water or new water 84–8; storm runoff sources compared 83, see also ephemeral pipes; perennial pipes
Mahaweli Ganga (River) 175–7
Mahaweli upper basin: changing runoff–rainfall ratios 178, 180; declining annual rainfall 178, 179; results of temporal analysis 178, 179–81; sources of data 177–8
Mancos Shale 29; drainage basins formed on 31, 32
Mandelbrot set 321
manifest underfits 49
material removal and storage, examined via sediment budget approach 335–6
maximum dynamic contributing area, pipeflow 75–7, 83–4
mean elevation curves 124–6, 128–9
meandering valleys 49; ingrowth of 48, 55
meanders: planform shape cf migration process, illustrating importance of textual detail 281–5; the underfit meander problem 46–59
measurement 269–73; transduction 270–2, 272; translation 269–70
Mercia Mudstones 227
Mesa Verde Group 29; drainage basins formed on 30, 31
Michigan Land Economic Survey 342
Milankovitch orbital effects 205
mined-land reclamation: Dave Johnston Mine 30, 33, 34; and drainage density 15–16; Jim Bridger Mine 30, 33, 34–5;

McKinley Mine 33, 34, 35; north-western Colorado, drainage density in 29–32
modelling: roles in scientific investigation 285–9, see also general circulation models; global climate models; hillslope evolution models
modes of enquiry 330, 330–1
moisture, recycling of over land areas 206
moisture balance/moisture budgets 193
morphologic analysis 351
morphological mapping 386–7
Morte Slates 220
Mosel River, middle reaches, terraces 53–4
mountain ranges 65, 66, 143; precipitation 198, 199, 200; and quasi-stationary waves 206
mountains, glaciated, process and form in erosion of 145–74

Napier, John, logarithms 257
necessity 324; and chaos theory 321–2
negative evidence, search for 274
New Red Sandstone 225, 226
New Zealand, glacial troughs 151–2
Newton, Isaac, interatomic force 254
Newtonian mechanics 314, 324
Nicholas, Tressilian 389
nivation 149–50
non-linear dynamic theory 322
North Africa, pluvial regimes 205
nutrient concentrations: and ephemeral pipes 99, 100; swales 104

oceans, evolution of thermal structure 195–6
on/off erosion switch 299–300; investigation of 300–8
open systems/closed systems 293–311
orographic forcing, and jet streams 201
overland flow 75, 82, 85, 183, 184, 322–3, see also Hortonian overland flow
oxygen isotope deep-sea stages 49

palaeochannels, glacial troughs as 151, 156
palaeoclimate: global, and unipolar glaciation 207; reconstructions see global climate models

palaeoclimatic studies: pre-Quaternary 195–200; Quaternary 200–5
Pangaea, simulation experiments, net effective precipitation 197–8
Paracelsian and Newtonian themes: geomorphology 257–60; physiology 252–7
parallel retreat 128
passive continental margins 141–3
peak lag time 78–80
pediment development/pediment gaps and passes 359–60
Penck, W. 121, 345, 358, 369; *Die morphologische Analyse* 348–9; *Primärrumpf* theory 355
Penckian analysis: use of and debate 352–8; Wooldridge's views 384, 386
peneplanes: 'summit' and 'primary', Sauer 353–4, 355, 356, *see also* Tertiary peneplanes
Peninsular Range, Sauer's fieldwork in 352–8
perennial pipes 83, 84, 85; and acid flushes 96; affecting peat thickness 101, *106*; and groundwater bogs 89; and stream rejuvenation 95
periglacial environment, Exmoor 232, 234
periglacial processes 160, 237, 298
pH and aluminium concentrations, Maesnant 96–7, *97*, *98*
phreatic surface, Maesnant *87*, 101, 104
physical geography, place-based methodology in research 279
physiography, taught by Sauer at Ann Arbor 341–2
Pickwell Down Sandstone 220
pipeflow 75, *76*; lag times 78, *79*; maximum dynamic contributing area 75–7; peak runoff rates 80, 82; runoff coefficients 77, *78*; sources of: old or new water 84–8
piping: affecting plant growth and soil profiles 99, 101, *102*, *103*, *104*; controls on distribution of in Britain 107–14; environmental impact of, Britain 95–104, *105*, *106*; and gully development 90–1; piping-gullying cycles 91
piston flow 88

Pitcairne, Archibald 253–4; hydraulic model 254, 255
plant ecology and soil development, Maesnant 99–104, *105*, *106*
plate tectonic processes–climate relationships 195
Playfair, John 61–2, 66–7
pluralism 328–9; need for 330–3; and social relevance of geomorphology 328
pluvial regimes, North Africa 205
pointed channel networks 91–2, *93*
Polacca Wash, Sauer 363–4, 368
polar and tropical continents, climatic effects compared 198
Porlock basin 225; deposits of 226
Portugal, southern: changed to a dry-valley environment 308, 309–10; concepts of geomorphological development 295–300; dating of valley fill 296, 298; Feio–Zbyszewski research 295, *297*; incision and fluvial deposition 298, 300, 301; investigation of the on/off erosion switch 300–8; reactivation of erosion system 304–5
positivism 268, 328, 333
post-mining stable topography, design of 36
Pozzuoli, Temple of Serapis 63
precipitation 34, 193; Pangaea, simulation experiments 197–8
precipitation, mean annual: and drainage density 16, *17*, *32*, 32; and valley drainage density 27, *28*
precipitation regimes: changed, and channel morphometry 301–2; present-day role of orography 198, *199*, 200
predictive models 285
pressure loss, due to friction (Young) 250–1
pressure release joints 165
probabilistic theory 315–16, 318–19
processes, diffusive and non-diffusive 122–8
proglacial lakes 232
Prosna River study 48

quantum theory 314
Quaternary: a chronology for southern Portugal *297*; implication of climate

shifts 187–8; palaeoclimatic studies 200–5; in West Somerset 231–4
radioactivity, and Earth's age estimates 62
rainfall, Mahaweli basin: data 177; decline in annual rainfall 178, *179*, 182
randomness: of chaotic motions 321; inherent and apparent 316, 318–19, 325
realism 266, 267–9, 273–4, 278, 280, 289, 333; basic tenets *267*; and modelling 285
reductionism 273
referential adequacy 274
relative denudation curve *124–6*, 128–9, *see also* denudation, relative
relief ratio 34
replication, multiplicative 274
research: field area/field work, precise roles of 280–5; intensive phase 278; interdisciplinary 274–7; place-based 278–9, 289; progressive design 277–8
resistance ratio (Young) 250
ribbon lakes 155–6
rilling, initiation of 90, 91, *92*
rillwash 132
roches moutonnées 154
rock avalanches 151; from cirque headwalls *168*; and stress release 165–7
rock basins 155
rock mass strength 152
rotation effects 135, 137
rotational cells, valley glaciers 160
runoff 41, 309; acid 96; deterministic chaos in 322–3; and drainage density on shale 33
runoff coefficients 77
runoff rates, peak 76–7, 80–3
runoff-rainfall ratios, changing, Sri Lanka 178, *180*
Russell, Richard J., bridge between Geology and Geography departments at Berkeley 344–5

Saharan dust, deposition in Atlantic 205
Salisbury, Rollin D., influence on Sauer 341
saturated hillslope flow 235
saturation overland flow 82
Sauer, Carl 340–79; before Berkeley

340–3; beyond the High Sierra 343–51; Chiricahua Mountains, his interpretation unsupported 358–60; doubting the Davisian system 345–6; 'Field project for slope-soil inquiry' 361–3; fieldwork style 349, 351; geomorphology and soil conservation 360–6; Handbook for Geomorphologists 366–8; 'Land resource and land use in relation to public policy' report 361; as member of Land-Use Committee 360–1; the Peninsular Range 352–8; Soil Conservation 370–2; and the Soil Conservation Service 363–8; 'Soil erosion and critical land margins' 361; 'The morphology of landscape' 351
scaling laws (Keill) 243, 244–52
scarps, antislope 166
Schumm, S.A. 121, 391
science, and geomorphology 336–7
scientific explanation, contrasting modes of 312–14
sediment budget, flexible concept of 334–7
sediment plumes 146
sediment queueing and threshold controls 299–300, 309
sediment transport 122, 132, 274; lost capacity, southern Portugal 306; and turbulent flow 271–2, *272*
sediment yields: basin 134; from drainage networks *40*, 40–1, *42*; glacial streams 148
seepage 101; landforms 88–95
Selworthy Beacon 216
semi-arid climate, drainage density in 20, 25
sensitivity analyses: spatially-distributed 288; use of models 285–6
Severn, River, inner valley at Shrewsbury 56–7
shale: drainage basins on 20, *21*, *24*, 25, *31*, 32, 41; drainage density on 32–3
Sharp, R.P., on Kirk Bryan 365–6
shear stress, deforming bed not banks 46–7
shingle movement 236
Shoalhaven River, NSW, thermoluminescence dates 51–2, 56
Shreve, R.L.: on probabilistic reasoning 315–16, 318; probabilistic-topological

approach 68, 70
simulation models, spatially distributed 286–8
simulations, of landform development 285
slip lines, in glaciers 153
slope evolution models *see* hillslope evolution models
slope-forming processes, West Somerset 235
slopes: reversed 154; slope process rates *127*; stability of and mountain glaciation 166–7
snow avalanche effects 150–1
snowline lowering 207; and lowered sea temperature 202–3
snowpatches 149–50
social change, impact of, southern Portugal 298, 302–4, 309–10
Soil Conservation Service 363–8; Sauer's contribution 368
soil development, and plant ecology, Maesnant 99–104, *105, 106*
soil erosion, and commercial exploitation 371
soil groups, Britain, distribution of piping in 112–13, *113*
soil loss, Sri Lanka 183
solute transport 273–4
Sparks, Bruce 389–90
spoils, reclaimed, infiltration rates changed 16
squared diameter law 245–6
Sri Lanka, central highlands, rainfall, stream flow and changing land use 182–3
Stamp, Professor Dudley 383–4
Stark, C.P., invasion–percolation model 68, 69–70
Steers, Alfred 7, 387–9; 1959 address to Institute of British Geographers 388–9
stochastic approaches 313–14; inherent indeterminacy 315, 317–18, 325; probabilistic theory 315–16, 318–19
stochastic complexity 323
Stogumber graben *see* Watchet–Stogumber basin
stormflow 78; in ephemeral pipes 101; in perennial pipes 84–5; yields 83–4
Strahler, Arthur 6, 390–1
Strahler ordering for streams 391

Strahler-ordered network analysis 243
strange attractors 189, 324
stream processes 143
stream profiles 139–41
stream-pinching, and 'flats' formation 229, *230*
stream/slope profiles 132, 138–41, 143
streamflow, Maesnant Basin 88
streamflow, Mahaweli basin: annual, increase in 182; data 177; seasonal trends in 178, *181*
streamhead pipes 92, *93*
streams: evolution of, West Somerset 229–31; first order channels 29, 41; meander wavelengths 47–8; subsurface, power of 113, *114*
stress release, and rock avalanches 165–7
structural controls, and subsurface networks 89
structural corroboration 274
structure, related to Earth processes 64–5
subsurface flow: contributing to stream runoff 82–3; and subsurface erosion 74–120
Sukkertoppen Ice Cap, calibration test 156–7
suspended sediment load, glacier fed streams 146
system interfacing 309–10

tectonic instability, Sauer in the Peninsular Range 352–6
tectonics, in geomorphological models 121–44
Tertiary peneplanes, West Somerset 227–31; sub–aerial and marine 228–9; an alternative view 229–31
textual detail, importance of 281–5
theoretical findings, validation of 273–7
thermoluminescence (TL) dating 51–2
thin elastic plate, two-dimensional deflection of 141–2
Thompson, D.W., *On Growth and Form* 259–60
Thornthwaite, C. Warren 366; became victim of Bryan's attacks 365; initiated Sauer's *Handbook for Geomorphologists* 367; and Polacca Wash studies 363

thresholds, rock, of fjords 155
throughflow 41, 75, *76* , 82; diffuse 108; parallel to pipes during storms 101, 104
Tibetan Plateau, effect of uplift 200
transduction 270–2, *272*
transient conditions 134; and characteristic form 129
translation 269–70
translational earthslides, Scotland, model–based assessment 285–6
turbulent flow, measurement problems 271–2, *272*

U-shaped valleys 151, 152
uncertainty, and stochastic reasoning 313–14
underfit, Osage-type 46–7
underfit meander problem 46–59
unifying concept and problem focus, need for 333–7
uplift 134–5, 142, 143, 200, 355; and rapid mountain building, global intervals of 195; tectonic, and steady downcutting 129, *131*, 132
USA, reclamation of mined land and drainage density 15–16

validation, consensual 274
valley bend, cutoff 56–7
valley bulging 165
valley drainage density 27, *28*, 28
valley meanders, wavelengths of 47–8
Valley of the Rocks, Lynmouth 232
valleys 65–6, 67–70; concept of stress concentration in 165; and streams, as a single set of phenomena 71
Variscan orogeny 221, *222*
vegetation: channel bank, effects of 305–6, 308, 309; and drainage density 34
velocity at different generations, formula for 247
'verrous'/'riegels' 154–5
Vexford breccias 227
von Bertalanffy, Ludwig 392–3

Watchet–Stogumber basin 225, 226, 227
water: increased demand for, southern Portugal 302–3; role in earth sculpture 61–2
water table: Algarve 303, 309; Maesnant Basin 84–5
water vapour–radiation feedback 190
water yield, deforested catchments 206
water–trees relationship, need for greater understanding of 185
watershed breaching, by ice sheets 159
weathering, chemical 149, 355–6
West Somerset landscape 216–18; current processes 234–6; effect of Devonian structures on coastal landforms 223; Permian and Triassic rocks of 225, *226*; primary framework 218–23; Quaternary facets 231–4; secondary basins 223–7; stream orientation and pattern, evidence on erosional history 229–31; Tertiary peneplanes 227–31
Willow Brook, alluvial fill 47
wind systems, global, under ice-age conditions 200–1, *202–3*
Windrush, River 57
Wooldridge, S.W. 6–7, 384–6
Wye, River, Redbrook and St Briavels 57

Young, Thomas, and the geometric series laws of arterial trees 250–2

Milton Keynes UK
Ingram Content Group UK Ltd.
UKHW031139141024
449569UK00024B/1208